Population, Regional Development and Transport

Series Editor

Pengjun Zhao, College of Urban and Environmental Sciences, Peking University, Beijing, China

This book series chiefly explores population change, regional development and transport in contemporary China. Its goal is to enhance our current understanding of population, regional development and sustainable transport in a context of rapid urbanization and transition – characterized by the shift from a centrally planned system to a market system, together with growing economic globalization and political decentralization.

The series will enrich the existing literature on population studies, regional development studies and transport studies. In particular, it highlights academic research on the interactions between population, regional development and transport. It will also shed new light on government practices with regard to regional development planning and management and transport investment.

Pengjun Zhao · Liangen Zeng

Transport Efficiency and Safety in China

Springer

Pengjun Zhao
College of Urban and Environmental Sciences
Peking University
Beijing, China

School of Urban Planning and Design
Peking University Shenzhen Graduate School
Shenzhen, China

Liangen Zeng
College of Urban and Environmental Sciences
Peking University
Beijing, China

This study was financially supported by the National Natural Science Foundation of China (41925003, 42130402), and Shenzhen Science and Technology Innovation Commission (JCYJ20220818100810024)

ISSN 2662-4613 ISSN 2662-4621 (electronic)
Population, Regional Development and Transport
ISBN 978-981-99-1054-0 ISBN 978-981-99-1055-7 (eBook)
https://doi.org/10.1007/978-981-99-1055-7

This Springer imprint is published by the registered company Springer Nature Singapore Pte Ltd.
The registered company address is: 152 Beach Road, #21-01/04 Gateway East, Singapore 189721, Singapore

Preface

The transport sector is one of the basic industries in the national economy of China, and it is the bond between logistics and the crowd in economic activities. Since the reform and opening-up, China has made great investment in transport and yielded fruitful results in transport construction; there has been a rapid growth in the length of highways of all classes, accompanied by a more mature network structure. However, the problem of over-investment in transport infrastructure soon manifested itself in some regions, and it is necessary to study the service efficiency of transport infrastructure in China. The transport sector is a major consumer of fossil energy and a significant source of CO_2 emissions. According to the International Energy Agency's World Energy Outlook in 2019, projections are that by 2030 the global transport sector will remain a major consumer of fossil energy, and that its fossil fuel consumption will reach 3,327 Mtoe in the Stated Policies Scenario, rising over 170% from 2000. Therefore, exploring the environmental impact of the transport sector on social development is of great relevance to global sustainable development. As a responsible developing country, China has pledged to peak CO_2 emissions as early as 2030. In this context, the transport sector, as the third-largest emitter of CO_2, is facing greater pressure to reduce CO_2 emissions. Transport safety is a common problem faced by countries all over the world, particularly emerging countries like China, where the number of traffic fatalities has ranked second in the world over the year. Road traffic accidents are the main type of traffic accidents. It is of great practical significance to study the current situation and influencing factors of road transport safety and further formulate effective measures.

In the book, transport efficiency and transport environmental efficiency in China are empirically studied using data envelopment analysis and the econometric modeling method, and the influencing factors of China's road traffic morality rate are empirically analyzed from socio-economic factors based on the Tobit model.

There are four innovative points regarding the research approach. Firstly, the epsilon-based measure (EBM) DEA model is applied to evaluate transport efficiency, and the EBM DEA model with undesirable outputs is applied to calculate transport environmental efficiency, which can address the defects of traditional DEA models and exhibit the potential for overestimation or underestimation of the efficiency value. Secondly, spatial autocorrelation is taken into account when analyzing the influencing factors of transport efficiency (highway and railway transport sectors) and transport environmental efficiency. Thirdly, in light of the actual situation in China, this book organically combines the advanced theory and method of environmental efficiency, and constructs a scientific evaluation theory system to calculate transport environmental efficiency of integrated transport systems, which is of great significance to improve the existing research on environmental efficiency, which will also provide a set of theoretical methods and a research framework for the green development accounting of transport sector. In the end, the regression analysis of influencing factors of road traffic mortality rate is based on macro socio-economic factors. Through analyzing the influence of specific variables of each factor on traffic accident casualties, this book puts forward corresponding suggestions and management measures for the development of road traffic safety in China, so as to provide theoretical basis and data support for improving road traffic safety in China.

The purpose of the book is to provide not only a theoretical base for transport planning, but also a numerical reference to various control measures in transport planning. It is hoped that transport planners and managers, governments, as well as students, researchers and consultants of transport planning, environmental management, energy and ecosystem management, will find this book useful.

Beijing, China Pengjun Zhao
Liangen Zeng

Acknowledgements We acknowledge the financial support of National Natural Science Foundation of China (41925003, 42130402), and Shenzhen Science and Technology Innovation Commission (JCYJ20220818100810024)

Contents

Chapter 1
Introduction

1.1 Research Background

The Intergovernmental Panel on Climate Change (IPCC) expected that transport-related CO_2 emissions would triple by 2100 if effective policies and measures were not implemented in the near future (IPCC, 2014). In this scenario, the average global temperature is expected to rise by over 4 °C above pre-industrial levels (IPCC, 2014). The transport sector is the third-largest source of CO_2 emissions in the world, ranking behind only the manufacturing sector and the electricity production sector in 2016 (Fig. 1.1).

According to the IEA's World Energy Outlook (WEO) in 2019, global transport energy consumption is projected to go up to 3327 Mtoe by 2030 in the Stated Policies Scenario, up 16.2% from 2018. Transport is a key sector for energy saving and emission reduction. Both developing an effective energy-saving means of transport and seeking scientific and reasonable carbon abatement policies are necessary choices for global sustainable transport development.

With the rapid development of China's economic reform, the imbalance between energy supply and demand, environmental pollution and ecological deterioration have become increasingly prominent. The growing transport sector is the main reason for rapidly increasing energy consumption and carbon emissions. In 2017, the final energy consumption by the transport sector in China was 325 Mtoe, which was over ten times the 1990 level, accounting for 15.8% of the total energy consumption, after energy consumption by the industry sector and residents (IEA, 2019). In China, energy saving and emission reduction in the transport sector has become a main approach to realize the harmonious development of energy, environment and economy. China pledges to peak carbon emissions by 2030 and achieve carbon neutrality before 2060 or sooner under the Paris Agreement to limit global warming to below 2 °C by the end of the century. To achieve carbon peaking and carbon neutrality goals, the transport sector should move towards low-carbon development as soon as possible and achieve net-zero carbon emission while building a transport power.

P. Zhao and L. Zeng, *Transport Efficiency and Safety in China*, Population, Regional Development and Transport, https://doi.org/10.1007/978-981-99-1055-7_1

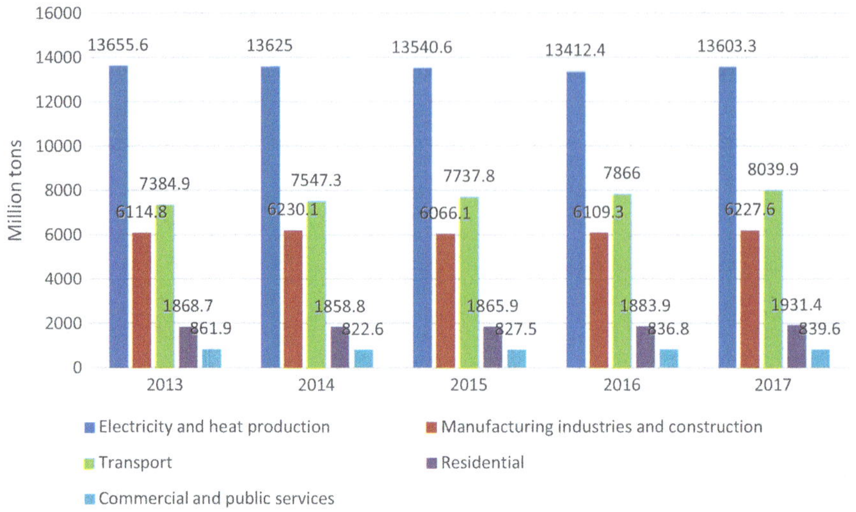

Fig. 1.1 CO_2 emissions from fuel combustion by sectors in the world from 2013 to 2017. *Data source* The authors, edited from CO_2 emissions from fuel combustion (IEA, 2015–2019)

In recent years, global road traffic fatalities have been rising. According to the "Global Status Report on Road Safety 2018", launched by the World Health Organization (WHO), there were 1.35 million road traffic deaths worldwide in 2016. Road traffic deaths are the eighth leading cause of death across all age groups (WHO, 2018). Without effective measures, road accidents will become the fifth leading cause of death worldwide (WHO, 2018). The report highlights that road accidents are the fifth leading cause of death for people aged 5 to 29 years, and that there is no reduction in the number of road traffic deaths in low-income countries, with road traffic safety being one of the most important public health issues worldwide (WHO, 2018).

Road accidents stand out as a threatening health problem. Without new or enhanced improvements, it is anticipated that with the increased use of vehicles, it will be the fifth leading cause of death by the end of 2030 (WHO, 2015).

With the implementation of "Road Traffic Safety Law of the People's Republic of China" in 2003, China has made significant achievements in transport safety. Data from the National Bureau of Statistics shows road traffic fatalities decreased from 93,853 in 2000 to 63,093 in 2016. However, at present road traffic safety in China leaves much to be desired, and there exist a few problems which are barely seen in European countries and the United States. For example, in 2016 the road traffic death rate per 10,000 vehicles was 3.4 in China, while the figure stood at 1.1, 0.6, and 0.8, respectively, in United States, Japan and Germany. China's overall road traffic safety needs to be improved.

"Road Safety Is No Accident" is the theme of World Health Day 2004. In October 2005, the WHO General Assembly adopted Resolution 60/5, which urged the member states of WHO to pay greater attention to road traffic injury prevention and invited the member states to set up the World Day of Remembrance for Road Traffic Victims that is on the third Sunday in November every year. In March 2010, the UN General Assembly adopted a draft resolution, proclaiming the period from 2005 to 2015 as the Decade of Action for Road Safety. The overall goal is to reduce transport accidents and casualties and save 5 million lives through carrying out more transport safety activities in countries and regions.

Since the reform and opening-up, China has made great achievements in transport construction. Transport infrastructure in China has experienced unprecedented development in the last decade due to strong government support for public infrastructure investment (Chen and Haynes, 2017). By the end of 2018, the length of railways in operation totaled 131,000 km, and electrified sections accounted for over 70%. The total length of highways reached 4,846,500 km, including 142,600 km of expressways, ranking first in the world. The length of navigable inland waterways in China amounted to 127,000 km. The number of civil airports in China reached 234 in 2018. Many of these indicators put China ahead of other countries in the world, and China has become a really big transport country. However, a transport power is not only embodied in the expansion of transport infrastructure, but also embodied in creating first-rate product techniques and services in comprehensive transport and infrastructure systems (Fig. 1.2).

Sustainable transport is the only way to realize the goal of building transport power. "Building China into a Country with a Strong transport Network" was released by the Central Committee of the Chinese Communist Party and the State Council of China in September 2019, which aims to raise its global competitiveness in the transport sector by setting up transport networks with wider coverage and higher speed, according to an official document. In the context of sustainable development,

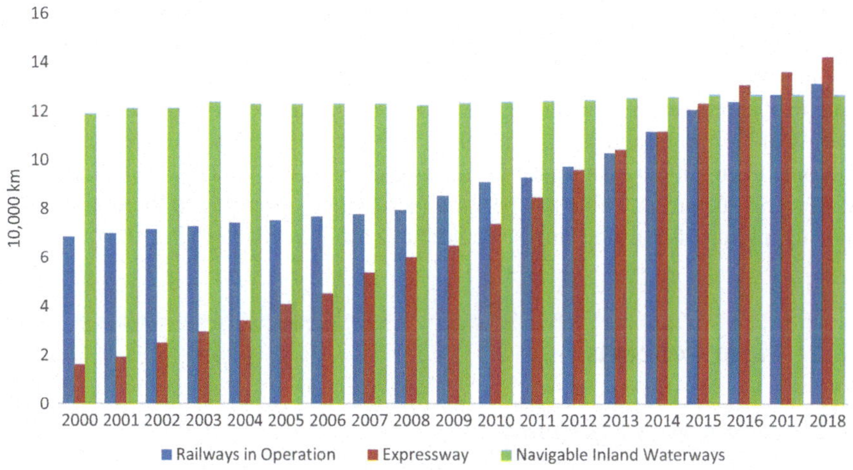

Fig. 1.2 Length of railways in operation, expressways, and navigable inland waterways in China from 2000 to 2018. *Data source* The authors, edited from China Statistical Yearbook (2019)

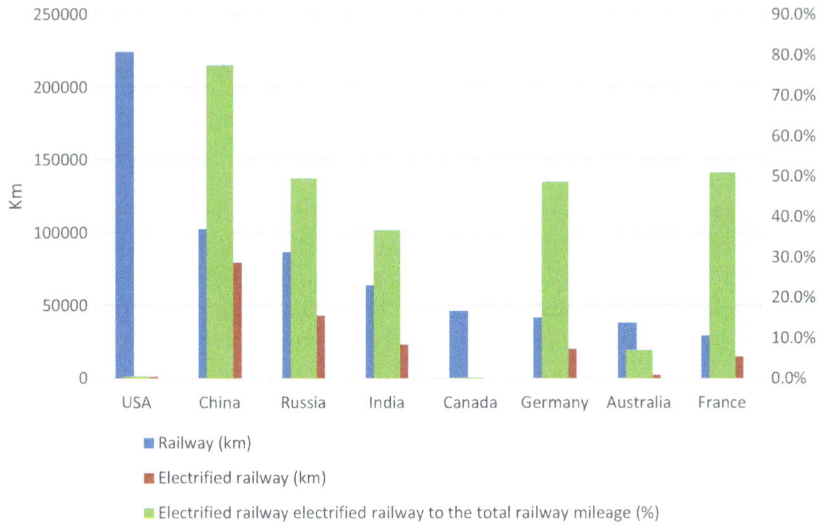

Fig. 1.3 Comparison of railway construction quality in major countries in 2016. *Data source* https://www.chyxx.com/industry/201709/559717.html

integrated transport systems aim to improve transport efficiency, reduce transport congestion and pollution, promote social equity, and save building maintenance costs.

After entering the twenty-first century, the transport sector in China has experienced healthy development. Transport network construction is in full swing based on the construction plans approved by China's State Council. For example, the total length of expressways increases at a speed of above 3000 km every year. Many construction plans of high-speed railways have come true. By the end of 2019, the length of electrified railways in operation reached 104,000 km, and its proportion to the total length of railways increased from 21.7% in 2000 to 74.3% in 2019; the length of expressways in operation reached 149,600 km, and its proportion to the total length of highways increased from 0.93% in 2000 to 3% in 2019. Expressways have become the main mode of transport for inter-city transport in China. Overall, transport infrastructure in China has become increasingly perfect, and the optimization and upgrading of transport infrastructure have been accelerated (Fig. 1.3).

1.2 Main Concepts

1.2.1 Transport Efficiency

The concept of "technical efficiency" was first put forward by Farrell in 1957. Technical efficiency relates the outputs (products or results) to the actual inputs (the resources consumed). This is the frontier of all production possibilities. This set summarizes all the technological possibilities of transforming inputs into outputs. A producer is technically inefficient, if his production is below the production possibility frontier. Technical efficiency indicates, for a given level of production, how a producer uses his resources in an optimal way (Coelli et al., 1998).

The technical efficiency of transport is usually called "transport efficiency". Different scholars have different definitions of transport efficiency; Costa and Markellos (1997) hold that transport efficiency is the ratio of a transport system's output to its input, and is an important measure to be assessed and monitored for governance and policymaking purposes. Kuang (2005) thinks that transport efficiency can be divided into three levels—macro, meso and micro—based on the role of transport production activities. The macro level refers to the overall technical efficiency of the transport industry in the national economy; the meso level refers to the allocation efficiency of transport resources among various transport sectors (such as highway, railway, aviation, and waterway transport); and the micro level refers to the technical efficiency of transport enterprises. Wu et al. (2013) divided transport efficiency into two levels: technical efficiency and economic efficiency. The former should be the integration of the technical efficiencies of various transport sectors and the allocation efficiencies of resources among different transport sectors. The latter refers to the proportion of the input of transport resources to the output in the form of economic currency, which focuses on the goal of traffic behavior, such as cost, tax or profit.

According to the above definitions of efficiency, in this research, transport efficiency can be understood from micro and macro perspectives. From the micro perspective, transport efficiency refers to the technical efficiency of traffic behavior, such as per unit of transport service output created by per unit of transport resource input. From the macro point of view, transport efficiency can be understood as the technical efficiency of the whole transport sector or specific transport departments (such as highway, railway, aviation, and waterway transport). In the following chapters, transport efficiency will be studied from the macro perspective. Transport efficiency is the ratio of a transport system's output to its input, and is an important measure to be assessed and monitored for governance and policymaking purposes (Costa & Markellos, 1997). A plethora of studies have constructed different measures to analyze the efficiency of various transport systems (Costa & Markellos, 1997, Fielding et al., 1978; Holmgren, 2013).

1.2.2 Transport Environmental Efficiency

The concept of "environmental efficiency" firstly appeared in 1992, defined as "ecological efficiency". It was mentioned by the World Business Council for Sustainable Development (WBCSD) in a business report on development and environment at the Earth Summit. The report points out that ecological efficiency can be used to evaluate the conformity between economic development and environmental protection. The concept of ecological efficiency emphasizes the creation of more goods and services while consuming fewer resources and producing less waste and pollution (Hinterberger & Schepelmann, 2001). This accords with the core idea of sustainable development, which fosters the harmonious development of the economy, resources, and environment. Reinhard and Knox Lovell (1999) and Montanar (2004) put forward the concept of environmental efficiency. Li et al., (2020) defined environmental efficiency as the ratio of the economic value added of production activities to the environmental pressure. It can be seen that ecological efficiency and environmental efficiency both aim at the minimization of environmental burden and the maximization of economic output, and emphasize that production activities can not only produce more and better output, but also reduce their negative impact on the environment to the greatest extent. At present, there is no strict distinction between the two concepts in academic circles, which needs further study.

In recognition of this, we define transport environmental efficiency (TEE) as comprehensive transport efficiency with which the transport sector achieves more transport outputs and creates much less pollution; the definition assumes a constant or decreasing input of the factors determining productivity in the transport sector (Cui & Li, 2015; Zhao et al., 2022).

1.3 Main Content of the Book

Transport efficiency, transport environmental efficiency and road transport safety in China are studied in this book. More specifically, using the theories of econometrics, input–output economics, and spatial economics, the book carries out a series of rigorous empirical analyses of transport efficiency in 31 provinces in mainland China from 2008 to 2016, transport environmental efficiency in 30 provinces in mainland China, excluding Tibet, from 2009 to 2016, and transport safety in 31 provinces in mainland China from 2003 to 2016. Therefore, based on the above research idea and logical framework, this book is divided into thirteen chapters.

This chapter is introduction. It mainly focuses on the research background, content, methods, conceptual framework, significance, and innovations of the book. Chapter 2 elaborates on the research methods. This chapter introduces the calculation method of transport efficiency and transport environmental efficiency, and the regression analysis methods for the empirical research on them. Chapter 3 introduces the research progress of transport efficiency. Chapter 4 is about the economic operation of the transport sector in China.

Chapters 5–9 present an empirical analysis of the transport efficiencies in highway, railway, airway, waterway, and urban transport sectors in China, respectively.

Chapter 10 contains an analysis of transport energy and climate change. In Sect. 10.1, transport energy consumption and climate change in the world are analyzed. Section 10.2 discusses energy consumption and CO_2 emissions in China. Chapter 11 gives an empirical analysis of transport environmental efficiency in China.

Chapter 12 presents an empirical analysis of transport safety in China. It first makes an overall analysis of road transport safety in China in Sect. 12.1, and then

Sect. 12.2 is mainly about the literature review on road transport safety. Section 12.3 analyzes the spatio-temporal characteristics of road traffic fatality rate in China, and Sect. 12.4 analyzes its influencing factors by establishing spatial econometric models. Section 12.5 reviews China's policies on road transport safety. In the end, the chapter gives policy proposals for promoting road transport safety.

1.4 Research Significance and Innovations of the Book

This book is of importance in the following aspects. Firstly, this book evaluates the input–output efficiency of the transport industry in China, and provides a reasonable reference for transport investment in the future. Transport investment can reduce the transport costs of residents, improve travel convenience and location accessibility, and bring about agglomeration and multiplier effects of economic production activities. In the 1990s, Aschauer (1990) carried out the research on the relationship between infrastructure investment and economic growth, and found that transport facilities have a significant positive impact on productivity improvement, and that the return rate of transport investment is highly elastic. Subsequently, Duffy-Deno and Eberts (1991) also confirmed the expansionary effect of transport investment on the economy.

In the early stage of the reform and opening-up in the 1980s and 1990s, compared with developed countries, China's transport investment still lagged behind as a whole, but the Chinese Government attached great importance to transport investment for a long time. Since the beginning of the twenty-first century, due to the needs of economic growth, social development and regional balance, China's investment in infrastructure, including transport, has increased rapidly. According to OECD statistics, in 2016, China's infrastructure expenditure was 2.5 times of the total infrastructure expenditure of G7 countries.

In 2020, the State Council of China issued the white paper "Sustainable Development of transport in China", which specified the transport development goals in 2035 and made it clear that China will continue to increase transport investment in the future. Therefore, in order to improve the efficiency of transport investment, it is necessary to comprehensively understand the recent transport input–output efficiency and evaluate the transport efficiency of various regions in China. That is why the in-depth analysis of the input–output efficiency of various transport departments has very important policy implications for China's comprehensive transport development.

Secondly, this book draws a clear map of energy use and environmental impacts of the transport system in China. With the development of the world economy, the environment and natural resources are severely damaged. Most developed countries experienced treatment after pollution and spent enormous amounts of money adjusting the relation between the economy and the environment. Over the years, many local governments in China have experienced this tortuous path. At present, environmental protection has received great attention in China; the protection of human life has become a top priority. The transport sector is one of the key sectors for energy conservation and emission reduction. Advocating a low-carbon transport system is way to achieve the sustainable development of China's future transport energy. In this era, the study of transport environmental efficiency is of great practical significance to the implementation of the basic national policy of energy conservation and emission reduction.

In the report on China's sustainable transport development, the Chinese government proposed to continuously optimize the freight transport structure through the policy of converting the bulk cargo transport mode from highway transport to railway transport or water transport. When analyzing the factors affecting transport environmental efficiency in this book, we take the transport freight structure as an important factor variable for regression analysis. We conclude that the larger the proportion of railway freight transport, the higher the level of transport environmental efficiency, which strongly supports the new policy of freight transport.

Thirdly, this book has important values for policy-making to improve global road traffic safety. The WHO launched the Decade of Action for Road Safety 2021–2030. The goal of this decade is to reduce the number of road traffic casualties by at least 50% by 2030. The road traffic deaths of China in 2018 ranked second in the world. Therefore, the improvement of China's traffic safety has a very significant impact on the realization of the goal of the Decade of Action for Road Safety 2021–2030. In the analysis of the influencing factors of road transport safety in this book, we construct a Tobit regression model to explore the socio-economic factors behind it, so as to provide macro traffic safety policies for traffic improvement in China.

This book also has important academic innovations. Firstly, the calculation methods of transport efficiency are innovative. The DEA and stochastic frontier analysis (SFA) methods are among the most widely used methods when measuring transport efficiency. SFA is a common parametric method, which can distinguish between noise and inefficiency while constructing the hypothesis test and confidence interval (Hjalmarsson et al., 1996). Some scholars use the SFA method to calculate technical efficiency in transport sectors or transport industries (Jarboui, 2016; Odeck, 2008). However, the SFA approach requires a prior specification on the functional form of the frontier, and the results of the efficiency calculation are easily influenced by subjective factors (Sun & Huang, 2021).

As an essential non-parametric measurement, DEA has two intrinsic advantages: It does not require a specified production function, and it is capable of calculating the relative efficiency of decision-making units (DMUs) with multiple inputs and outputs (Lampe & Hilgers, 2015). Therefore, the DEA method has been widely applied in calculating transport efficiency (Agarwal et al., 2010; Bandyopadhayay et al., 2016; Fitzová et al., 2018; Kabasakal et al., 2005). The DEA method can be divided into two basic models: radial and non-radial DEA models, such as CCR, BCC and SBM models. However, both radial and non-radial analysis models have inherent shortcomings. The main shortcoming of the radial model is that it neglects non-radial slacks. This can lead to a biased measure while evaluating the efficiency of the DMUs. Furthermore, radial models require input or output variables to change proportionally, which cannot cope with such cases properly (Tone & Tsutsui, 2010;

Zeng, 2022). In contrast, non-radial models directly capture the non-radial slacks not considered in radial models and may lose the original proportionality, which is inappropriate for efficiency analysis. Hence, it is necessary to compile the radial model and the non-radial model into a composite model to measure efficiency in a more reasonable way.

This study is different from others in two aspects: (1) It applies an epsilon-based measure (EBM) model to evaluate the transport efficiency of highways, railways, airways, and waterways across provinces of China. The EBM model is applied to evaluate transport efficiency, because it has the advantages of both radial and non-radial DEA methods, and can generate precise numerical results. (2) It utilizes the EBM DEA model with undesirable outputs to measure transport environmental efficiency at the provincial level; this method can address the defects of radial DEA and non-radial DEA, and consider undesirable outputs, which exhibit the potential for overestimation or underestimation of transport environmental efficiency (Yang et al., 2018; Zeng et al., 2022b). We can therefore obtain more accurate and scientific calculated results for transport environmental efficiency.

Secondly, the spatial Durbin model is adopted to analyze the influencing factors of transport efficiency (of highways and railways) and transport environmental efficiency. Many mathematical models have been extended to examine the interaction mechanism between transport efficiency and related determinants, mostly via a regression model. The regression can either be a non-spatial regression or a spatial regression. For non-spatial regression approaches, the Tobit regression (Alam et al., 2020; Asmild et al., 2009; Fitzová et al., 2018; Kutlar et al., 2013) and panel regression models (Karlaftis & McCarthy, 1998) are widely used in previous studies related to the influencing factors of transport efficiency. Everything is related to everything else, but near things are more related to each other (Tobler, 1970), even transport activities. However, the ubiquitous spatial effect, which mainly involves spatial dependence and spatial heterogeneity, was frequently overlooked by non-spatial regression approaches. The estimated coefficients of non-spatial econometric methods will be biased or inconsistent (Lesage & Pace, 2009) when independent variables have significant spatial correlation characteristics.

The panel Durbin model is adopted to analyze the influencing factors of transport efficiency (highway and railway transport sectors) and transport environmental efficiency, which takes spatial autocorrelation into account and makes the measurement system more accurate.

Thirdly, it reveals the major factors influencing road traffic safety. At present, scholars have embarked upon the study of micro-level influencing factors of road transport safety, and achieved very great success. Micro-level influencing factors mainly include traffic participants, vehicles, roads, climatic environment (Chang & Wang, 2006; Eboli et al., 2020; Kopelias et al., 2007), and traffic laws and regulations, and the goal is to put forward targeted improvement measures for specific traffic scenarios. In contrast, macro socio-economic factors are paid little attention to. Although some scholars' studies found that macro socio-economic factors have an impact on traffic accident casualties, there is no effective empirical analysis of the influencing factors of the road traffic mortality rate (RTMR) in China from a macro perspective.

Based on the panel data of 31 provinces in China from 2003 to 2018, we analyze the main influencing factors of the RTMR in China from a socio-economic perspective using a Tobit panel model. As the value of RTMR is limited in a certain range, the use of traditional regressions, such as ordinary least square, cannot lead to consistent estimators, due to truncated data. The Tobit regression model based on the principle of maximum likelihood estimation can effectively avoid problems such as inconsistency and bias in parameter estimation (Zeng et al., 2022a), and more reasonable parameter estimation results are obtained.

To sum up, the theoretical analyses and discussions in this book will enhance our existing knowledge of the change characteristics and driving factors of the transport system's efficiency in a context of rapid urbanization, industrialization and marketization in China. The findings of the existing policy evaluation will bring fresh evidence for transport policy performances to both scholars and politicians. In particular, it shows policymakers how to create an efficient transport system in order to save energy, reduce GHGs emissions and improve social security.

References

Agarwal, S., Yadav, S. P., & Singh, S. P. (2010). Based estimation of the technical efficiency of state transport undertakings in India. *Opsearch, 47*, 216–230. https://doi.org/10.1007/s12597-011-0035-4

Alam, K. M., Xuemei, L., Baig, S., et al. (2020). Analysis of technical, pure technical and scale efficiencies of Pakistan railways using data envelopment analysis and Tobit regression model. *Networks and Spatial Economics, 20*, 989–1014. https://doi.org/10.1007/s11067-020-09510-9

Aschauer, D. A. (1990). Why is infrastructure important? In A. H. Munnell (Ed.), *Is there a shortfall in public capital investment?* Federal Reserve Bank of Boston. https://xueshu.baidu.com/usercenter/paper/show?paperid=1c0v0v703p5808g0mg6m0eh03u352041&site=xueshu_se

Asmild, M., Holvad, T., & Hougaard, J. L., et al. (2009). Railway reforms: do they influence operating efficiency? *Transportation, 36*, 617–638. https://doi.org/10.1007/s11116-009-9216-x

Bandyopadhayay, A., Banerjee, A., & Abidi, N. R. A. (2016). Measuring routes efficiency of Kolkata bus transport: A modified DEA approach. *Annals of Data Science, 3*, 305. https://doi.org/10.1007/s40745-016-0074-z

Chang, L. Y., & Wang, H. W. (2006). Analysis of traffic injury severity: An application of non-parametric classification tree techniques. *Accident, Analysis and Prevention, 38*, 1019–1027. https://doi.org/10.1016/j.aap.2006.04.009

Chen, Z., & Haynes, K. E. (2017). Transport infrastructure and economic growth in China: A meta-analysis. In *Socioeconomic environmental policies and evaluations in regional science* (vol. 24). Springer. https://doi.org/10.1007/978-981-10-0099-7_18

China Statistical Yearbooks. (2019). China Statistical Publishing House. https://data.cnki.net/yearbook/Single/N2020100004

Coelli, T. J., Rao, D. S. P., Battese, G. E. (1998). *An introduction to efficiency and productivity analysis*. Kluwer Academic Publishers. SpringerLink. https://xueshu.baidu.com/usercenter/paper/show?paperid=9ffa0b479d531860a54c98b08e22f629

Costa, A., & Markellos., R. N. (1997). Evaluating public transport efficiency with neural network models. *Transportation Research Part C: Emerging Technologies, 5*(5), 301–312. https://doi.org/10.1016/S0968-090X(97)00017-X

Cui, Q., & Li, Y. (2015). An empirical study on the influencing factors of transport carbon efficiency: Evidences from fifteen countries. *Applied Energy, 141*, 209–217. https://doi.org/10.1016/j.apenergy.2014.12.040

Duffy-Deno, K. T., & Eberts, R. W. (1991). Public infrastructure and regional economic development: A simultaneous equation approach. *Journal of Urban Economics, 30*, 329–343. https://doi.org/10.1016/0094-1190(91)90053-A

Eboli, L., Forciniti, C., & Mazzulla, G. (2020). Factors influencing accident severity: an analysis by road accident type. *Transport Research Procedia, 47*, 449–456. https://doi.org/10.1016/j.trpro.2020.03.120

Fielding, G. J., Glauthier, R. E., & Lave, C. A. (1978). Performance indicators for transit management. *Transport, 7*, 365–379. https://doi.org/10.1007/BF00168037

Fitzová, H., Matulová, M., & Tomeš, Z. (2018). Determinants of urban public transport efficiency: Case study of the Czech Republic. *European Transport Research Review, 10*, 42. https://doi.org/10.1186/s12544-018-0311-y

Guido, F., Marc, I., & Cathering, V. (2009). Railway (De)Regulation: A European efficiency comparison. https://doi.org/10.1111/j.1468-0335.2008.00739.x

Hinterberger, F. K., & Schepelmann, F. (2001). *Eco-efficiency of regions: Toward reducing total material input*. Sustainable Europe Research Institute. Available online: https://www.researchgate.net/profile/Friedrich_Hinterberger/publication/228597679_Eco-Efficiency_of_Regions_Toward_Reducing_Total_Material_Input/links/00463519fb6a1d7570000000.pdf

Hjalmarsson, L., Kumbhakar, S. C., & Heshmati, A. (1996). DEA, DFA and SFA: A comparison. *Journal of Productivity Analysis, 7*, 303–327. https://doi.org/10.1007/BF00157046

Holmgren, J. (2013). The efficiency of public transport operations—An evaluation using stochastic frontier analysis. *Research in Transportation Economics, 39*(1), 50–57. https://doi.org/10.1016/j.retrec.2012.05.023

Intergovernmental Panel on Climate Change (IPCC). (2014). Working Group III—Mitigation of Climate Change, Chapter 8: Transport, 117. https://www.researchgate.net/publication/262069271_Intergovernmental_Panel_on_Climate_Change_Working_Group_III-_Mitigation_of_Climate_Change_Chapter_8_Transport

International Energy Agency (IEA). (2015–2019). *CO_2 emissions from fuel combustion (2015–2019).* https://www.oecd-ilibrary.org/fr/energy/co2-emissions-from-fuel-combustion_20783426

International Energy Agency (IEA). (2019). World Energy Outlook (WEO) 2019. https://doi.org/10.1787/caf32f3b-en

Jarboui, S. (2016). Managerial psychology and transport firms efficiency: A stochastic frontier analysis. *Review of Managerial Science, 10*, 365–379. https://doi.org/10.1007/s11846-014-0149-1

Kabasakal, A., Kutlar, A., & Sarikaya, M. (2005). Efficiency determinations of the worldwide railway companies via DEA and contributions of the outputs to the efficiency and TFP by panel regression. *Central European Journal of Operations Research, 23*, 69–88. https://doi.org/10.1007/s10100-013-0303-x

Karlaftis, M. G., & McCarthy, P. (1998). Operating subsidies and performance in public transit: An empirical study. *Transport Research Part A: Policy and Practice, 32*(5), 359–375. https://doi.org/10.1016/S0965-8564(98)00002-0

Kopelias, P., Papadimitriou, F., Papandreou, K., et al. (2007). Urban freeway crash analysis. *Transportation Research Board, 2015*, 123–131. https://doi.org/10.3141/2015-14

Kuang, M. (2005). *On the transport efficiency: The theory and practice of optimal allocation of transport resources.* China Railway Publishing House.

Kutlar, A., Kabasakal, A., & Sarikaya, M. (2013). Determination of the efficiency of the world railway companies by method of DEA and comparison of their efficiency by Tobit analysis. *Quality & Quantity, 47*, 3575–3602. https://doi.org/10.1007/s11135-012-9741-0

Lampe, H. W., & Hilgers, D. (2015). Trajectories of efficiency measurement: A bibliometric analysis of DEA and SFA. *European Journal of Operational Research, 240*(1), 1–21. https://doi.org/10.1016/j.ejor.2014.04.041

LeSage, J., & Pace, R. K. (2009). *Introduction to spatial econometrics*. CRC Press. https://doi.org/10.1007/BF03354894

Li, D. S., Zhao, Y. W., & Li, L. Y. C. (2020). Change of environmental efficiency and environmental productivity of coal cities: Based on panel data of 11 cities in Shanxi province. *Journal of Natural Resources, 36*(3), 618–633. https://t.cnki.net/kcms/detail?v=3uoqIhG8C46NmWw7YpEsKMypi3qVj28LEUDxQXHYyS3fbyEOrIfBzPW0Yvxeh1vAHoPsaXmTaqUf5LZbnoAoESz_jWgj4Hxd&uniplatform=NZKPT

Montanari, R. (2004). Environmental efficiency analysis for enel thermo-power plants. *Journal of Cleaner Production, 12*, 403–414. https://doi.org/10.1016/S0959-6526(03)00015-5

Odeck, J. (2008). Efficiency measurement of ferries serving road networks in Norway: A stochastic frontier approach. *Maritime Economics and Logistics, 10*, 409–428. https://doi.org/10.1057/mel.2008.14

Reinhard, C. A., & Knox Lovell, G. T. (1999). Econometric estimation of technical and environmental efficiency: An application to Dutch dairy farms. *American Journal of Agricultural Economics, Agricultural and Applied Economics Association, 81*(1), 44–60. https://ideas.repec.org/a/oup/ajagec/v81y1999i1p44-60.html

Sun, S. F., & Huang, C. C. (2021). Energy structure evaluation and optimization in BRICS: A dynamic analysis based on a slack based measurement DEA with undesirable outputs. *Energy, 216*, 119251. https://doi.org/10.1016/j.energy.2020.119251

Tobler, W. R. (1970). A computer movie simulating urban growth in the Detroit region. *Economic Geography, 46*, 234–240. https://www.jstor.org/stable/143141

Tone, K., & Tsutsui, M. (2010). An epsilon-based measure of efficiency in DEA—A third pole of technical efficiency. *European Journal of Operational Research, 207*, 1554–1563. https://doi.org/10.1016/j.ejor.2010.07.014

World Health Organization (WHO). (2015). *Global status report on road safety 2018*. WHO Press. https://www.who.int/roadsafety/events/2015/Appendix_14.pdf

World Health Organization (WHO). (2018). *Global status report on road safety*. WHO Press. https://www.who.int/publications/i/item/9789241565684

Wu W, Cao Y H, Liang, S.B. (2013). Review of transport efficiency and its research trends from the transport geography perspectives. *Progress in Geography, 32*(2), 243–250. http://www.progressingeography.com/CN/abstract/abstract13661.shtml

Yang, L., Wang, K. L., & Geng, J. C. (2018). China's regional ecological energy efficiency and energy saving and pollution abatement potentials: An empirical analysis using epsilon-based measure model. *Journal of Cleaner Production, 194*, 300–308. https://doi.org/10.1016/j.jclepro.2018.05.129

Zeng, L. (2022). The driving mechanism of urban land green use efficiency in China based on the EBM model with undesirable outputs and the Spatial Durbin model. *International Journal of Environmental Research and Public Health, 2022*(19), 10748. https://doi.org/10.3390/ijerph191710748

Zeng, L., Li, H., Lao, X., Hu, H., Wei, Y., Li, C., Yuan, X., Guo, D., & Liu, K. (2022a). China's road traffic mortality rate and its empirical research from socio-economic factors based on the Tobit model. *Systems, 10*, 122. https://doi.org/10.3390/systems10040122

Zeng, L., Li, P., Yu, Z., Nie, Y., Li, S., Gao, G., & Huang, D. (2022b). Spatiotemporal characteristics and influencing factors of water resources' green utilization efficiency in China: Based on the EBM model with undesirable outputs and SDM model. *Water, 2022*(14), 2908. https://doi.org/10.3390/w14182908

Zhao, P., Zeng, L., Li, P., et al. (2022). China's transport sector carbon dioxide emissions efficiency and its influencing factors based on the EBM DEA model with undesirable outputs and Spatial Durbin model. *Energy, 238*, 121934. https://doi.org/10.1016/j.energy.2021.121934

Chapter 2
Research Methods

2.1 The Calculation Method of Transport Efficiency

This book applies the EBM model to calculate transport efficiency (in highway and railway transport sectors). In this section we introduce CCR and SBM models as representative radial and non-radial measures of efficiency respectively and point out their shortcomings, and then introduce the advantages of the EBM method.

There are N decision-making units (DMUs) ($j = 1, \ldots, N$), which have M inputs ($i = 1, \ldots, M$) and S outputs ($i = 1, \ldots, S$). The input matrices are denoted by $X = \{x_{ij}\} \in R^{M \times N}$, and in response to this the output matrices are denoted by $Y = \{y_{ij}\} \in R^{S \times N}$. The method assumes $X > 0$ and $Y > 0$. N, S, M stand for the number of DMUs, the outputs and the inputs, respectively. S^- is the input slacks, x_{ij} and y_{rj} stand for the ith input and the rth output of DMU_j, and k is the intensity variable. k represents the intensity vector.

The radial input-oriented CCR model for measuring the efficiency γ^* of DMU_0 can be expressed in the form (Tone & Tsutsui, 2010; Yang et al., 2018):

$$[\text{CCR - I}]$$

$$\gamma^* = \min_{\gamma^*, \lambda, S^-} \gamma \tag{2.1}$$

$$s.t. \begin{cases} \lambda x_o = X\lambda + S^- \\ y_o \leq Y\lambda \\ \lambda \geq S_i^- \geq 0 \end{cases} \tag{2.2}$$

For solving [CCR-I], we shall first solve [CCR-I] and obtain γ^* (weak efficiency), and then, we maximize $\sum_i^m \frac{S_i^-}{xio}$ in terms of λ and S^-, subject to (2.2) and $\gamma = \gamma^*$. The CCR model has its own problem because it neglects non-radial slacks S^- in the efficiency score γ^*. In solving this problem, Tone (2001) proposed slacks-based measure (SBM) models. The formula of the SBM model under the constant-returns-to-scale assumption is as follows:

P. Zhao and L. Zeng, *Transport Efficiency and Safety in China*, Population, Regional Development and Transport, https://doi.org/10.1007/978-981-99-1055-7_2

$$\kappa = \min\left(1 - \frac{1}{M}\sum_{i=1}^{M}\frac{S_i^-}{x_{io}}\right);$$

$$s.t.\begin{cases} \sum_{j=1}^{N} \lambda j x_{ij} = x_{io} - S_i^- \quad i = 1, 2, \ldots, M \\ \sum_{j=1}^{N} \lambda j y_{ij} \geq y_{io} \quad i = 1, 2, \ldots, S \\ \lambda j \geq 0(\forall j), S_i^- \geq 0(\forall i) \end{cases} \tag{2.3}$$

where κ^* is the efficiency of the non-radial input-oriented SBM model. The non-radial input slacks vector is $S^- = \left(S_1^-, \cdots, S_M^-\right)^T$. Let an optimal parameter of [SBM-I-C] be (λ^*, S^{-*}), the objective function can also be simplified as

$$\kappa^* = \frac{1}{M}\sum_{i=1}^{M}\frac{x_{io} - S_i^-}{x_{io}} \tag{2.4}$$

where the SBM score κ^* is the average of the component-wise reduction rates which may vary from one input to another. The major drawback of the BCC model is the neglect of the radial factors. In the BCC model, S^- is not necessarily proportional to x_0. The projected DMU may lose the proportionality in the original (Cui & Li, 2015).

As the previous section illustrated, both CCR and SBM models have their pros and cons. Tone and Tsutsui (2010) thus proposed the EBM model, which can combine radial and non-radial DEA models in a unified framework. Based on the assumption of constant returns to scale, the input-oriented EBA model is:

[EBM - I - C]

$$\beta^* = \min_{\gamma,\lambda,S^-}\left(\gamma - \varepsilon x \sum_{i=1}^{M}\frac{w_i^- S_i^-}{x_{io}}\right) \tag{2.5}$$

$$s.t\begin{cases} \gamma \mathrm{x}_{\mathrm{o}} - \mathrm{X}\lambda - \mathrm{S}^- = 0 \\ \mathrm{Y}\lambda \geq \mathrm{y}_{\mathrm{o}} \\ \lambda \geq 0 \\ S^- \geq 0 \end{cases} \tag{2.6}$$

[Dual]

$$\beta^* = \max_{\mathrm{v,u}} \mathrm{uyo} \tag{2.7}$$

$$s.t\begin{cases} \mathrm{vxo} = 1 \\ -\mathrm{vX} + \mathrm{uY} \leq 0 \\ vi \geq \frac{\varepsilon x w_i^-}{xio}(i = 1, ..., m) \\ \mathrm{u} \geq 0 \end{cases} \tag{2.8}$$

Here w_i stands for the relative importance of input i and satisfies $\sum w_i^{-1} = 1(\forall i w_i^- \geq 0)$. ε_x is a parameter which combines the radial γ^* and non-radial slacks terms. $w^- = (w_1^-, \ldots, w_M^-)$ and ε_x must be determined in advance before measuring the efficiency. From the term $\frac{w_i^- S_i^-}{x_{io}}$ in Eq. (2.5), $\frac{S_i^-}{x_{io}}$ is units-invariant and so w_i^- should be a units-invariant value reflecting the relative importance of resource i. The succeeding discussion about this subject refers to the study of Tone and Tsutsui (2010).

2.2 The Calculation Method of Transport Environmental Efficiency

Because Tone and Tsutsui's EBM did not consider any undesirable factors, Li and Chiu et al. combined the EBM-DEA and an undesirable factor into the EBM model with undesirable outputs. The book applies it for the evaluation of transport environmental efficiency of 30 provinces across mainland China from 2009 to 2016.

Suppose n DMU$_k$ ($k = 1, 2, \ldots, $ n) and m type inputs X_j $(X_{1j}, X_{2j}, \ldots, X_{mj})$ can produce s type outputs Y$_j$ $(Y_{1j}, Y_{2j}, \ldots, Y_{mj})$. Li and Chiu et al.'s non-oriented EBM model with undesirable outputs is as follows:

$$\pi^* = \min_{0,\lambda,s^-,s^g,s^b} \left(\frac{\theta - \varepsilon x \sum_{i=1}^{m} \frac{\omega_i^- s_i^-}{x_{i0}}}{\zeta + \varepsilon y \sum_{i=1}^{s1} \frac{\omega_i^{+s1} s_i^g}{y_{i0}} + \varepsilon y \sum_{i=1}^{s2} \frac{\omega_i^{-s2} s_i^b}{y_{i0}}} \right)$$

$$s.t \begin{cases} X\lambda - \theta X_0 + s^- = 0 \\ Y^g\lambda - \zeta Y_0 - s^g = 0 \\ Z^b\lambda - \zeta Y_0 - s^b = 0 \\ \lambda_1 + \cdots + \lambda_n = 1 \\ \lambda \geq 0, s^- \geq 0, s^g \geq 0, s^b \geq 0, \theta \geq 1, \theta \leq 1 \end{cases} \tag{2.9}$$

where s_i^- stands for the slack variable of input; s^g and s^b are the slacks of desired output and undesired output, respectively; and w_i^- is the weight of input i, which satisfies $\sum \omega_i^{-1} = 1(\forall i \omega_i^- \geq 0)$. w_i^{+s1} and w_i^{-s2} indicate the weights of the desired output i and the undesired output i, respectively, which satisfy $\sum \omega_i^{+s1} + \sum \omega_i^{-s2} = 1(\forall i \omega_i^+ \geq 0)$. ε_x represents the combination of radial θ and non-radial slack, and ε_y denotes the combination of radial φ and non-radial slack. γ^*, which is the optimal solution in the EBM model, stands for the efficiency value of the DMU. With the value range between 0 and 1, the DMU is in the efficient state (if $\gamma^* = 1$) or the inefficient state (if $\gamma^* < 1$). An inefficient DMU can reach the production frontier by reducing inputs and undesirable outputs or expanding desirable outputs.

Using the Charnes-Cooper transformation method, (2.9) is transformed into linear programming, as shown in (2.10) (Wu et al., 2019):

$$\upsilon^* = \min\left(\frac{\varphi - \varepsilon x\sum_{i=1}^{m}\frac{\omega_i^- s_i^-}{xi0}}{\psi + \varepsilon y\sum_{i=1}^{s1}\frac{\omega_i^{+s1} s_i^g}{y_{i0}} + \varepsilon y\sum_{i=1}^{s2}\frac{\omega_i^{-s2} s_i^b}{y_{i0}}}\right)$$

$$s.t\begin{cases} X\eta - \varphi X_0 + s^- = 0 \\ Y^g\eta - \psi Y_0 - s^g = 0 \\ Y^b\eta - \psi Y_0 - s^b = 0 \\ \lambda_1 + \cdots + \lambda_n = 1 \\ \lambda \geq 0, s^- \geq 0, s^g \geq 0, s^b \geq 0, \geq 1, \theta \leq 1 \end{cases} \tag{2.10}$$

The best solution for this formula is $(t^*, \upsilon^*, \psi^*, \varphi^*, \eta*, s^{-*}, s^{*+g}, s^{*-b},)$ and the optimal solution for the EBM is defined as:

$$\begin{aligned} \pi^* = \upsilon^*, \theta^* = \varphi^*/t^*, \lambda^* = \eta^*/t^*, s^{-*} = S^{-*}/t^* \\ s^{*+g} = S^{*+g}/t*, s^{*-b} = S^{*-b}/t^* \end{aligned} \tag{2.11}$$

$\pi^* = 1$ indicates that the non-oriented EBM is the most efficient and the DMU is the most efficient; if other DMUs seek to attain the highest efficiency, the following adjustments are needed:

$$\begin{aligned} X_0^* &= X\lambda^* = \theta^* X_0 - s^{-*} \\ Y_0^{*+g} &= X^{*+g}\lambda^* = \eta^* y_0 + s^{+g} \\ X_0^{*-b} &= X^{*-b}\lambda^* = \eta^* y_0 - s^{-b} \end{aligned} \tag{2.12}$$

2.3 Regression Analysis Method

2.3.1 *Spatial Autocorrelation Analysis*

According to the First Law of Geography, everything is related to everything else, but near things are more related than distant things (Tobler, 1970). Global spatial autocorrelation indexed by Moran's I is commonly used to measure the degree of general spatial clustering of attribute variables, and it is calculated as follows:

$$It = \frac{N\sum_{i=1}^{N}\sum_{j=1}^{N} W_{i,j}\left(X_{i,t} - \overline{X}_t\right)\left(X_{j,t} - \overline{X}_t\right)}{\sum_{i=1}^{N}\sum_{j=1}^{N} W_{i,j}\left[\sum_{i=1}^{N}\left(X_{i,t} - \overline{X}_t\right)^2\right]} \tag{2.13}$$

In Eq. (2.13), $X_{i,t}$ indicates the value of X for province i in year t. $\overline{X}$ represents the average of X for all provinces in year t. It indicates the value of Global Moran's I in year t, and the scale ranges from − 1 to 1. If the result is greater than 0 and reaches a significant level, the TLEE has an obviously positive spatial correlation. If the value is less than 0 and reaches a significant level, a negative spatial correlation is implied. If the value is 0 or if the significance test fails, no spatial relationship is shown. N identifies the total number of provinces. W is the spatial matrix. i and j represent two adjacent provinces. If province i borders on j, $W_{i,j}$ equals 1; otherwise, $W_{i,j}$ equals 0.

The significance of Global Moran's I can be examined by Z-tests, which is calculated as follows:

$$Z = \frac{I - E(I)}{\sqrt{Var(I)}} \tag{2.14}$$

where E(I) and V(I) stand for the expected value and variance of Global Moran's I, respectively. $E(I) = \frac{-1}{N-1}$ and $V(I) = E(I^2) - E(I)^2$. If $Z_I \geq 1.96$ and $p < 0.05$, there is a significant positive spatial autocorrelation.

2.3.2 *The Spatial Durbin Model (SDM)*

This book analyzes the influencing factors of transport efficiency (in highway and railway transport sectors) and transport environmental efficiency in China with the spatial Durbin model (SDM). In this section we introduce SLM and SEM models, summarize their shortcomings, and then point out the rationality in applying SDM.

As the spatial lag model (SLM) is similar to the spatial autoregressive model (SAR) in time series, SLM is also called SAR, which is expressed as follows (Zhao et al., 2017):

$$Y = \rho WY + \beta X + \varepsilon \tag{2.15}$$

where *Y* represents the dependent variable. ρ stands for the spatial autocorrelation coefficient, and *W* is the spatial weight matrix, which denotes the adjacent spatial relationship between regions. Additionally, *WY* indicates the spatial lag of the dependent variable. Moreover, β denotes the spatial regression coefficient, and *X* represents the matrix of control variables. Finally, ε indicates random errors and belongs to *N* (0, $\sigma^2 I_N$). In the SLM, the dependent variable (*WY*) is only explained by a spatially lagged explained variable, and is incapable of analyzing the influence of explanatory variables in the adjoining areas (Manski, 1993).

The spatial error model (SEM) is given as:

$$\begin{aligned} Y &= \beta X + \varepsilon \\ \varepsilon &= \lambda W\varepsilon + \mu \end{aligned} \tag{2.16}$$

where λ is the spatial error coefficient of the dependent variable vector, and μ stands for the random error vector of normal distribution. SEM only emphasizes the spatially autocorrelated error term, and neglects the spatial lag of the explained variable (Manski, 1993). All other variables in Eq. (2.15) have the meaning defined in Eq. (2.15).

The spatial Durbin model (SDM) can test the influence of explanatory variables in local areas on explained variables, as well as both explained variables and explanatory variables in neighboring areas (Lesage & Pace, 2009; Yu et al., 2013; Zeng et al., 2022). The basic form of SDM can be described as follows:

$$Y = aI_N + \rho WY + \beta X + \theta WX + \varepsilon \tag{2.17}$$

where *WX* indicates the spatial lag of the spatially lagged explanatory variables; *a*, θ and β are vectors of regression coefficient estimates (Yu et al., 2013); I_N is an N-order identity matrix. All other variables in Eq. (2.18) have the same connotation in Eqs. (2.15) and (2.16). The SDM not only allows one to capture the spatial correlations between dependent variables (*WY*) and spatial spillover effects of independent variables (*WX*), but also considers the interactive effects of endogenic, exogenic, and autocorrelated terms. If the lag effect from the independent variables (exogenous interaction) is not present, SDM will degenerate into SLM. If the lag effect from the dependent variables (endogenous interaction) is not present, SDM will not be simplified to SEM (Lesage & Pace, 2009; Wang et al., 2019).

References

Cui, Q., & Li, Y. (2015). An empirical study on the influencing factors of transportation carbon efficiency: evidences from fifteen countries. *Applied Energy, 141*, 209–217. https://doi.org/10.1016/j.apenergy.2014.12.040

Lesage, J. P., & Pace, P. K. (2009). *Introduction to spatial econometrics*. CRC Press Taylor & Francis Group. https://doi.org/10.1007/BF03354894. Accessed November 7, 2019.

Manski, C. F. (1993). Identification of endogenous social effects: The reflection problem. *Review of Economic Studies, 60*(3), 531–554. https://doi.org/10.2307/2298123

Tobler, W. R. (1970). A computer movie simulating urban growth in the Detroit region. *Economic Geography, 46*, 234–240. https://www.jstor.org/stable/143141

Tone, K. (2001). Slacks-based measure of efficiency in data envelopment analysis. *European Journal of Operational Research, 130*(3), 498–509. https://doi.org/10.1016/S0377-2217(99)00407-5

Tone, K., & Tsutsui, M. (2010). An epsilon-based measure of efficiency in DEA: a third pole of technical efficiency. *European Journal of Operational Research, 207*, 1554–1563. https://doi.org/10.1016/j.ejor.2010.07.014

Wang, L., Huang, J. X., Cai, H. Y., Liu, H. Z., Lu, J. M., & Yang, L. S. (2019). A study of the socioeconomic factors influencing migration in Russia. *Sustainability, 11*(6), 1650–1663. https://doi.org/10.3390/su11061650

Wu, P., Wang Y. Q., Chui, Y. H., Li, Y., & Lin, T. Y. (2019). Production efficiency and geographical location of Chinese coal enterprises-undesirable EBM DEA. *Resource Policy, 64*, 101527. https://doi.org/10.1016/j.resourpol.2019.101527

Yang, L., Wang, K. L., & Geng, J. C. (2018). China's regional ecological energy efficiency and energy saving and pollution abatement potentials: An empirical analysis using epsilon-based measure model. *Journal of Cleaner Production, 194*, 300–308. https://doi.org/10.1016/j.jclepro.2018.05.129

Yu, N. N., Jong, M. D., Storm, S., & Mi, J. N. (2013). Spatial spillover effects of transport infrastructure: Evidence from Chinese regions. *Journal of Transport Geography, 28*, 56–66. https://doi.org/10.1016/j.jtrangeo.2012.10.009

Zhao, L. S., Sun, C. Z., & Liu, F. C. (2017). Interprovincial two-stage water resource utilization efficiency under environmental constraint and spatial spillover effects in China. *Journal of Cleaner Production, 164*, 715–725. https://doi.org/10.1016/j.jclepro.2017.06.252

Zeng, L., Li, C., Liang, Z., Zhao, X., Hu, H., Wang, X., Yuan, D., Yu, Z., Yang, T., Lu, J., Huang, Q., & Qu, F. (2022). The carbon emission intensity of industrial land in China: Spatiotemporal characteristics and driving factors. *Land, 11*, 1156. https://doi.org/10.3390/land11081156

Chapter 3
Research Progress of Transport Efficiency

3.1 Measurements Methods

In the literature, we can find different methods for measuring efficiency of the transport sector. The typical methodologies for transport efficiency assessment include index analysis method, stochastic frontier analysis (SFA), and Data envelopment analysis (DEA). The index analysis method computes the ratio of transport-related indicators. DEA can be defined as a nonparametric method to measure the efficiency of the decision-making unit (DMU) with multiple inputs and multiple outputs. SFA, as an alternative approach to DEA, supposes a parametric function between inputs and outputs. In the following, we give a brief description of each approach.

3.1.1 Index Analysis Method

Typical index analysis methods include single index analysis and multi-index analysis. The single index method computes the ratio of certain transport outcomes, such as passengers transported, to certain resources utilized in generating the outcomes such as the number of vehicles. Fielding et al. (1978) applied the multi-index method to evaluate the transport efficiency of California's urban bus system from the aspects of efficiency and benefits, respectively. Martland (1997) used the single index method to evaluate railway transport efficiency from 1966 to 1995 in the United States. Barrett (1995) used the multi-index method to evaluate the production efficiency of Irish railways. Waters et al. (1997) also used the multi-index method to analyze the total factor productivity of railways in Canada from 1956 to 1991.

Wu (2001) constructed the evaluation index system of the transport efficiency by six index layers: transport labor productivity, transport production technology, transport, transport resources utilization transport environmental impact, and transport safety. The index analysis is most widely used due to its simplicity of implementation

P. Zhao and L. Zeng, *Transport Efficiency and Safety in China*, Population, Regional Development and Transport, https://doi.org/10.1007/978-981-99-1055-7_3

and ease of interpretation. However, it has been criticized for generating inconsistent results.

3.1.2 Data Envelopment Analysis (DEA)

DEA was first put forward by Charnes et al. in 1978. Therefore, the original DEA model was also called the Charnes-Cooper-Rhodes (CCR) model. In contrast to parametric methods, DEA is a non-parametric approach, which has superiority in estimating the validity of multiple inputs and multiple outputs, and does not need any presupposition about the production functional form on the frontier (Dong et al., 2014). Hence, the DEA method has been widely used in many domains, particularly in the evaluation of energy, ecological, and environmental efficiency.

In 1983 Banker et al. developed a modified DEA method for realizing the separation of technical and scale efficiencies, which is called the Banker-Charnes-Cooper (BCC) model. Both CCR and BCC models are based on the proportional reduction (enlargement) of input (output) vectors and neglect slacks (Tone, 2001), which have been extensively adopted to evaluate the efficiency of the transport sectors in recent years. Kutlar et al. (2013) used CCR and BCC models to analyze the technical efficiency and allocation efficiency of 31 freight railway companies and passenger railway companies around the world from 2000 to 2019. Cowie (1999) used the BCC model to calculate the technical efficiency, management efficiency and organizational efficiency of Spanish private railway company and state-owned railway company. Chapin and Schmidt (1999) used the BCC model to evaluate the production efficiency of American railway companies. The main shortcoming of CCR and BCC models is that they require the total input to be proportional to the total output. This neglects the effects of non-radial slacks and cannot realize the decomposition of factors in evaluating efficiency (Tone, 2001). To resolve their shortcoming, Tone (2001) proposed a non-radial DEA method, namely the SBM model, which can directly handle input or output slacks and eliminate the radial and oriented deviation. The SBM model has been widely used to calculate transport sustainable efficiency later. For example, Bai et al. (2020) builds a dynamic network SBM model to measure the efficiency value of China's railway passenger transport sector.

3.1.3 Stochastic Frontier Analysis (SFA)

Stochastic frontier analysis (SFA) was proposed by Aigner et al. (1977) and Meeusen and van den Broeck (1977). A positive characteristic of SFA is that it allows for the decomposition of the residual in two terms: the statistical noise and the inefficiency effect. Thus, SFA outputs a stochastic frontier (Silva et al., 2017). The SFA method is widely used in the field of transport efficiency calculation because it can overcome the errors caused by stochastic errors in deterministic models. Gathon and Pestieau

(1995) established a logarithmic stochastic frontier production model to measure the technical efficiency of 19 European railway companies from 1961 to 1988. Cantos and Maudos (2000) used the SFA approach to calculate the productivity, efficiency, and technical change for 15 European railways, concluding that the companies with higher degrees of autonomy have the higher efficiency. Cantos and Maudos (2001) calculated the both cost efficiency and revenue efficiency of 16 European railways by SFA method, and they agreed that the existence of inefficiency could cause by the strong policy of regulation and intervention. However, the shortcomings of the SFA method are also obvious.. However, the shortcomings of the SFA method are also obvious. It requires a specific functional form a priori, which determines the shape of the efficient frontier, and a probability distribution for the DMU's efficiency levels. The selected functional form introduces inductive bias in the stochastic process and may lead to severe degradation of the results when the shape is not consistent with the data (Silva et al., 2017).

In a word, the DEA method is a nonparametric method, which can evaluate the technical efficiency of multi-input and multi-output DMUs simultaneously, and there is no need to preset a strict production function. As for the SFA method, the advantage is that it eliminates the influence of external environmental factors on performance evaluation. The disadvantage of SFA is that it can only consider the situation of multiple inputs and a single output, and that a strict production function needs to be preset. In a word, DEA and SFA have their own advantages and disadvantages.

3.2 Influencing Factors

To sum up, the research mainly focuses on the impact of ownership, financial subsidies, management and operation, transport scale, external environment, urbanization and other factors on transport efficiency.

3.2.1 Privatization Policy

Most empirical studies show that privatization policy has a significant impact on transport efficiency. However, scholars have not yet reached a consensus on the impact of privatization policy on transport efficiency. Some scholars believe that privatization policy can promote transport efficiency. For example, Gathon and Pestieau (1995) found that the giving of management autonomy to state-owned railway enterprises will improve the operational efficiency of railway transport enterprises. Oum et al. (2008) found that airport privatization is conducive to improving the operating efficiency of civil airports. Cantos et al. (1999) believes that enterprise autonomy has positive effects on the efficiency of railway transport enterprises. Cowie et al. (2003) analyzed the influencing factors of railway transport efficiency, and found that privatization has a positive impact on the improvement of railway transport efficiency. Cowie (2002) found that deregulation and privatization have improved the efficiency of firms, and that productivity and efficiency are higher in the private sector than in the public sector. According to Roy (2007), private operators surpass public ones via cutting-edge techniques; this result was confirmed later by Sakai and Takahashi (2013) in the Japanese context. Other scholars, such as Perry and Babitsky (1986), Kerstens (1996) and Cullinane et al. (2006), have also verified the positive impact of private ownership on transport efficiency.

But some scholars are skeptical about privatization policy. Vicente and Lourdes (2001) found that the technical efficiency of the private transport sector is basically equivalent to that of the public transport sector, and that the privatization policy of the transport industry should be treated with caution. Else and James (1995) studied the impact of market structure on railway transport efficiency, and considered that the privatization policy in the railway sector should be treated with caution. Pina and Torres (2001) analyzed the efficiency of urban transport services in Spain, and concluded that privatization has a bad impact on the efficiency of urban public transport services. Dalen and Gómez-Lobo (2003) believe that there is no significant difference between public and private firms in technical efficiency.

3.2.2 Management Modes

Advanced and scientific transport production and management modes play a significant role in promoting transport efficiency. Sampaio et al. (2008) made a comparative analysis of the public transport systems in Europe and Brazil. He believes that the adoption of flexible tax policies and equal management organizations will improve the efficiency of the public transport system. Cantos et al. (1999) concluded that management specialization is of positive significance for the efficiency of railway transport enterprises. Li (2008) believes that the levels of railway operation and management are important factors significantly affecting the efficiency of railway production. Gao et al. (2011) believes that strengthening management will improve enterprise efficiency.

Multimodal transport refers to the flexible combination of different modes of transport, which is an important mode of transport management. Xiong et al. (2006) and Ye et al. (2019) believe that the development of intermodal transport is an important measure to promote transport efficiency. Ji and Yan (2014) believe that the Intermodal Surface transport Efficiency Act of 1991, a United States law, can reduce freight transport costs and improve logistics efficiency.

3.2.3 Technical Progress

With the continuous development of science and technology, technical progress plays an increasingly important role in the field of transport. Therefore, the technical factors used in a transport system is also one of the important factors affecting its efficiency. With the application of information technology in urban transport, means of transport are becoming more and more diversified, comfortable and convenient. The higher the degree of informatization, the more easily traffic congestion can be solved. transport efficiency can thus be improved. Fujdiak et al. (1995) believes that an intelligent transport management system can bring smoother transport for the city and solve the existing urban transport problems of high emissions, long traffic delays and a high traffic accident rate. Stathopoulos and Karlaftis (2003) thinks that an intelligent transport system (ITS) can effectively deal with urban congestion and improve transport efficiency.

3.2.4 Government Subsidies

Some scholars, for example, Sakano and Obeng (1995), believe that government subsidies inhibit public transport efficiency. Some scholars also discussed the impact of government subsidies at different administrative levels on transport efficiency. Beria and Grimaldi (2010) proposed that transport subsidies and ownership are two major factors affecting the efficiency of public transport in Sicily of Italy. Teye et al. (2017) found that government subsidies and road pricing policies improve the cargo flow of intermodal transport, as well as the transport efficiency of ports and surrounding areas. Karlaftis and McCarthy (1999) used panel data analysis to study the relationship between financial subsidies for public transport and public transport efficiency, but they found that financial subsidies have no significant impact on public transport efficiency. Fitzová and Matulova (2020) found that the proportion of subsidies in revenue has a bad effect on the urban public transport efficiency of the Czech Republic.

3.2.5 Enterprise Size

Many empirical studies show that enterprise size in the transport industry has a positive impact on transport efficiency. Barros and Dieke (2007) found that the transport efficiency of large airports is generally higher than that of small and medium-sized airports due to the existence of economies of scale. Odeck (2008) made an empirical analysis of Norwegian automobile transport companies, and found that enterprise mergers promote enterprise productivity. Hirschhausen and Cullmann (2010) found that the merger and reorganization of public transport companies will improve the efficiency of German public transport companies. Jarboui et al.,(2013) measured the technical efficiency of 64 public road transport operators in 18 countries and found that that large-size operators with more investment capacity tend to be more technically efficient than small-size operators. Chang and Tovar (2014) agreed that the size of port positively affect the technical efficiency of port terminals.

3.3 Research Regions

Research on transport efficiency is conducted across four regions: the European region, the American region, the Asian region, and other regions.

In the European region, Hirschhausen and Cullmann (2010) analyzed efficiency of 179 communal public transport bus companies in Germany during 1990 to 2014 by DEA method, and they find that the average technical efficiency of German bus companies is relatively low. Hilmola (2007) used DEA to analyze the efficiency and productivity of railway freight transport in different European countries from 1980 to 2003. Furthermore, Savolainen and Hilmola (2008) used DEA to evaluate the relative technical efficiencies of three European transport systems: railway, maritime and air transport. Results show that privately owned airline companies are significantly more efficient in passenger services.

In the American region, Chapin and Schmidt (1999) measure the efficiency of US Class I railroad companies since deregulation by applying the DEA approach, agreeing that their efficiency had been improved since deregulation, but not because of mergers by regression analysis. Barnum et al. (2008) used DEA to compare the performance of 46 bus routes of an urban transit agency in the United States, treating each route as a DMU. Karlaftis (2004) employed DEA in evaluating the efficiency of 256 US transit systems over a five-year period from 1990 to 1994. Boame (2004) studied the technical efficiency of Canadian urban transit systems by bootstrap DEA; he found that the average technical efficiency is 78%. According to the comparative efficiency study of Sampaio et al. (2008), only 14.3% of Brazilian systems analyzed are efficient, accounting for 5.3% of all systems analyzed. And only 25% of European systems are inefficient.

In the Asian region, Liu et al., (2018) study the spatio-temporal evolution of China's railway transportation efficiency during 2005–2013 using the ultra-efficient SBM model and Malmquist index method. Hahn et al. (2013) developed a network DEA model for evaluating the efficiency scores of 58 bus companies in Seoul in 2009, and they thought that the current evaluation system has difficulty in motivating the inefficient companies to take actions to improve their efficiency since it focuses only on profit sharing and does not impose mandatory penalties. Alam et al. (2020) applied DEA to evaluate the technical, pure technical and scale efficiencies of Pakistan railways from 1950 to 2016 and the super efficiency model was employed to rank efficient DMUs. Agarwal (2009) examined differences in technical efficiency and scale efficiency of 29 state transport undertakings in India. Movahedi et al. (2007) evaluated the efficiency of Iranian railways from 1971 to 2004 using DEA. To calculate the efficiency of each year, he used six input and three output variables.

3.4 Summary

The research on transport efficiency in Western countries started earlier. Foreign scholars have established a lot of evaluation systems for the efficiency evaluation model, including index analysis methods, SFA, and DEA. In contrast to parametric methods, DEA has superiority in estimating the validity of multiple inputs and multiple outputs, and does not need any presupposition about the production functional form on the frontier. Hence, the DEA method has been widely used in transport efficiency assessment. Foreign scholars have discussed the influencing factors of transport efficiency, and attempted to verify the effectiveness of transport planning schemes or transport policies through empirical research on the influencing factors of transport efficiency. To sum up, the research mainly focuses on the impact of ownership, financial subsidies, management and operation, transport scale, external environment, and other factors on transport efficiency.

As for China, the research started relatively late and is at an empirical stage—foreign measurement models are used to study different regional scales in China. Current domestic research is mostly on transport efficiency based on spatio-temporal characteristics using DEA and SFA. An increasing number of spatial dimensions in domestic research have been observed. However, the analysis of driving factors of transport efficiency lacks research on spatial mechanisms, with limited application of spatial econometric models.

References

Aigner, D. J., Lovell, C. A. K., & Schmidt, P. (1977). Formulation and estimation of stochastic frontier production function models. *Journal of Econometrics, 6*(1), 21–37. https://doi.org/10.1016/0304-4076(77)90052-5

Agarwal, S. (2009). Measuring the efficiency of public transport sector in India: an application of data envelopment analysis. In: *Proceedings of the conference in Philadelphia, USA.* http://astro.temple.edu/banker/dea2009/paper/Agarwal.pdf

Alam, K. M., Xuemei, L., Baig, S., et al. (2020). Analysis of technical, pure technical and scale efficiencies of Pakistan railways using data envelopment analysis and Tobit regression model. *Networks and Spatial Economics, 20,* 989–1014. https://doi.org/10.1007/s11067-020-09510-9

Bai, X., Jin, Z., & Chiu, Y. H. (2020). Performance evaluation of China's railway passenger transportation sector. *Research in Transportation Economics., 90*(100859), 0739–8859. https://doi.org/10.1016/j.retrec.2020.100859

Banker, R. D., Charnes, A., & Cooper, W.W. (1983). Some models for estimating technical and scale inefficiencies in data envelopment analysis. *Management Science, 30,* 1078–1092. http://www.jstor.org/stable/2631725

Barrett, S. D. (1995). Measuring railway productivity in Ireland: a comment. *Economic & Social Review, 26*(3), 323–325. http://edepositireland.ie/bitstream/handle/2262/64181/26%20apr%2095%20barrett.pdf?sequence=1&isAllowed=y

Barnum, D. T., Tandon, S., & Mcneil, S. (2008). Comparing the performance of bus routes after adjusting for the environment using data envelopment analysis. *Journal of Transportation Engineering, 134*(2), 77–85. https://www.nstl.gov.cn/paper_detail.html?id=24afa1253aa0b1d5c9800085fbcb81ff

Barros, C. P., & Dieke, P. U. C. (2007). Performance evaluation of Italian airports: A data envelopment analysis. *Journal of Air Transport Management, 13*(4), 184–191. https://doi.org/10.1016/j.jairtraman.2007.03.001

Beria, P., & Grimaldi, R. (2010). Unconventional factors of efficiency in public transport. A case study and theory. MPRA Paper 29234, University Library of Munich, Germany. https://mpra.ub.uni-muenchen.de/29234/1/MPRA_paper_29234.pdf

Boame, A. K. (2004). The technical efficiency of canadian urban transit systems. *Transportation Research Part E, 40*(5), 401–416. https://www.sciencedirect.com/science/article/pii/S1366554503000784

Cantos, P., & Maudos, J. (2000). Efficiency, technical change and productivity in the European rail sector: A stochastic frontier approach. *International Journal of Transport Economics, 27*, 55–76. http://worldcat.org/issn/03918440

Cantos, P., & Maudos, J. (2001). Regulation and efficiency: The case of European railways. *Transportation Research, 35A*, 459–472. https://www.sciencedirect.com/science/article/pii/S0965856400000070

Cantos, P., Pastor, J. M., & Serrano, L. (1999). Productivity, efficiency and technical change in the European railways: A non-parametric approach. *Transportation, 26*, 337–357. https://doi.org/10.1023/A:1005127513206

Chang, V., & Tovar, B. (2014). Efficiency and productivity changes for Peruvian and Chilean ports terminals: A parametric distance functions approach. *Transport Policy, 31*(Jan), 83–94. https://www.sciencedirect.com/science/article/pii/S0967070X1300173X

Chapin, A., & Schmidt, S. (1999). Do mergers improve efficiency? Evidence from deregulated rail freight. *Journal of Transport Economics and Policy, 33*(2), 147–162. http://www.jstor.org/stable/20053802

Charnes, A., Cooper, W. W., & Rhodes, E. (1978). Measuring the efficiency of decision making units. *European Journal of Operational Research, 2*(6), 429–444. https://doi.org/10.1016/0377-2217(78)90138-8

Cowie, J. (1999). The technical efficiency of public and private ownership in the rail industry. *Journal of Transport Economics and Policy, 33*(3), 241–251. https://www.jstor.org/stable/20053814

Cowie, J. (2002). Acquisition, efficiency and scale economies: an analysis of the British bus industry. *Transport Reviews, 22*(2), 147–157. https://ideas.repec.org/a/taf/transr/v22y2002i2p147-157.html

Cowie, J. (2003). The technical efficiency of public and private ownership in the rail industry. *Journal of Transport Economics and Policy, 33*(3), 241–252. https://xueshu.baidu.com/usercenter/paper/show?paperid=197c0ru0ju1a02g0bc2g0aw02w257665&site=xueshu_se

Cullinane, K., Wang, T., & Song, D., et al. (2006). The technical efficiency of container ports: Comparing data envelopment analysis and stochastic frontier analysis. *Transportation Research Part A: Policy and Practice, 40*(4), 354–374. https://doi.org/10.1016/j.tra.2005.07.003

Dalen, D. M., & Gómez-Lobo, A. (2003). Yardsticks on the road: Regulatory contracts and cost efficiency in the Norwegian bus industry. *Transportation, 30*, 371–386. https://doi.org/10.1023/A:1024784517628

Dong, Y., Hamilton, R., & Tippett, M. (2014). Cost efficiency of the Chinese banking sector: A comparison of stochastic frontier analysis and data envelopment analysis. *Economic Modelling, 36*, 298–308. https://doi.org/10.1016/j.econmod.2013.09.042

Else P.K., James T.J. (1995). Privatisation and the quality of rail services. *Transportation Research, 35*, 387–400. https://ideas.repec.org/a/eee/transa/v29y1995i6p387-400.html

Fitzová, H., & Matulova, M. (2020). Comparison of urban public transport systems in the Czech Republic and Slovakia: Factors underpinning efficiency. *Research in Transportation Economics, 81*(SI), 1–9. https://www.sciencedirect.com/science/article/pii/S0739885920300135

Fielding, G. J., Glauthier, R. E. & Lave, C. A. (1978). Performance indicators for transit management. *Transportation, 7*, 365–379. https://doi.org/10.1007/BF00168037

Fujdiak, R., Masek, P., Mlynek, P., et al. (1995). Advanced optimization method for improving the urban traffic management. *Journal of the Operational Research Society, 47*(10), 962–962. https://doi.org/10.1109/FRUCT-ISPIT.2016.7561507

Gao, Y., Li, W. D., You, X. Y. (2011). Research on the efficiency evaluation of China's railway transport enterprises with network DEA. *China Soft Science, 245*(05), 176–182. https://t.cnki.net/kcms/detail?v=3uoqIhG8C44YLTlOAiTRKgchrJ08w1e7tvjWANqNvp-75z6vSxsXez1kLbgKD5flTtw8OM1HS18I4aalnPb08hUh0eVvVLKs&uniplatform=NZKPT

Gathon, H. J., & Pestieau, P. (1995). Decomposing efficiency into its managerial and its regulatory components: The case of European railway. *European Journal of Operational Research, 80*, 500–507. https://doi.org/10.1016/0377-2217(94)00133-W

Hahn, J. S., Kim, D. K., Kim, H. C., et al. (2013). Efficiency analysis on bus companies in Seoul city using a network DEA model. *KSCE Journal of Civil Engineering, 17*, 1480–1488. https://doi.org/10.1007/s12205-013-0467-x

Hilmola, O. P. (2007). European railway freight transportation and adaptation to demand decline: Efficiency and partial productivity analysis from period of 1980–2003. *International Journal of Productivity and Performance Management, 56*(3): 205–225. https://doi.org/10.1108/17410400710731428/full/html

Hirschhausen, C. V., & Cullmann, A. (2010). Questions to airport benchmarkers—Some theoretical and practical aspects learned from benchmarking other sectors. https://xueshu.baidu.com/usercenter/paper/show?paperid=db3898640c1ad3b47c34b90347fa108e&site=xueshu_se

Jarboui, S., Forget, P., & Boujelbene, Y. (2013). Public road transport efficiency: A stochastic frontier analysis. *Journal of Transportation Systems Engineering and Information Technology, 13*(5), 64–71. https://doi.org/10.1016/S1570-6672(13)60123-3

Ji, X. L., & Yan, L. (2014). Urban logistics cost control anti economic structure optimized: The international experience and references: Theory & practice. *Shanghai Journal of Economics, 2014*(07), 90–96. https://doi.org/10.19626/j.cnki.cn31-1163/f.2014.07.010

Karlaftis, M. G. (2004). A DEA approach for evaluating the efficiency and effectiveness of urban transit systems. *European Journal of Operational Research, 152*, 154–164. https://www.sciencedirect.com/science/article/pii/S0377221703000298

Karlaftis, M., & Mccarthy, P. (1999). The effect of privatization on public transit costs. *16*(1), 27–44. https://doi.org/10.1023/A:1008179629733

Kerstens, K. (1996). Technical efficiency measurement and explanation of French urban transit companies. *Transportation Research Part a: Policy & Practice, 30*(6), 431–452. https://doi.org/10.1016/0965-8564(96)00006-7

Kutlar, A., Kabasakal, A., & Sarikaya, M. (2013). Determination of the efficiency of the world railway companies by method of DEA and comparison of their efficiency by Tobit analysis. *Quality & Quantity, 47*(6), 3575–3602. https://doi.org/10.1007/s11135-012-9741-0

Li, L. B. (2008). Efficiency analysis of the regional railway in China: Based on the application of DEA-Tobit approach. *Soft Science, 22*(4), 58–63. https://xueshu.baidu.com/usercenter/paper/show?paperid=55eeb9f01ce31399b784b791bf36b828&site=xueshu_se

Liu, B. Q., Wu, W., Su, Q., & Zhang, L. L. (2018). Spatio-temporal evolution characteristics and mechanism of railway efficiency in China. *Geographical Research, 37*(3), 512–526. http://www.geogsci.com/EN/abstract/abstract40181.shtml

Martland, C. (1997). Sources of financial improvement in the U.S. rail industry (1966–1995). In Transportation Research Forum, 39th annual meeting.

Meeusen, W., & van den Broeck, J. (1977). Efficiency estimation from cobb-Douglas production functions with composed error. *International Economic Review, 18*, 435–444. https://doi.org/10.2307/2525757

Movahedi, M. M., Saati, S., & Vahidi, A. R. (2007). Iranian railway efficiency (1971–2004): an application of DEA. *International Journal of Contemporary Mathematical Sciences, 2*(32), 1569–1579. http://www.m-hikari.com/ijcms-password2007/29-32-2007/lotfiIJCMS29-32-2007.pdf

Odeck, J. (2008). The effect of mergers on efficiency and productivity of public transport services. *Transportation Research Part A: Policy and Practice, 42*(4), 696–708. https://www.sciencedirect.com/science/article/pii/S0965856408000207

Oum, T. H., Yan, J., & Yu, C. (2008). Ownership forms matter for airport efficiency: A stochastic frontier investigation of worldwide airports. *Transportation Research Forum, 64*(2), 422–435. https://doi.org/10.1016/j.jue.2008.03.001

Perry, J. L., & Babitsky, T. T. (1986). Comparative performance in urban bus transit: Assessing privatization strategies. *Public Administration Review, 46*(1), 57–66. https://www.docin.com/p-1739048923.html

Pina, V., & Torres, L. (2001). Analysis of efficiency of local government services delivery: An application to urban public transport. *Transportation Research Part A, Policy and Practice, 35*(10), 924–994. https://www.sciencedirect.com/science/article/pii/S0965856400000331

Roy, W. (2007). Ownership, contractual practices and technical efficiency: The case of urban public transport in France. *Journal of Transport Economics and Policy,41*(2), 257–282 .https://www.xueshufan.com/publication/2902635193

Sakano, R., & Obeng, K. (1995). Re-examination of inefficiencies in urban transit systems: A stochastic frontier approach. *Logistics and Transportation Review, 31*(4), 377. https://www.proquest.com/citedby/MSTAR_197442995/CCF7E6B555854692PQ/1?accountid=13151

Sakai, H., & Takahashi, Y. (2013). Ten years after bus deregulation in japan: an analysis of institutional changes and cost efficiency. *Research in Transportation Economics, 39*(MAR.), 215-225. https://doi.org/10.1016/j.retrec.2012.06.016

Savolainen, V., & Hilmola, O. (2008). Does privatisation matter in the efficiency of airline freight transports? Evidence from the European market. *World Review of Intermodal Transportation Research, 2*(1), 9–29. https://xueshu.baidu.com/usercenter/paper/show?paperid=1a46bcb530606145ed8f1adc7b6c0f03

Sampaio, B. R., Neto, O. L., & Sampaio, Y. (2008). Efficiency analysis of public transport systems: Lessons for institutional planning. *Transportation Research Part A: Policy and Practice*, 445–454 https://ageconsearch.umn.edu/record/207952/files/2006_2A_EffAnalysis_paper.pdf

Silva, T. C., Tabak, B. M., Cajueiro, D. O., et al. (2017). A comparison of DEA and SFA using micro-and macro-level perspectives: Efficiency of Chinese local banks. *Physical A, 2017*(469), 216–223. https://doi.org/10.1016/j.physa

Stathopoulos, A., & Karlaftis, M. G. (2003). A multivariate state space approach for urban transportation flow modeling and prediction. *Transportation Research Part C Emerging Technologies, 11*(2), 121–135. https://doi.org/10.1016/S0968-090X(03)00004-4

Teye, C., Bell, M. G. H., & Bliemer, M. C. J. (2017). Entropy maximising facility location model for port city intermodal terminals. *Transportation Research Part e: Logistics and Transportation Review, 100*, 1–16. https://doi.org/10.1016/j.tre.2017.01.006

Tone, K. (2001). Slacks-based measure of efficiency in data envelopment analysis. *European Journal of Operational Research, 130*(3), 498–509. https://doi.org/10.1016/S0377-2217(99)00407-5

Vicente, P., & Lourdes, T. (2001). Analysis of the efficiency of Local government services delivery: A application to Urban public transport. *Transportation Research Part A, 35*, 924–944. https://doi.org/10.1016/S0965-8564(00)00033-1

Waters,W. G., Tretheway, M. W., & Fok, A. K. (1997). The total factor productivity of the Canadian railways. *Journal of Transport Economics and Policy, 39*(4), 93. https://xueshu.baidu.com/usercenter/paper/show?paperid=1n5q04p00x0208p08e5t0pr0eu019633&site=xueshu_se

Wu, W. H. (2001). Study on the evaluation system of China transportation efficiency. *Comprehensive Transportation, 2*, 37–39.

Xiong, C. J., Ning, X. X., & Pang, Y. L. (2006). Evaluation of coordinated development between Chinese comprehensive transportation modes. *Systems Engineering, 24*(6), 1–7. https://t.cnki.net/kcms/detail?v=3uoqIhG8C44YLTlOAiTRKgchrJ08w1e7eWoVfj7plMy3_Zth0dEIBOtQKp0GZhvEfxIMfq5OWxp0g5ig6Sgrrmm0k4UJvCOp&uniplatform=NZKPT

Ye, C. H., Ge, Y. Y., Wang, J. L., & Zhang, J. F. (2019). Research on the impact of China's transportation industry on the promotion of export trade——Empirical analysis based on Zhejiang panel data. *Price: Theory & Practice, 12*, 67–70. https://doi.org/10.19851/j.cnki.CN11-1010/F.2019.12.025

Chapter 4
The Economic Operation of the Transport Industry

4.1 Fixed Asset Investment in the Transport Industry

4.1.1 The Investment Scale is Relatively Large, with Its Growth Rate Slowing Down in Recent Years

Transport investment plays an important role in promoting economic growth. Transport infrastructure construction projects have a long industrial chain, which can fully drive the growth of the steel, cement, building materials, machinery and other industries while stimulating consumption. Transport investment in central and western China is of great significance to promoting balanced development among regions.

Investment, as part of the "troika" that drives GDP growth, plays a very important role in the macro economy of a country. In China, transport infrastructure construction has always occupied an important position in investment. According to the National Bureau of Statistics of China, the fixed asset investment in the transport industry increased from RMB 1.7 trillion in 2008 to RMB 5.39 trillion in 2016. From 2008 to 2016, the annual average proportion value of fixed asset investment in transport to the whole industry was 9.4%. Since the implementation of the new economic growth concept in 2013, the growth of fixed asset investment in transport has slowed down, but its proportion has remained at a high level; from 2013 to 2016, its proportion was about 8.58% (Fig. 4.1).

China's fixed asset investment in transport grew rapidly from 2008 to 2010, especially in 2009, with a nominal growth rate of 46.7%, which was mainly due to the positive effect of the 4 trillion-yuan fiscal policy. At the end of 2008, the Chinese government introduced an economic stimulus package of RMB 4 trillion to deal with the global financial crisis. According to Ping Zhang, chairman of the National Development and Reform Commission at that time, approximately RMB 1.5 trillion would be invested in infrastructure, including roads, railroads, airports, and seaports (Qin, 2016). The fixed asset investment in transport in 2011 showed a negative growth of 5.93% compared with that in 2010, which was mainly caused by the decrease of

P. Zhao and L. Zeng, *Transport Efficiency and Safety in China*, Population, Regional Development and Transport, https://doi.org/10.1007/978-981-99-1055-7_4

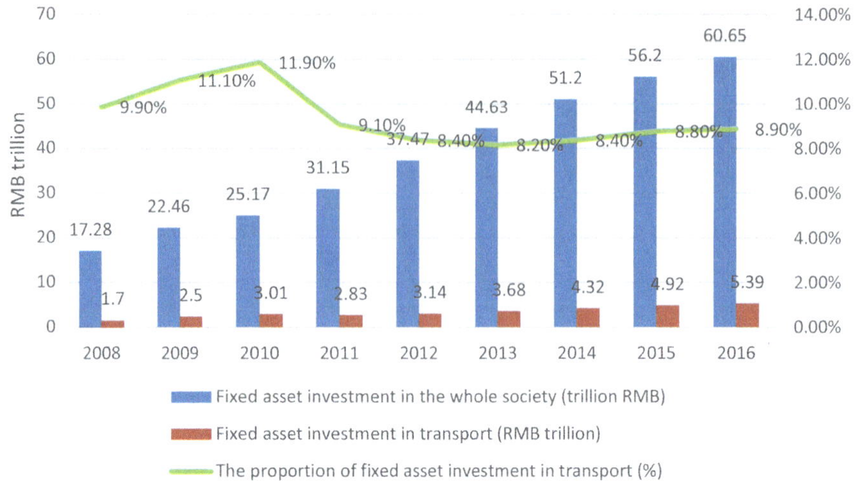

Fig. 4.1 Fixed asset investment in transport in China from 2008 to 2016 (2000–2017). *Data source* The authors, edited from China Statistical Yearbook (2017)

investment in railways. After 2012, the growth slowed down, and the average annual growth rate of fixed asset investment in transport was maintained at 13.8%, lower than the growth rate from 2008 to 2010 (Table 4.1; Fig. 4.1).

The large-scale investment in transport infrastructure has fundamentally changed the Chinese economy and society. For one thing, the improved freight and passenger mobility has facilitated regional trade flows and increased the allocative efficiency of factor endowments, which has in turn improved the productivity of economic activities. For another, transport bottleneck constraints due to a shortage of infrastructure supply have been alleviated as new infrastructures are built. Meanwhile, a reduction of transport costs is expected to boost travel demand, which in turn will further promote economic growth (Chen & Haynes, 2017).

4.1.2 Highway Transport is the Most Important Investment Sector, and Investment in Air Transport is Growing Rapidly

From the perspective of investment structure, there is a big difference in the amount of investment among different transport sectors. According to the Statistical Yearbook of the Chinese Investment in Fixed Assets (2017), transport investment in China mainly flowed to highway transport in 2016, and the proportion of highway transport investment in the entire transport investment showed a rapid upward trend, rising from 47.2% in 2008 to 65.9% in 2016. In China, with the sustained growth of urban industry and population size, highway transport needs have been on the rise.

Table 4.1 Fixed asset investment in transport in China from 2008 to 2016

Year	Fixed asset investment in the whole society (trillion RMB)	Fixed asset investment in transport (trillion RMB)	The proportion of fixed asset investment in transport (%)	The nominal growth rate of fixed asset investment in transport (%)
2008	17.28	1.70	9.9	20.28
2009	22.46	2.50	11.1	46.70
2010	25.17	3.01	11.9	20.42
2011	31.15	2.83	9.1	− 5.93
2012	37.47	3.14	8.4	11.15
2013	44.63	3.68	8.2	17.00
2014	51.20	4.32	8.4	17.47
2015	56.20	4.92	8.8	13.85
2016	60.65	5.39	8.9	9.53

Data source The authors, edited from Statistical Yearbook of the Chinese Investment in Fixed Assets (2009–2013, 2014–2017) and statistical yearbooks of provinces over the years.

Railway transport and water transport are the transport sectors with the second and third largest fixed asset investment, respectively, but the proportion of fixed asset investment in these two transport sectors showed a downward trend, from 25.9% and 7.7% in 2008 to 14.4% and 4% in 2016, respectively. China Railway Corporation is mainly responsible for railway operation and construction in China. At present, there are great debt pressure and huge liabilities. During the 12th Five-Year Plan period from 2010 to 2015, the total railway investment in China was RMB 3.5 trillion and its liabilities were RMB 3.94 trillion, with an annual interest of more than RMB 180 billion. The heavy financial burden restrained railway investment, which was between RMB 770 billion and RMB 774.8 billion from 2014 to 2016.

During the study period, except for 2016 (4.1%), the proportion of air transport investment in the entire transport investment fluctuated between 3 and 4%. The proportion of pipeline transport investment was below 1% all the year round (Tables 4.2 and 4.3).

4.1.3 Western China's Share of Transport Investment is Growing

Furthermore, the communications and transport system in western China is far behind that in the coastal region of China (Ramesh, 2017). In 1999, the Western Development Program (WDP) was implemented by the Chinese government. Its primary task is to build a good transport infrastructure. This is not only the need of economic development, but also the premise of attracting investment in the western region. In recent years, benefiting from the strategy of western development, the total investment in

Table 4.2 Fixed asset investment in different transport sectors in China from 2008 to 2016

Year	The entire transport industry	Railway transport	Highway transport	Water transport	Air transport	Pipeline transport
2008	1570.0	407.3	741.2	120.4	59.0	13.9
2009	2327.1	666.1	1055.8	167.1	60.5	7.3
2010	2788.3	762.2	1276.4	208.0	89.3	9.5
2011	2776.6	591.5	1385.6	192.2	83.6	14.8
2012	3088.1	612.9	1746.6	200.8	112.4	20.5
2013	3632.9	669.1	2050.3	212.3	131.4	37.4
2014	4289.0	770.7	2451.3	243.5	143.0	31.5
2015	4897.5	773.0	2861.4	235.2	184.0	29.9
2016	5362.8	774.8	3293.7	216.3	222.0	26.3

Data source The authors, edited from Statistical Yearbook of the Chinese Investment in Fixed Assets (2009–2013, 2014–2017) and statistical yearbooks of provinces over the years

Table 4.3 Proportion of fixed asset investment in each transport sector

Year	Railway transport (%)	Highway transport (%)	Water transport (%)	Air transport (%)	Pipeline transport (%)
2008	25.9	47.2	7.7	3.8	0.9
2009	28.6	45.4	7.2	2.6	0.3
2010	27.3	45.8	7.5	3.2	0.3
2011	21.3	49.9	6.9	3.0	0.5
2012	19.8	56.6	6.5	3.6	0.7
2013	18.4	56.4	5.8	3.6	1.0
2014	18.0	57.2	5.7	3.3	0.7
2015	15.8	58.4	4.8	3.8	0.6
2016	14.4	61.4	4.0	4.1	0.5

Data source The authors, edited from Statistical Yearbook of the Chinese Investment in Fixed Assets (2009–2013, 2014–2017) and statistical yearbooks of provinces over the years

transport in the western region has increased year by year, the proportion of which in the whole country has increased steadily, as shown in Fig. 4.2. In the WDP, the Chinese government pays attention to the construction of rural transport infrastructure. In 2002, China started a road-building program to access western counties. The proposed program included 252 building plans in over 15 regions that will increase the number of highways to 26,000, a significant component of the government's strategy to encourage tourism and economic growth in western China (Dimitriou, 2019).

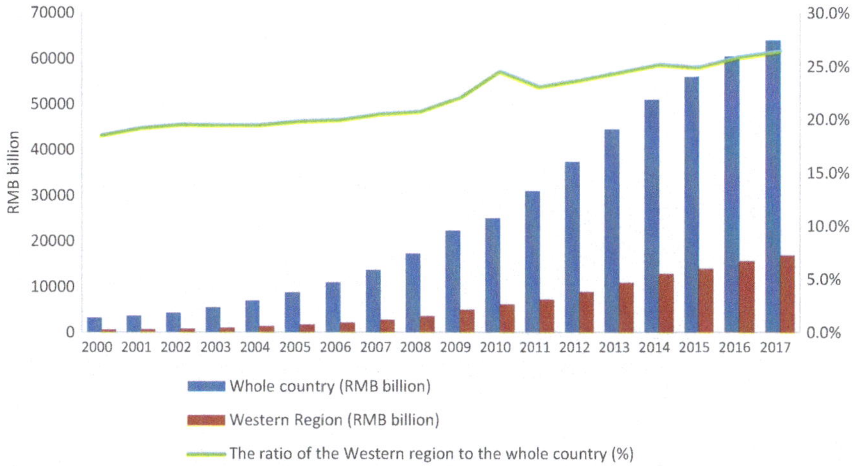

Fig. 4.2 The growth trend of fixed asset investment in transport in western China. *Data source* The authors, edited from China Statistical Yearbook (2020)

The increasing infrastructure investment in the western region is meant to promote regional economic growth and reduce regional differences. But some scholars question the effectiveness of WDP policies. Ramesh (2017) analyzed the economic data of both western and coastal regions and found that the western region's share of national GDP was 17.2% in 2000 and it fell to approximately 13% in 2014. According to Shi and Huang (2014), the western region overinvested in infrastructure after 2008 due to China's fiscal stimulus.

4.2 Economic Characters of the Transport Industry

4.2.1 The Economic Scale of the Transport Industry is at a High Level, and Its Growth Has Slowed Down in Recent Years

The overall economic scale of the transport industry in China grew steadily from 2008 to 2016. The added value of transport increased from RMB 1636.76 trillion in 2008 to RMB 3302.87 billion in 2016, which more than doubled in nine years (Table 4.4; Fig. 4.3). This shows that there is a strong demand for transport service and a broad space for transport development. The average annual growth rate was about 9.2%, and the annual growth rate peaked in 2011; the growth slowed down after 2011.

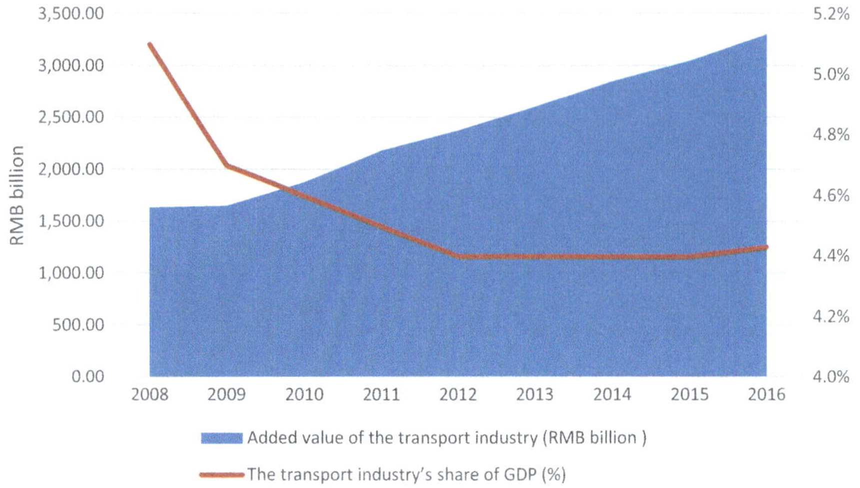

Fig. 4.3 The added value of the transport industry in China (2008–2016). *Data source* the authors, edited from China Statistical Yearbook (2017)

Table 4.4 The economic scale of the transport industry in China (2008–2016)

Year	Added value of the transport industry (million RMB)	The transport industry's share of GDP (%)	The growth rate of the added value of the transport industry (%)
2008	16,367.6	5.1	12.1
2009	16,522.4	4.7	0.9
2010	18,783.6	4.6	13.7
2011	21,842	4.5	16.3
2012	23,763.2	4.4	8.8
2013	26,042.7	4.4	9.6
2014	28,534.4	4.4	9.6
2015	30,519.5	4.4	7.0
2016	33,028.7	4.43	6.96

Data source The authors, edited from China Statistical Yearbook (2017)

4.2.2 The Contribution Rate of the Transport Industry to Economic Growth is Declining

During the study period, the contribution rate of the transport industry to the whole economic scale showed a downward trend, and the proportion of the added value of the transport industry in GDP decreased from 5.1% in 2008 to 4.4% during the period from 2012 to 2016. This may be due to the improvement of logistics technology and the decrease of domestic logistics costs during the study period.

Internationally, logistics efficiency is evaluated by the proportion of the total logistics costs to GDP. The lower the proportion of that, the higher the logistics efficiency and the more developed the logistics industry. From 2008 to 2016, the average proportion of that was 15.53%. During this period, the proportion of that decreased slowly, which indicates that the overall logistics efficiency did not improve much. The logistics efficiency significantly improved in 2016, and the proportion of that dropped to 14.9%, which is the lowest value in the study period. (Table 4.5; Fig. 4.4).

The decline of the ratio of the total social logistics costs to GDP is affected by many factors. Generally speaking, the demand for logistics of primary and secondary industries is larger than that of the tertiary industry, so the logistics costs of primary and secondary industries are also higher than the logistics cost of the tertiary industry. In recent years, with the transformation of the economic growth mode and the continuous optimization of industrial structure, in China, the proportion of the tertiary industry in GDP has increased, and therefore the ratio of logistics costs to GDP is decreasing. Moreover, the Chinese government has deepened the supply-side reform to promote the steady improvement of logistics management and operation efficiency. In the end, the degree of logistics marketization in China has been gradually improved .

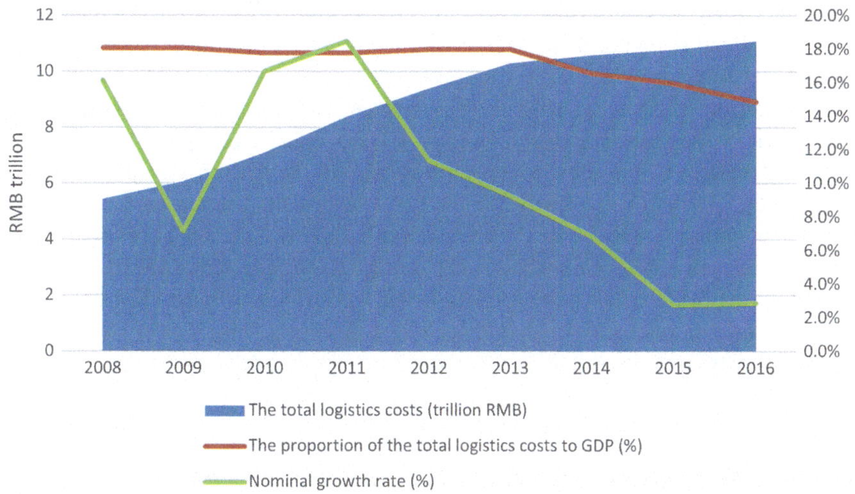

Fig. 4.4 The total logistics costs in China (2008–2016). *Data source* The authors, edited from China Logistics Yearbook (2017)

Table 4.5 The total logistics costs in China (2008–2016)

Year	The total logistics costs (trillion RMB)	The proportion of the total logistics costs to GDP (%)	Nominal growth rate (%)
2008	5.45	18.1	16.2
2009	6.08	18.1	7.2
2010	7.1	17.8	16.7
2011	8.4	17.8	18.5
2012	9.4	18.0	11.4
2013	10.3	18.0	9.3
2014	10.6	16.6	6.9
2015	10.8	16.0	2.8
2016	11.1	14.9	2.9

Data source The authors, edited from China Logistics Yearbook (2017)

4.2.3 *The Total Logistics Costs Are High*

Although the proportion of China's total social logistics costs in GDP has been at a low level in recent years, compared with the United States, the proportion is still too high. In 2016, the logistics costs of China accounted for 14.9% of GDP, which were significantly higher than the level of 7.2% in the United States. Transport costs are the largest component of logistics costs in China and the United States. In 2016, the proportion of transport costs in the logistics costs of China was about 47.6%, while that of the United States in 2015 was about 63.2%. In addition, the per unit logistics costs in China show an upward trend, and the per ton logistics costs rose from RMB 211 in 2008 to RMB 253 in 2016 (Table 4.6). The reason for this rise is that the economic structure of China has been adjusted. In order to promote sound economic development, China has tried to reduce the transport volume of many bulk commodities, which makes the growth of the total freight volume slow down after 2012. Its growth rate is lower than that of logistics costs, which makes the per unit logistics costs rise.

4.2.4 *The Proportion of Railway Freight Transport is Obviously Low*

From 2008 to 2016, the growth of the freight turnover in China was relatively stable. The freight turnover increased from 11,030 billion ton-km in 2008 to 186,629.48 billion ton-km in 2016, an increase of 69.2% in nine years. This shows that the demand for freight transport is strong. The average annual growth rate was about 7.23%, with the highest growth rate in 2010; the growth slowed down after 2011, and was even negative in 2015. That was driven largely by the decline in demand for

Table 4.6 The per unit logistics costs in China (2008–2016)

Year	Freight volume (billion)	Growth rate of freight volume (%)	Per unit logistics costs (RMB/t)
2008	25.86	13.6	211
2009	28.25	9.3	215
2010	32.42	14.7	219
2011	36.97	14.0	227
2012	41.00	10.9	229
2013	40.99	– 0.04	251
2014	41.67	1.7	254
2015	41.76	0.2	259
2016	43.87	5.0	253

Data source The authors, edited from China Logistics Yearbook (2017) and China Statistical Yearbook (2017)

smuggled goods. Since 1990, bulk goods and raw material movements (especially coal) have accounted for about 75% of total freight transport (Cui et al., 2021). Coal transport accounted for 53% of total railway freight transport in China in 2015 (Zhang et al., 2020). Since 2010, the Chinese economy is shifting from the high-speed growth period to the moderate-speed growth period; the growth of fixed asset investment is slowing down, and the growth of added value in major industries such as coal, metallurgy, power, and chemical industries is receding. The railway transport market of smuggled goods, typified by coal, has been weak for some time; the railway freight turnover shows a downward trend since 2012. But in general, the basic demand for transporting bulk goods such as coal, petroleum and smelting materials will not suffer any fundamental change. Grain, fertilizer, pesticide, and cotton are materials providing a guaranteed demand for rail transport. As a result, during the 13th Five-Year Plan period, transport of bulk goods will remain the principal market of railway freight (Xiao, 2016).

From the perspective of freight transport structure, railway, highway and water freight accounts for 17.42%, 31.89% and 48.58% of the total freight turnover, respectively. In the nine years, the market share of water freight increased steadily, the market share of rail freight continued to decline, and the market share of highway freight increased first and then decreased (Fig. 4.5; Tables 4.7 and 4.8).

Compared with the United States, the proportion of railway transport in freight transport in China is obviously lower. In 2016, railway transport only accounted for 12.75% of the total freight turnover, while the proportion of railway transport in the United States remained between 33 and 35% from 2012 to 2015.

According to the Ministry of Ecology and Environment of China, building a medium- and long-distance passenger and freight transport system with electrified railways and clean ships as the main part, as well as a short-distance passenger and freight transport system with low emission vehicles and new energy vehicles as the main part, is a fundamental measure for the prevention and control of motor vehicle pollution in the future. In addition, it is necessary to establish and implement the strictest environmental supervision system for motor vehicles, especially high emission diesel trucks, to ensure that the proportion of railway transport in freight transport is significantly increased.

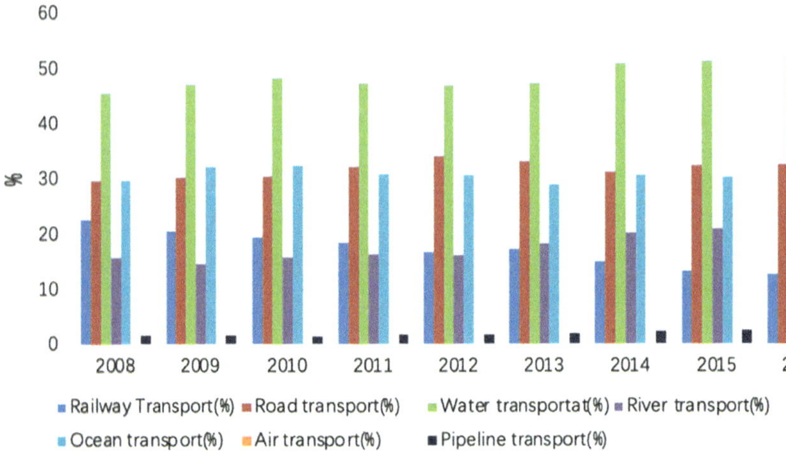

Fig. 4.5 The market share of different transport sectors (2008–2016). *Data source* The authors, edited from China Statistical Yearbook (2017)

Table 4.7 The freight turnover in different transport sectors (2008–2016)

Year	The total freight turnover (100 million ton-km)	Railway transport (100 million ton-km)	Highway transport (100 million ton-km)	Water transport (100 million ton-km)	River transport (100 million ton-km)	Ocean transport (100 million ton-km)	Air transport (100 million ton-km)	Pipeline transport (100 million ton-km)
2008	110,300	25,106.28	32,868.19	50,262.7	17,411.7	32,851	119.6	1944.03
2009	122,133.31	25,239.17	37,188.82	57,556.67	18,032.67	39,524	126.23	2022.42
2010	141,837.42	27,644.13	43,389.67	68,427.53	22,428.53	45,999	178.9	2197.19
2011	159,323.6	29,465.79	51,374.74	75,423.84	26,068.84	49,355	173.91	2885.44
2012	173,804.46	29,187.09	59,534.86	81,707.58	28,295.58	53,412	163.89	3211.04
2013	168,013.8	29,173.89	55,738.08	79,435.65	30,730.65	48,705	170.29	3495.89
2014	181,667.69	27,530.19	56,846.9	92,774.56	36,839.56	55,935	187.77	4328.28
2015	178,355.9	23,754.31	57,955.72	91,772.45	37,536.45	54,236	208.07	4665.35
2016	186,629.48	23,792.26	61,080.1	97,338.8	39,263.8	58,075	222.45	4195.87

Data source The authors, edited from China Statistical Yearbook (2017)

Table 4.8 The market share of different transport sectors (2008–2016)

Year	Railway transport (%)	Highway transport (%)	Water transport (%)	River transport (%)	Ocean transport (%)	Air transport (%)	Pipeline transport (%)
2008	22.76	29.80	45.57	15.79	29.78	0.11	1.76
2009	20.67	30.45	47.13	14.77	32.36	0.10	1.66
2010	19.49	30.59	48.24	15.81	32.43	0.13	1.55
2011	18.49	32.25	47.34	16.36	30.98	0.11	1.81
2012	16.79	34.25	47.01	16.28	30.73	0.09	1.85
2013	17.36	33.17	47.28	18.29	28.99	0.10	2.08
2014	15.15	31.29	51.07	20.28	30.79	0.10	2.38
2015	13.32	32.49	51.45	21.04	30.41	0.12	2.62
2016	12.75	32.73	52.16	21.04	31.12	0.12	2.25
Mean	17.42	31.89	48.58	17.74	30.84	0.11	2.00

Data source The authors, edited from China Statistical Yearbook (2017)

4.2.5 Highway Transport is the Most Important Way of Passenger Transport

From 2008 to 2012, the growth of the passenger turnover in China was relatively stable. The figure decreased in 2013, and it showed an upward trend after 2013. The 2013 figures about passenger and freight traffic are calculated with data from the 2013 survey of transport economics, and have different coverages. The same applies to the following tables (NBSC, 2021). Railway transport and highway transport are the main modes of passenger transport, while water transport has a very low share of the passenger transport market.

It is worth noting that the proportion of highway transport in the passenger transport market was maintained at above 53%, and the proportion of railway transport showed a downward trend. After 2012, the proportion of highway transport showed a downward trend, while the proportion of railway transport showed an upward trend (Fig. 4.6; Tables 4.9 and 4.10). With the large-scale construction of high-speed railways in various regions and the six railway speed-up campaigns, many passenger transport modes have changed into high-speed railway transport. China's high-speed rail system has taken shape (Wang et al., 2015). High-speed rail is defined as rail services operating at a speed above 200 km per hour. High-speed railway transport is more time-efficient than highway transport. With the further expansion of the high-speed rail network, it is likely that high-speed rail will increase its share in passenger traffic (Qin, 2016).

During the study period, the proportion of air transport showed an upward trend, which indicated that economic development promoted the passenger demand for air transport (Wu & Man, 2018). Connectivity by air transport has been significantly improved over the past years. The number of airports increased from 135 in 2005 to

202 in 2014. The number of domestic and international routes more than doubled during the same period (Qin, 2016). Moreover, Chinese people's desire for efficiency drives air transport development. Nevertheless, despite the rapid growth of air transport in China, it is not as developed as that in other high-income countries. Currently, the average number of civil airports per million people in China is only 0.35, which is considerably lower than those in the United States (2.02) and Japan (1.38) in 2006 (Mao et al., 2009).

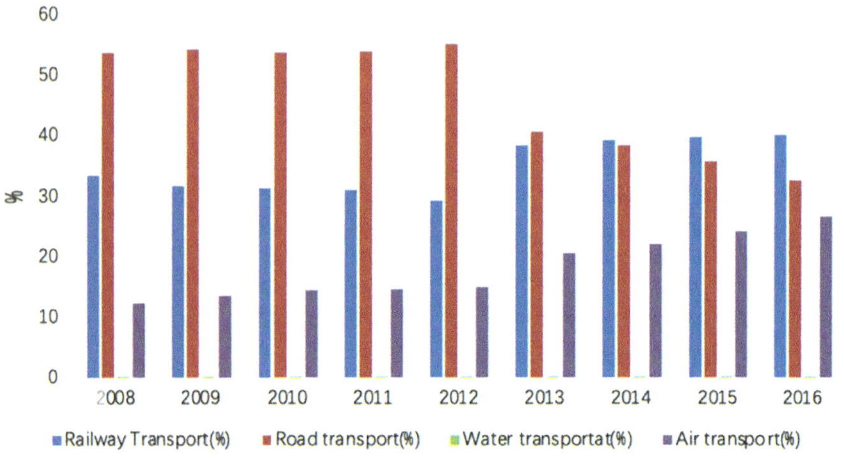

Fig. 4.6 The market share of different transport sectors (2008–2016). *Data source* The authors, edited from China Statistical Yearbook (2017)

Table 4.9 The passenger turnover in different transport sectors (2008–2016)

Year	The total passenger turnover (100 million person-km)	Railway transport (100 million person-km)	Highway transport (100 million person-km)	Water transport (100 million person-km)	Air transport (100 million person-km)
2008	23,196.70	7778.60	12,476.11	59.18	2882.80
2009	24,834.94	7878.89	13,511.44	69.38	3375.24
2010	27,894.26	8762.18	15,020.81	72.27	4039
2011	30,984.03	9612.29	16,760.25	74.53	4536.96
2012	33,383.09	9812.33	18,467.55	77.48	5025.74
2013	27,571.65	10,595.62	11,250.94	68.33	5656.76
2014	28,647.13	11,241.85	10,996.75	74.34	6334.19
2015	30,058.90	11,960.60	10,742.66	73.08	7282.55
2016	31,258.46	12,579.29	10,228.71	72.33	8378.13

Data source The authors, edited from China Statistical Yearbook (2017)

Table 4.10 The market share of different transport sectors (2008–2016)

Year	Railway transport (%)	Highway transport (%)	Water transport (%)	Air transport (%)
2008	33.53	53.78	0.26	12.43
2009	31.73	54.40	0.28	13.59
2010	31.41	53.85	0.26	14.48
2011	31.02	54.09	0.24	14.64
2012	29.39	55.32	0.23	15.05
2013	38.43	40.81	0.25	20.52
2014	39.24	38.39	0.26	22.11
2015	39.79	35.74	0.24	24.23
2016	40.24	32.72	0.23	26.80

Data source The authors, edited from China Statistical Yearbook (2017)

4.3 Transport Infrastructure Investment and Economic Growth

Transport is vital to the operation of society, and transport investment has been widely debated from both academic and policy perspectives. A large body of studies on the macro-level analysis of the economic impact of transport investment have been conducted since the 1980s. Although consensus has not been reached yet, it is generally believed that transport infrastructure investment facilitates economic growth (Bhatta & Drennan, 2003; Boarnet, 1998; Cadot et al., 2006; Chen & Haynes, 2017). In this chapter, we construct a time series model to analyze the impact of transport infrastructure investment on economic growth in China.

4.3.1 Data Source and Method

The official statistics agency of China did not release the data of capital stock in the transport industry. This study uses the perpetual inventory method (Goldsmith, 1951) to estimate the capital stock. This study takes 2003 as the base year and uses the panel data of 31 provinces (except Hong Kong, Macao and Taiwan) in China from 2003 to 2018. The data comes from China Statistical Yearbook (2004–2019) and internal data of the Ministry of transport of China. This book uses the perpetual inventory method of Goldsmith (1951) to measure the fixed capital stock, and refers to the method of Zhang et al. (2004) to measure the capital stock of the base year, which uses the fixed capital investment in 2003 divided by 10%. For the depreciation rate of the fixed capital stock of the transport industry, the authors refer to the research results of Li and Zhang (2016), taking 8.76%, while for the fixed capital stock of other industries, the authors refer to the research results of Zhang et al. (2004), taking 9.6%.

In this study, two models are used to reflect the effect of transport infrastructure on economic growth: basic production function model and time lag model.

The basic production function model is as follows:

$$Log(Y_{it}) = \beta_0 + \beta_1 \log(L_{it}) + \beta_2 \log(TK_{it}) + \beta_3 \log(NTK_{it}) + \varepsilon_{it} \quad (4.1)$$

where i is the ith province and t stands for the year; L is the number of employees; *TK* is the fixed capital stock of transport; and *NTK* is the fixed capital stock of other industries. In order to study the time lag effect of the fixed capital of transport infrastructure on economic growth, the following model is introduced:

$$Log(Y_{it}) = \beta_0 + \beta_1 \log(L_{it}) + \beta_2 \log(TK_{it-n}) + \beta_3 \log(NTK_{it}) + \varepsilon_{it} \quad (4.2)$$

Since it takes at least one to four years for the transport project to be completed and put into operation, the lag years $n = 1, 2, 3, 4, 5$ are selected. In order to reflect the impact of transport infrastructure on economic growth in different periods, the two models are estimated according to different periods.

4.3.2 Results

4.3.2.1 Results of the Basic Model

From the regression results of the basic model (Table 4.11), the elasticity of the fixed capital of transport to GDP was 0.167 from 2003 to 2018, which means that if fixed capital investment in transport increases by 1%, GDP will increase by 0.167%. However, we can find that the elasticity of the fixed capital of transport to GDP first decreased and then increased. The elasticity of the fixed capital of transport to GDP from 2009 to 2012 was significantly higher than that from 2003 to 2008 and from 2013 to 2018. The transport infrastructure in China has basically been able to meet the needs of social and economic development, resulting in the decline of marginal efficiency of investment. Meanwhile, in some areas, the transport network is not reasonable enough, and transport construction is somewhat excessive. This makes the transport infrastructure not to play a greater role in improving economic growth.

4.3.2.2 Results of the Time Lag Model

We find that the elasticity of the fixed capital of transport to GDP basically remained unchanged in different lag periods, which indicates that transport investment has an effect on economic growth, not only during the investment period but for years to come.

Table 4.11 Results of the basic model

	2003–2018	2003–2008	2009–2012	2013–2018
ln(L)	0.179831***	0.0261787	– 0.0274749	0.22397***
ln(TK)	0.166838***	0.2639191***	0.3556143***	0.2192342***
ln(NTK)	0.760957***	0.9060848***	0.825802 ***	0.6690321***

Note: *** indicates significance at the 1% level.
Data source Edited by the authors

According to Tables 4.12, 4.13, 4.14, and 4.15, the elasticity of capital stock of transport to GDP shows an increasing trend during the whole study period of the lag phase I (0.1565115**), lag phase II (0.1793539**), lag phase III (0.2201451**) and lag phase IV (0.2683532**), respectively. It reflects that the fixed capital investment in transport infrastructure will have a growing positive effect on economic growth year by year. This result requires that policy makers should pay attention to the impact of the improved transport network and accessibility on economic development after transport infrastructure is completed and put into use.

Table 4.12 Results of lag phase I

	2004–2018	2004–2008	2009–2012	2013–2018
ln(L)	0.20373***	0.0395682	– 0.0159378	0.2291251***
ln(TK)	0.1565115***	0.1815369***	0.3539402***	0.2776419***
ln(NTK)	0.738465***	0.9461151 ***	0.8176241***	0.6182199***

Note: *** indicates significance at the 1% level.
Data source Edited by the authors

Table 4.13 Results of lag phase II

	2005–2018	2005–2008	2009–2012	2013–2018
ln(L)	0.221649***	0.0432086	– 0.0134856	0.2273731***
ln(TK)	0.1793539 ***	0.1419886 ***	0.3304846***	0.3512635***
ln(NTK)	0.6980806***	0.9647483 ***	0.8343119 ***	0.5619264***

Note: *** indicates significance at the 1% level.
Data source Edited by the authors

Table 4.14 Results of lag phase III

	2006–2018	2006–2008	2009–2012	2013–2018
ln(L)	0.2346904***	0.0393112	− 0.0227811	0.2218881***
ln(TK)	0.2201451***	0.1320793***	0.3009147***	0.4268999***
ln(NTK)	0.6498925***	0.9669557 ***	0.8665931***	0.5085713***

Note: *** indicates significance at the 1% level.
Data source Edited by the authors

Table 4.15 Results of lag phase IV

	2007–2018	2007–2008	2009–2012	2013–2018
ln(L)	0.2442628***	0.0295683	− 0.0320266	0.2240137***
ln(TK)	0.2683532***	0.1425264*	0.2709933***	0.4717478***
ln(NTK)	0.6013115 ***	0.9595924 ***	0.8982111***	0.4729722 ***

Note: *** and * indicate significance at the 1% and 10% levels, respectively
Data source Edited by the authors

4.4 Conclusions

4.4.1 Characteristics of Transport Investment

Firstly, the scale of fixed asset investment in the transport industry is relatively large, and its growth has slowed down in recent years. The large-scale investment in transport infrastructure has fundamentally changed the Chinese economy and society. Secondly, highway transport is the most important investment sector, and the investment in air transport is growing rapidly. Railway transport and water transport are the transport sectors with the second and third largest fixed asset investment, respectively, but the proportion of fixed asset investment in these two transport sectors shows a downward trend. Thirdly, western China's share of transport investment is growing, benefiting from the strategy of western development.

4.4.2 Characteristics of the Transport Economy

Firstly, the economic scale of the transport industry is at a high level, and its growth has slowed down in recent years. The contribution rate of the transport industry to economic growth is declining. Secondly, the total logistics costs are high, and the per unit logistics costs in China show an upward trend. Thirdly, from 2008 to 2016, the growth of the freight turnover in China was relatively stable, but the proportion of railway freight transport was obviously low. It is worth noting that the proportion of highway transport in the passenger transport market was maintained at above 53%, and the proportion of railway transport showed a downward trend from 2008 to 2016.

4.4.3 The Influence of Transport Investment on Economy

In the empirical research on the influence of transport investment on economic growth in China, we found that transport investment has an effect on economic growth, not only during the investment period but for years to come. The elasticity of fixed transport capital to GDP first decreased and then increased from 2003 to 2018. Policy makers should pay attention to the impact of the improved transport network and accessibility on economic development after transport infrastructure is completed and put into use.

References

Bhatta, S. D., & Drennan, M. P. (2003). The economic benefits of public investment in transportation. *Journal of Planning Education and Research, 22*(3), 288–296. https://xueshu.baidu.com/usercenter/paper/show?paperid=6100119c210557f61729d09842142381&site=xueshu_se

Boarnet, M. G. (1998). Spillovers and the location effect of public infrastructure. *Journal of Regional Science, 38*(3), 381–400. https://doi.org/10.1111/0022-4146.00099

Cadot, O., Röller, L. H., & Stephan, A. (2006). Contribution to productivity or pork barrel? The two faces of infrastructure investment. *Journal of Public Economics, 90*, 1133–1153. https://www.doc88.com/p-8952142139163.html?r=1

Chen, Z., & Haynes, K. E. (2017). Transportation infrastructure and economic growth in China: A meta-analysis. In H. Shibusawa, K. Sakurai, T. Mizunoya, S. Uchida (Eds.), *Socioeconomic environmental policies and evaluations in regional science. New frontiers in regional science: Asian perspectives* (vol. 24). Springer. https://doi.org/10.1007/978-981-10-0099-7_18

China Logistics Yearbook. (2017). China Statistical Publishing House. https://data.cnki.net/yearBook/single?id=N2017110176

China Statistical Yearbooks. (2017). China Statistical Publishing House. https://data.cnki.net/yearbook/Single/N2020100004

China Statistical Yearbooks. (2020). China Statistical Publishing House, Beijing. https://data.cnki.net/v3/yearbook/Single/N2020100004

Cui, S. N., Pittman, R., & Zhao, J. A. (2021). Restructuring the Chinese freight railway: Two scenarios. *Asia and the Global Economy, 1*(1), 100002. https://doi.org/10.1016/j.aglobe.2021.100002

Dimitriou, H. T. (2019). Urban public transport and economic development. In *Sustainable approaches to urban transport* (pp. 187–192). https://xueshu.baidu.com/usercenter/paper/show?paperid=1c2c0pm07q0404d07k3602204d534867&site=xueshu_se

Goldsmith, R. M. (1951). A perpetual inventory of national wealth. *Studies in Income and Wealth, 14*, 5–73. http://www.nber.org/books/unkn51-2

Li, J. W., & Zhang, G. Q. (2016). Estimation of capital stock and capital return rate of China's transportation infrastructure. *Contemporary Finance & Economics* 6, 3–14. (In Chinese). https://doi.org/10.13676/j.cnki.cn36-1030/f.2016.06.001

Mao, B. H., Sun, Q. X., & Chen, S. Q. (2009). Structural analysis on 2008 intercity transport system of China. *Journal of Transportation Systems Engineering and Information Technology, 9*, 10–18. https://doi.org/10.1016/S1570-6672(08)60043-4

National Bureau of Statistics of China (NBSC). (2021). http://data.stats.gov.cn/easyquery.htm?cn=E0103

Qin, Y. (2016). China's transport infrastructure investment: Past, present, and future. *Asian Economic Policy Review., 11*(2), 199–207. https://doi.org/10.1111/aepr.12135

Ramesh, S. (2017). Transportation infrastructure and spatial development in China. In *China's lessons for India: Volume I*. Palgrave Macmillan. https://doi.org/10.1007/978-3-319-58112-5_7

Shi, H., & Huang, S. (2014). How much infrastructure is too much? A new approach and evidence from China. *World Development, 56*, 272–286. https://doi.org/10.1016/j.worlddev.2013.11.009

Statistical Yearbook of the Chinese Investment in Fixed Assets (2009–2013, 2015–2017), China Statistical Publishing House. https://data.cnki.net/yearBook/single?id=N2019030174

Wang, J., Jiao, J., Du, C., et al. (2015). Competition of spatial service hinterlands between high-speed rail and air transport in China: Present and future trends. *Journal of Geographical Sciences, 25*, 1137–1152. https://doi.org/10.1007/s11442-015-1224-5

Wu, X., & Man, S. (2018). Air transportation in China: Temporal and spatial evolution and development forecasts. *Journal of Geographical Sciences, 28*, 1485–1499. https://doi.org/10.1007/s11442-018-1557-y

Xiao, J. (2016). Development of railway logistics in China. In: L. Wang, S. Lee, P. Chen, X. Jiang, B. Liu (Eds.), *Contemporary logistics in China. Current Chinese Economic Report Series*. Springer. https://doi.org/10.1007/978-981-10-1052-1_7

Zhang, J., Wu, G. Y., & Zhang, J. P. (2004). The estimation of China's provincial capital stock: 1952–2000. *Economic Research Journal, 10*, 35–44. http://en.cnki.com.cn/Article_en/CJFDTOTAL-JJYJ200410004.htm

Zhang, Q., Kennedy, C., Wang, T., Wei, W., Li, J., & Shi., L. (2020). Transforming the coal and steel nexus for China's ecocivilization: Interplay between rail and energy infrastructure. *Journal of Industrial Ecology, 24*(6), 1352–1363. https://ideas.repec.org/a/bla/inecol/v24y2020i6p1352-1363.html

Chapter 5
Highway Transport Efficiency

5.1 Background and Methods

5.1.1 Background

Against the backdrop of sustainable economic development and industrial transformation and upgrading, the modern transport system of China has been continuously optimized and improved. With the guidance and support of the Chinese government through a series of policies, remarkable achievements have been made in the highway transport sector. From 2013 to 2017, the total length of highways in China grew at an annual rate of 25%. By the end of 2017, the total length of highways in China reached 4.7735 million km (NBSC, 2021).

With the continuous improvement of infrastructure, the highway transport sector is playing an increasingly significant role in social and economic development. From 2014 to 2018, the annual average ratios of passenger volume and freight volume to the total transport volume were 81.12% and 76.06%, respectively (NBSC, 2021). It can be seen that the research on highway transport efficiency has a far-reaching impact on the development of the modern comprehensive transport system.

In order to accelerate the supply-side reform and promote sustainable economic development, it is necessary to scientifically evaluate the highway passenger transport efficiency (HPTE) and highway freight transport efficiency (HFTE) in China, analyze their regional characteristics, and give targeted suggestions for improving the efficiency of different regions.

5.1.2 Methods

Considering the accessibility of the statistics, the study will cover 31 provincial-level administrative regions in mainland China from 2008 to 2016. The areas will be divided into eight groups in conformity with the regional economic division of

P. Zhao and L. Zeng, *Transport Efficiency and Safety in China*, Population, Regional Development and Transport, https://doi.org/10.1007/978-981-99-1055-7_5

Table 5.1 Eight economic zones of China

Economic zones	Constitution
Northern coast	Beijing, Tianjin, Hebei, Shandong
Eastern coast	Shanghai, Jiangsu, Zhejiang
Southern coast	Fujian, Guangdong, Hainan
Northeast	Liaoning, Jilin, Heilongjiang
Middle reaches of the Yellow River	Shanxi, Inner Mongolia, Henan, Shaanxi
(Middle Yellow River)	
Middle reaches of the Yangtze River	Anhui, Jiangxi, Hubei, Hunan
(Middle Yangtze River)	

Data source NBSC (2021)

China's State Council to study the differences of HPTE and HFTE in these areas (Table 5.1; Fig. 5.1).

The EBM model can take radial and non-radial factors into account simultaneously, which can compensate for the weakness of traditional DEA methods (Yang et al., 2018). The research calculates the HPTE in 31 Chinese provinces from 2008 to 2016 using the EBM model.

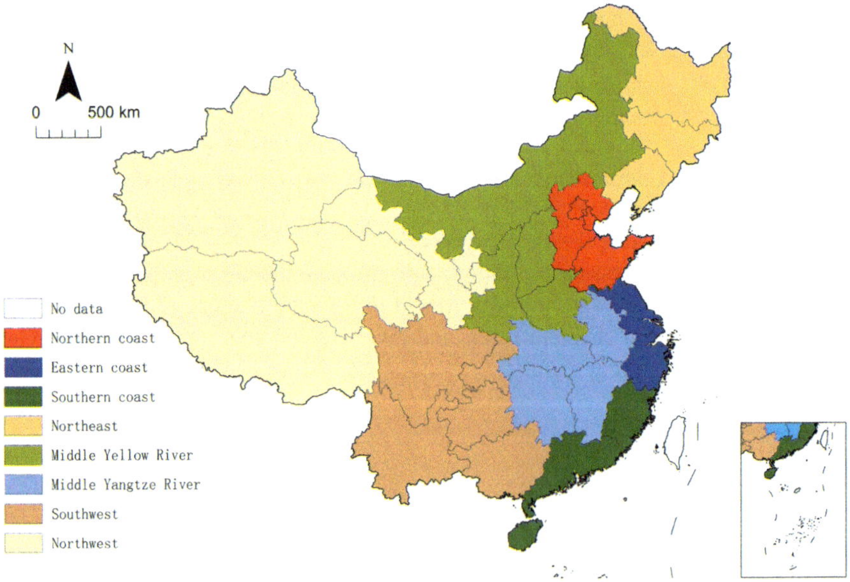

Fig. 5.1 Eight economic zones of China. *Data source* China's National Bureau of Statistics (2021)

The input indicators for the calculation of HFTE include the capital stock of the highway transport sector, the number of operating freight vehicles on highways, and the length of highways. The output indicator is the highway freight turnover (Table 5.2).

This study estimates the capital stock of the highway transport sector using the perpetual inventory method (Goldsmith, 1951). The basic formula is: $K_{it} = I_{it} + (1 - \delta)K_{it-1}$. $K_{i,t}$ represents the capital stock of the highway transport sector in t year of I province, $I_{i,t}$ represents the total fixed capital investment in the highway transport sector in t year of I province, and δ represents the capital depreciation rate of the highway transport sector. Based on the research results of Zhang et al. (2004), the capital stock of the highway transport sector in the base period is estimated by dividing the total fixed capital investment in highway transport in 2008 by 10%. The annual depreciation rate δ is 12.5% based on the research results of Liu and Liu (2007).

The data come from China Statistical Yearbook (2009–2017), Statistical Yearbook of the Chinese Investment in Fixed Assets (2009–2013, 2015–2017), China Transport Statistical Yearbook (2009–2017) and statistical yearbooks of provinces over the years (Table 5.2).

Table 5.2 The measurement index system of highway transport efficiency

	Primary indexes	Secondary indexes
Highway passenger transport efficiency	Inputs	The capital stock of the highway transport sector (billion RMB)
		The length of highways (kilometers)
		Operating passenger vehicles on highways (unit)
	Output	Highway passenger turnover (passenger-kilometer)
Highway freight transport efficiency	Inputs	The capital stock of the highway transport sector (billion RMB)
		The length of highways (kilometers)
		Operating freight vehicles on highways (unit)
	Output	Highway freight turnover (ton-kilometer)

Data source China Statistical Yearbook (2009–2017), Statistical Yearbook of the Chinese Investment in Fixed Assets (2009–2013, 2015–2017), China Transport Statistical Yearbook (2009–2017) and statistical yearbooks of provinces over the years

5.2 Measurement Results

The HPTE and HFTE of 31 provinces from 2008 to 2016 in China are calculated using the EBM model and MaxDEA 7.9 Ultra. The results are listed in Tables 5.3 and 5.4. Figs 5.2 and 5.3 are the spatial distribution maps of both HPTE and HFTE, respectively.

5.2.1 The Overall Characteristics

From 2008 to 2016, the annual average values of HPTE and HFTE in all provinces of China were 0.564 and 0.417, respectively, which means the overall level of HPTE and HFTE across China was low. As shown in Fig. 5.4, there are obvious spatial differences for HPTE and HFTE.

Generally speaking, the HPTE in coastal areas (Eastern coast, Northern coast and Southern coast) was higher than that in the middle reaches (Middle reaches of the Yangtze River, Middle reaches of the Yellow River and the Northeast), while the HPTE in the middle reaches was significantly higher than that in inland areas (Southwest and Northwest). (For the convenience of description, Northeast is temporarily included in Middle reaches.) The HFTE in Middle reaches of the Yangtze River, Middle reaches of the Yellow River and Northern coast was higher than the national level during the study period, followed by Eastern coast, Northeast, and Northwest. Southern coast and Eastern coast had the lowest HFTE (Table. 5.5).

5.2.2 The National HPTE and HFTE First Decreased and then Increased

As shown in Fig. 5.4, the mean values of both HPTE and HFTE of all provinces in China overall showed a U-shaped trend, and the lowest level was observed in 2012 and 2011, respectively. In response to the global financial crisis in 2008, China implemented a loose fiscal and monetary policy (Li et al., 2015), and many regions began to increase investment in highway infrastructure construction. The total investment in the highway transport sector increased from RMB 741.151 billion in 2008 to RMB 1385.635 billion in 2011 (Statistical Yearbook of the Chinese Investment in Fixed Assets, 2009, 2012). The continuous large-scale investment in highway construction brought about repeated construction and other problems of overcapacity, inhibiting highway transport efficiency. The Chinese government stopped the loose fiscal policy and adopted a prudent fiscal policy from 2011. As a result, the investment increment and stock of the highway transport sector were optimized, and then the overall HPTE and HFTE of China showed a fluctuating upward trend rather than falling after 2011 (Figs. 5.5 and 5.6; Tables 5.5 and 5.6).

Table 5.3 HPTE of 31 provinces in China from 2008 to 2016

Provinces	2008	2009	2010	2011	2012	2013	2014	2015	2016	Mean
Beijing	1	1	1	1	1	0.764	0.691	0.728	0.731	0.879
Tianjin	1	1	0.805	0.734	0.714	0.790	0.657	0.823	0.839	0.818
Hebei	0.330	0.318	0.339	0.344	0.312	0.341	0.266	0.392	0.345	0.332
Shanxi	0.687	0.589	0.301	0.277	0.261	0.448	0.304	0.422	0.343	0.404
Inner Mongolia	0.232	0.222	0.362	0.368	0.340	0.419	0.297	0.460	0.445	0.350
Liaoning	0.381	0.388	0.419	0.368	0.334	0.571	0.469	0.616	0.679	0.470
Jilin	0.348	0.311	0.470	0.429	0.376	0.421	0.348	0.552	0.507	0.418
Heilongjiang	0.191	0.189	0.251	0.264	0.246	0.398	0.343	0.585	0.533	0.333
Shanghai	0.565	0.551	0.640	0.623	0.636	1	1	1	1	0.779
Jiangsu	0.894	0.943	0.859	0.813	0.744	0.908	0.694	1.000	0.978	0.870
Zhejiang	0.969	0.904	0.809	0.724	0.624	0.817	0.608	0.917	0.772	0.794
Anhui	0.823	0.820	0.876	0.954	1	1	1	1	0.821	0.921
Fujian	0.618	0.639	0.397	0.351	0.314	0.592	0.445	0.580	0.536	0.497
Jiangxi	0.690	0.669	0.452	0.393	0.366	0.625	0.533	0.723	0.715	0.574
Shandong	0.796	0.766	0.884	0.867	0.781	0.600	0.483	0.680	0.654	0.723
Henan	0.827	0.782	0.668	0.761	0.751	0.746	0.876	1.000	1.000	0.824
Hubei	0.559	0.536	0.402	0.387	0.380	0.377	0.353	0.538	0.516	0.450
Hunan	0.845	0.854	0.389	0.361	0.343	0.583	0.527	0.675	0.613	0.577
Guangdong	0.912	0.985	1	1	1	1	1	1	1	0.989
Guangxi	0.933	0.925	0.707	0.664	0.616	0.609	0.459	0.670	0.638	0.691
Hainan	1	1	1	1	1	0.885	0.697	0.782	0.724	0.899
Chongqing	0.521	0.527	0.381	0.371	0.362	0.584	0.435	0.695	0.628	0.500
Sichuan	0.699	0.684	0.408	0.371	0.339	0.394	0.301	0.504	0.435	0.459
Guizhou	0.572	0.551	0.272	0.285	0.300	0.490	0.411	0.612	0.600	0.455
Yunnan	0.379	0.360	0.225	0.243	0.242	0.312	0.263	0.380	0.341	0.305
Tibet	0.214	0.187	0.102	0.085	0.078	0.201	0.166	0.178	0.154	0.152
Shaanxi	0.474	0.488	0.309	0.312	0.292	0.421	0.348	0.472	0.435	0.395
Gansu	0.395	0.335	0.392	0.453	0.442	0.513	0.464	0.671	0.621	0.476
Qinghai	0.180	0.173	0.207	0.215	0.248	0.339	0.317	0.506	0.485	0.297
Ningxia	0.354	0.343	0.339	0.326	0.322	0.434	0.411	0.604	0.551	0.409
Xinjiang	0.475	0.376	0.289	0.310	0.325	0.645	0.518	0.543	0.463	0.438
China	0.608	0.594	0.515	0.505	0.487	0.588	0.506	0.655	0.616	0.564

Data source Edited by the authors

Table 5.4 HFTE of 31 provinces in China from 2008 to 2016

Provinces	2008	2009	2010	2011	2012	2013	2014	2015	2016	Mean
Beijing	0.154	0.140	0.127	0.140	0.134	0.180	0.161	0.178	0.164	0.153
Tianjin	0.546	0.480	0.412	0.393	0.426	0.498	0.457	0.505	0.465	0.465
Hebei	0.648	0.669	0.705	0.761	0.798	1.000	0.859	1.000	1.000	0.827
Shanxi	0.344	0.247	0.217	0.214	0.233	0.258	0.237	0.369	0.391	0.279
Inner Mongolia	0.581	0.641	0.635	0.658	0.735	0.453	0.456	0.783	0.696	0.627
Liaoning	0.509	0.520	0.522	0.518	0.542	0.642	0.589	0.714	0.765	0.591
Jilin	0.265	0.254	0.239	0.244	0.257	0.345	0.313	0.409	0.421	0.305
Heilongjiang	0.305	0.202	0.172	0.163	0.166	0.209	0.189	0.256	0.260	0.213
Shanghai	0.805	0.642	0.555	0.511	0.465	0.691	0.485	0.513	0.422	0.565
Jiangsu	0.242	0.232	0.216	0.204	0.206	0.296	0.282	0.439	0.425	0.282
Zhejiang	0.405	0.376	0.323	0.298	0.289	0.295	0.274	0.463	0.522	0.361
Anhui	1	1	1	1	1	1	1	1	1	1
Fujian	0.241	0.229	0.220	0.234	0.252	0.309	0.301	0.456	0.425	0.296
Jiangxi	0.751	0.679	0.632	0.603	0.661	0.720	0.707	1	1	0.750
Shandong	1	1	0.877	0.750	0.656	0.573	0.518	0.866	0.866	0.789
Henan	0.632	0.711	0.751	0.741	0.709	0.534	0.533	0.816	0.842	0.696
Hubei	0.265	0.294	0.275	0.297	0.327	0.459	0.461	0.670	0.566	0.402
Hunan	0.293	0.304	0.334	0.398	0.468	0.534	0.543	0.740	0.633	0.472
Guangdong	0.254	0.279	0.252	0.262	0.271	0.382	0.348	0.547	0.579	0.353
Guangxi	0.364	0.352	0.349	0.379	0.426	0.438	0.409	0.633	0.625	0.442
Hainan	0.280	0.247	0.228	0.186	0.173	0.137	0.127	0.159	0.126	0.185
Chongqing	0.211	0.202	0.199	0.234	0.240	0.261	0.251	0.354	0.312	0.251
Sichuan	0.189	0.167	0.157	0.162	0.180	0.220	0.222	0.316	0.259	0.208
Guizhou	0.153	0.134	0.124	0.134	0.158	0.255	0.260	0.339	0.289	0.205
Yunnan	0.141	0.129	0.112	0.106	0.109	0.163	0.149	0.221	0.215	0.149
Tibet	0.164	0.134	0.106	0.086	0.072	0.216	0.169	0.222	0.139	0.145
Shaanxi	0.452	0.401	0.364	0.373	0.410	0.460	0.427	0.528	0.498	0.435
Gansu	0.370	0.337	0.264	0.273	0.324	0.303	0.297	0.380	0.333	0.320
Qinghai	0.373	0.325	0.278	0.262	0.252	0.231	0.215	0.274	0.251	0.273
Ningxia	0.868	0.781	0.672	0.588	0.583	0.465	0.394	0.670	0.690	0.635
Xinjiang	0.292	0.246	0.215	0.192	0.188	0.226	0.210	0.328	0.369	0.252
China	0.423	0.398	0.372	0.367	0.378	0.411	0.382	0.521	0.501	0.417

Data source Edited by the authors

Fig. 5.2 The mean value of HPTE intervals in 31 provinces from 2008 to 2016. *Data source* Edited by the authors

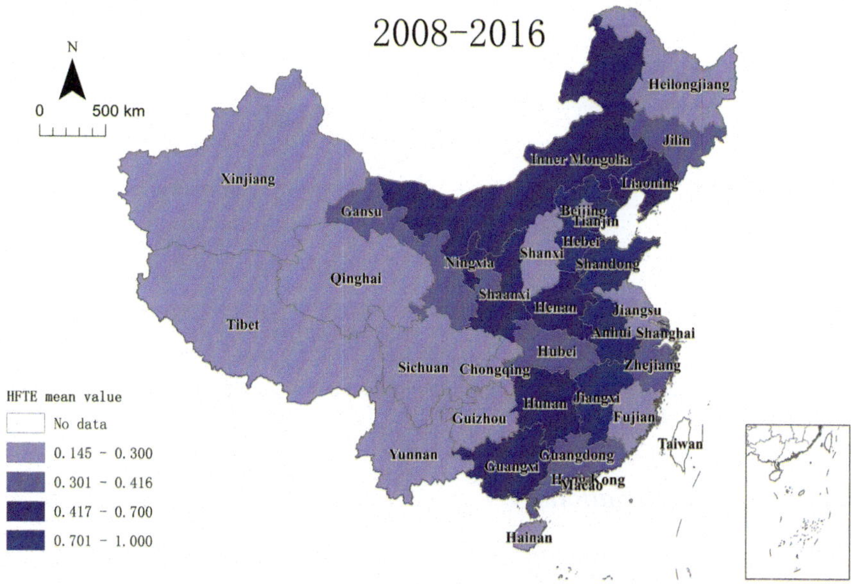

Fig. 5.3 The mean value of HFTE intervals in 31 provinces from 2008 to 2016. *Data source* Edited by the authors

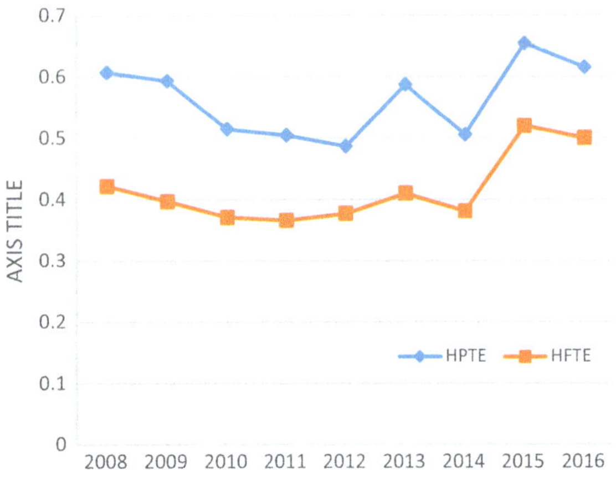

Fig. 5.4 The trend of national HPTE and HFTE from 2008 to 2016. *Data source* Edited by the authors

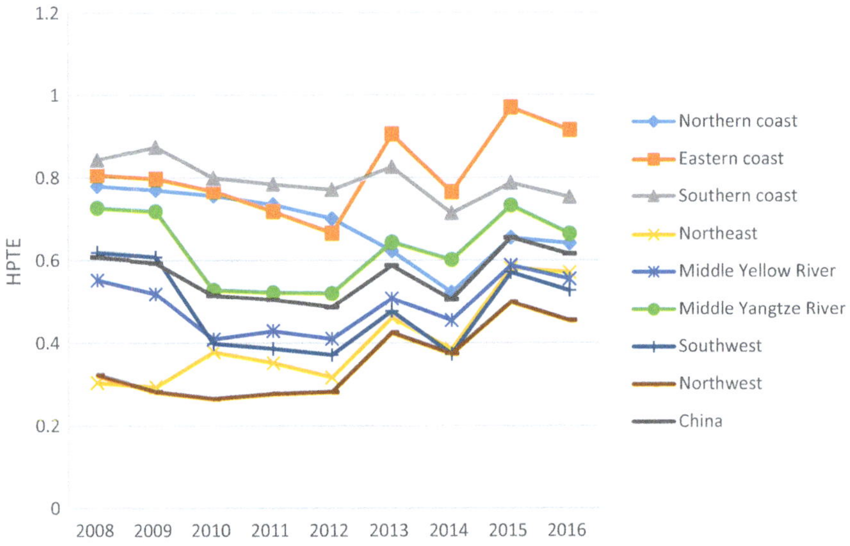

Fig. 5.5 The trend of HPTE nationwide and in eight economic zones from 2008 to 2016. *Data source* Edited by the authors

5.2.3 Spatial Variations

5.2.3.1 The Northern Coast

During the study period, in the northern coast, the annual average values of HPTE and HFTE in each province were 0.688 and 0.559, respectively, higher than the national average of 0.124 and 0.142 in the same period, indicating that the efficiency of the

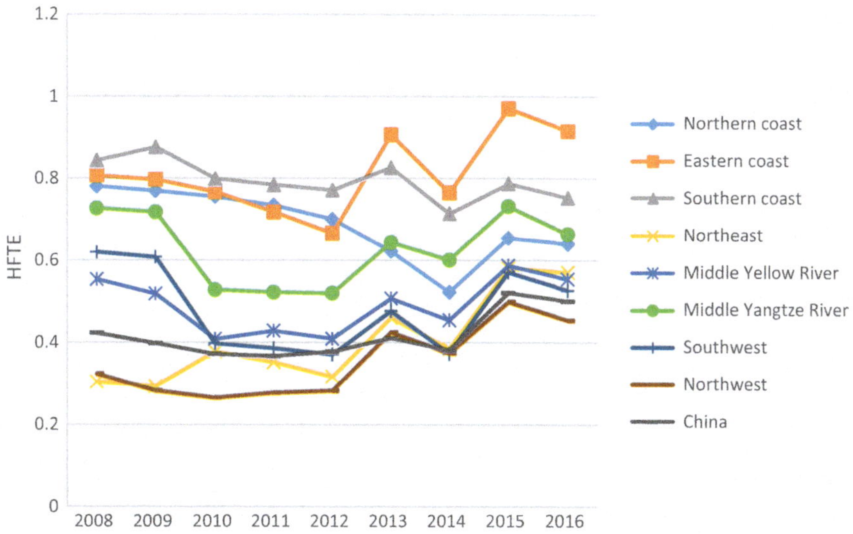

Fig. 5.6 The trend of HFTE nationwide and in eight economic zones from 2008 to 2016. *Data source* Edited by the authors

Table 5.5 HPTE of eight economic zones from 2008 to 2016

Economic zones	2008	2009	2010	2011	2012	2013	2014	2015	2016	Mean
Northern coast	0.782	0.771	0.757	0.736	0.702	0.624	0.524	0.656	0.642	0.688
Eastern coast	0.809	0.799	0.769	0.720	0.668	0.908	0.767	0.972	0.917	0.814
Southern coast	0.843	0.875	0.799	0.784	0.771	0.826	0.714	0.787	0.753	0.795
Northeast	0.307	0.296	0.380	0.354	0.319	0.463	0.387	0.584	0.573	0.407
Middle Yellow River	0.555	0.520	0.410	0.430	0.411	0.509	0.456	0.589	0.556	0.493
Middle Yangtze River	0.729	0.720	0.530	0.524	0.522	0.646	0.603	0.734	0.666	0.631
Southwest	0.621	0.609	0.399	0.387	0.372	0.478	0.374	0.572	0.528	0.482
Northwest	0.324	0.283	0.266	0.278	0.283	0.426	0.375	0.500	0.455	0.354
China	0.608	0.594	0.515	0.505	0.487	0.588	0.506	0.655	0.616	0.564

Data source Edited by the authors

highway system was at a high level. HPTE and HFTE first declined and then rose, with the lowest point in 2014.

The northern coast has a relatively comprehensive transport system, as well as an advanced industrial structure and a high technical level. The region is one of the most powerful high-tech R&D and manufacturing centers (Sun and Huo, 2019). The region is densely populated. Core cities in this region, such as Beijing, Tianjin and

Table 5.6 HFTE of eight economic zones from 2008 to 2016

Economic zones	2008	2009	2010	2011	2012	2013	2014	2015	2016	Mean
Northern coast	0.587	0.572	0.530	0.511	0.504	0.563	0.499	0.637	0.624	0.559
Eastern coast	0.484	0.417	0.365	0.338	0.320	0.427	0.347	0.472	0.456	0.403
Southern coast	0.258	0.252	0.233	0.227	0.232	0.276	0.259	0.387	0.377	0.278
Northeast	0.360	0.325	0.311	0.308	0.322	0.399	0.364	0.460	0.482	0.370
Middle Yellow River	0.502	0.500	0.492	0.497	0.522	0.426	0.413	0.624	0.607	0.509
Middle Yangtze River	0.577	0.569	0.560	0.575	0.614	0.678	0.678	0.853	0.800	0.656
Southwest	0.212	0.197	0.188	0.203	0.223	0.267	0.258	0.373	0.340	0.251
Northwest	0.413	0.365	0.307	0.280	0.284	0.288	0.257	0.375	0.356	0.325
China	0.423	0.398	0.372	0.367	0.378	0.411	0.382	0.521	0.501	0.417

Data source Edited by the authors

Qingdao, have attracted a huge flow of people and logistics from the surrounding regions, driving the improvement of regional highway transport efficiency.

In terms of provincial differences, the HPTEs in Beijing, Tianjin and Shandong were higher than that in Hebei. The HFTEs in Hebei and Shandong were higher than those in Tianjin and Beijing. Beijing and Tianjin were the growth poles in the region, attracting a large number of people in the surrounding areas (Li and Luo, 2017), and the highway passenger turnover was at a high level.

Hebei and Shandong have a high highway network density. Hebei is the province with the highest HFTE in China, and its highway freight turnover is leading the country all the year round (NBSC, 2021). Hebei undertakes the industry dispersion from Beijing and Tianjin, which brings huge freight demand to Hebei. The highway passenger turnover in Shandong is second only to Hebei all the year round, conducive to a high level of HPTE. The HFTE in Beijing was the lowest not only in the region, but also across China, with an annual average value of less than 0.2. The non-capital function distribution and transport control on the highway passage time segments and carrying capacity of freight vehicles lead to a lower freight turnover in Beijing (Zhao et al., 2020) (Figs. 5.7, 5.8, 5.9 and 5.10).

5.2.3.2 The Eastern Coast

The eastern coast had a high level of HPTE, with an annual average value of 0.8143, which was the highest among the eight regions, while the regional HFTE was little lower than the national average. The HPTE and HFTE first declined and then gyrated up, and the lowest points of regional HPTE and HFTE were in 2012.

In terms of provincial differences, the annual average values of HPTEs of three provinces were high, which were all above 0.7, and there was little difference among the three provinces, while the HFTE of Shanghai was significantly higher than that

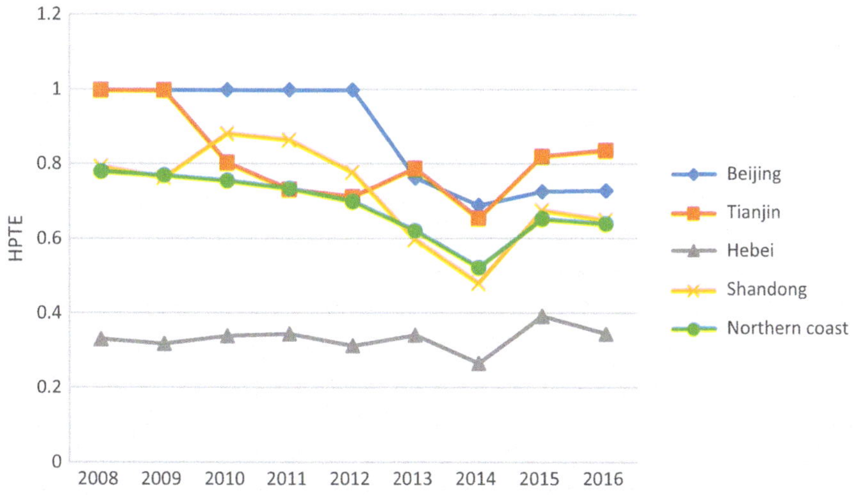

Fig. 5.7 The trend of HPTE in the northern coast from 2008 to 2016. *Data source* Edited by the authors

Fig. 5.8 The mean value of HPTE in the northern coast from 2008 to 2016. *Data source* Edited by the authors

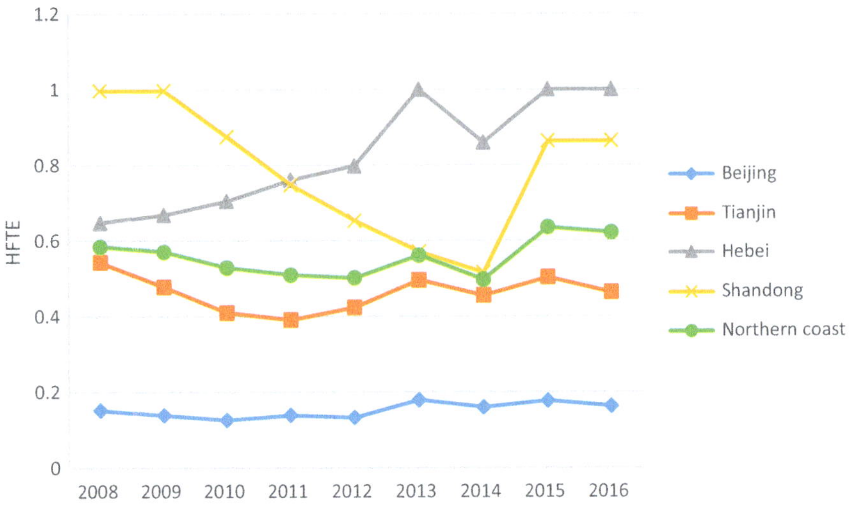

Fig. 5.9 The trend of HFTE in the northern coast from 2008 to 2016. *Data source* Edited by the authors

Fig. 5.10 The mean value of HFTE in the northern coast from 2008 to 2016. *Data source* Edited by the authors

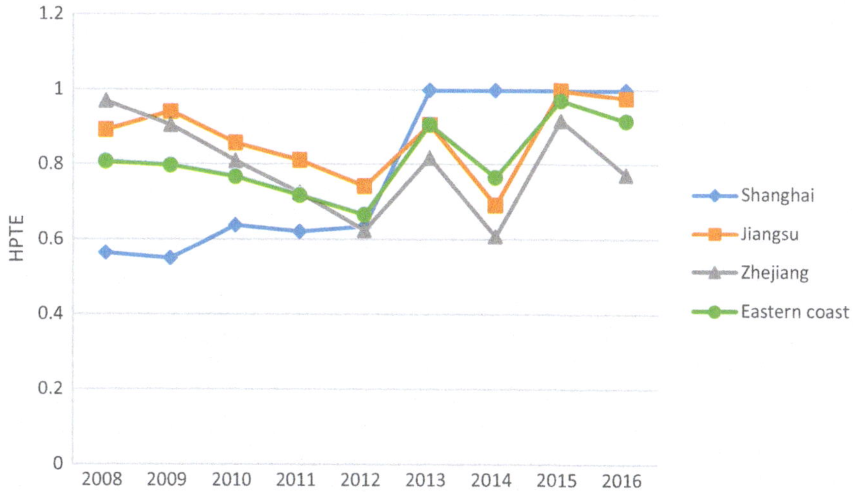

Fig. 5.11 The trend of HPTE in the eastern coast from 2008 to 2016. *Data source* Edited by the authors

of Jiangsu and Zhejiang. The HPTE of Shanghai significantly improved during the study period, and reached the production frontier surface from 2013 to 2016. The HPTEs of Jiangsu and Zhejiang first declined and then gyrated up, while the HFTE of these two provinces showed a trend of first declining and then rising (Figs. 5.11, 5.12, 5.13 and 5.14).

In the past century, the eastern coast has been China's economic center, with abundant capital and technology resources. It boasts a more reasonable industrial system and perfect transport infrastructure (Sun, 2021; Zhao et al., 2020), as well as strong economic attractiveness. At the same time, the region is an important scenic area in China, thus forming a huge highway passenger flow.

5.2.3.3 The Southern Coast

During the whole study period, in the southern coast, the annual average values of HPTE and HFTE of the provinces in the region were 0.795 and 0.278, respectively, indicating that the regional HPTE was at a relatively high level compared with the rest of the country, while the regional HFTE was lower than the national average level. The HPTE showed a downward trend, while the HFTE first decreased and then showed a fluctuating upward trend.

The southern coast is one of the earliest regions in China open to the outside world, and has accumulated abundant capital, advanced technology and rich management experience. The Pearl River Delta region in Guangdong is an important processing and production base for export goods in China, which has attracted a large number of migrant workers from the surrounding areas (Zhao et al., 2020). With the deepening

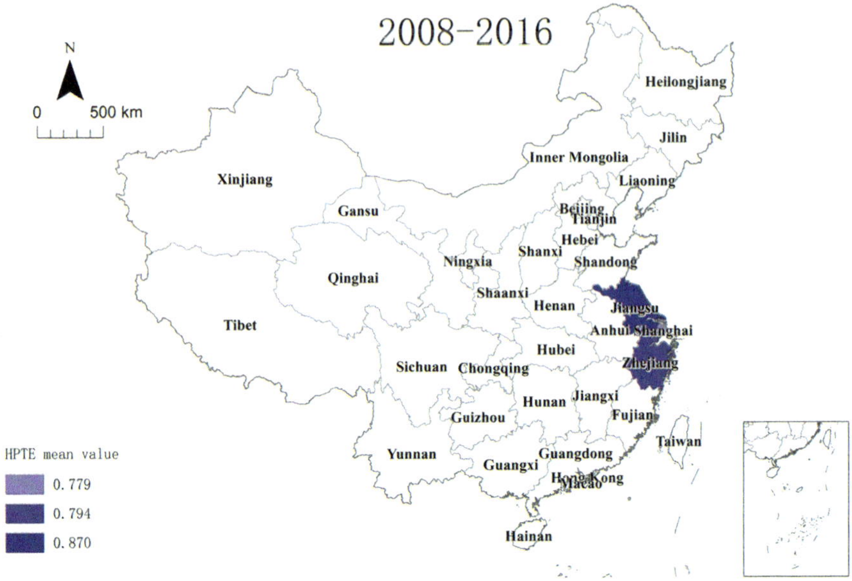

Fig. 5.12 The mean value of HPTE in the eastern coast from 2008 to 2016. *Data source* Edited by the authors

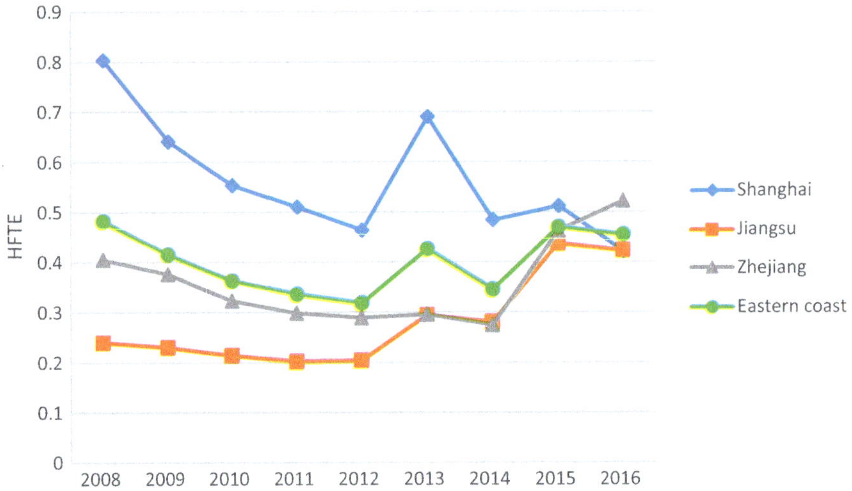

Fig. 5.13 The trend of HFTE in the eastern coast from 2008 to 2016. *Data source* Edited by the authors

Fig. 5.14 The mean value of HFTE in the eastern coast from 2008 to 2016. *Data source* Edited by the authors

of reform and industrial optimization and upgrading, the southern coast keeps introducing new technologies, a constant improvement in regional economic development is witnessed, and the economic attractiveness continues to strengthen, forming a huge highway passenger flow.

The HPTE in Hainan was in the production frontier surface from 2008 to 2012, but it showed a downward trend after 2012, while the HFTE showed a downward trend during the whole study period. Hainan is an island with a small population. Its geographical location largely limits the improvement of its HFTE (Figs. 5.15, 5.16, 5.17 and 5.18).

5.2.3.4 The Northeast

The annual average values of HPTE and HFTE in the Northeast during the study period were 0.407 and 0.3697, respectively, which were lower than the national average level, indicating that the regional highway transport efficiency was at a low level. The HPTE showed a fluctuating upward trend, while the HFTE first declined and then rose, with the lowest level in 2011.

Due to its geographical location, the Northeast features widespread distribution of frozen earth and the low density of highways and railways, which are not conducive to transport development. As an old industrial base, it struggles with concentrated

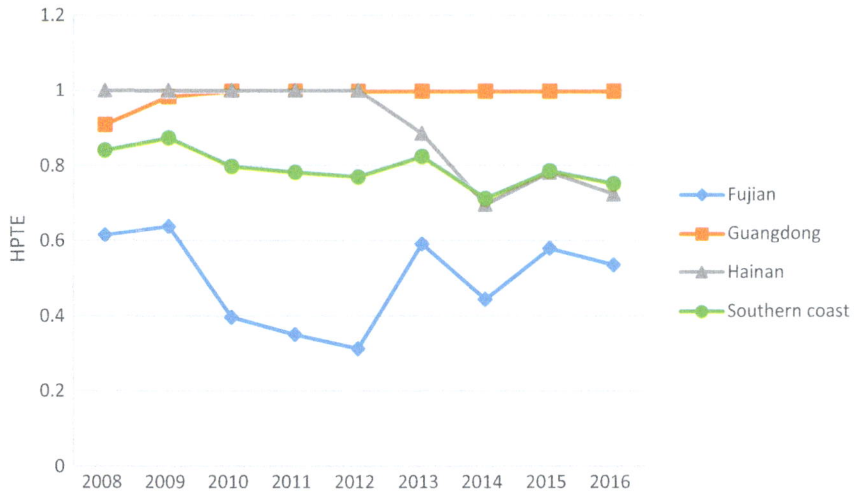

Fig. 5.15 The trend of HPTE in the southern coast from 2008 to 2016. *Data source* Edited by the authors

Fig. 5.16 The mean value of HPTE in the southern coast from 2008 to 2016. *Data source* Edited by the authors

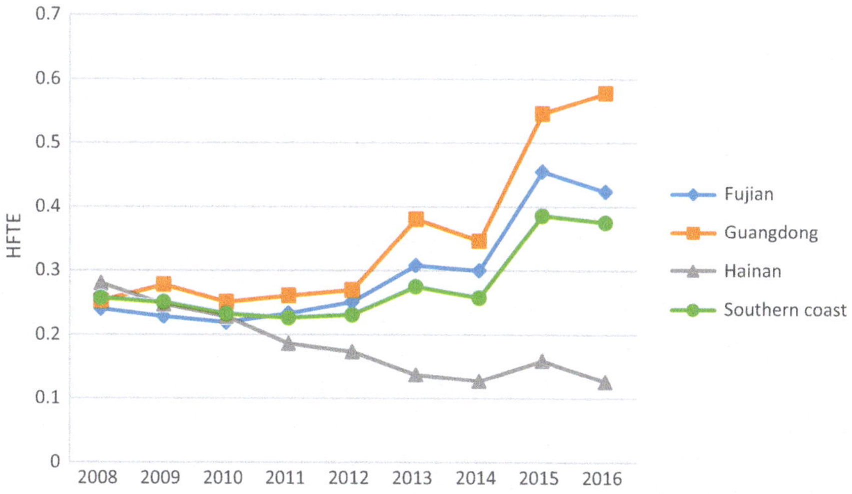

Fig. 5.17 The trend of HFTE in the southern coast from 2008 to 2016. *Data source* Edited by the authors

Fig. 5.18 The mean value of HFTE in the southern coast from 2008 to 2016. *Data source* Edited by the authors

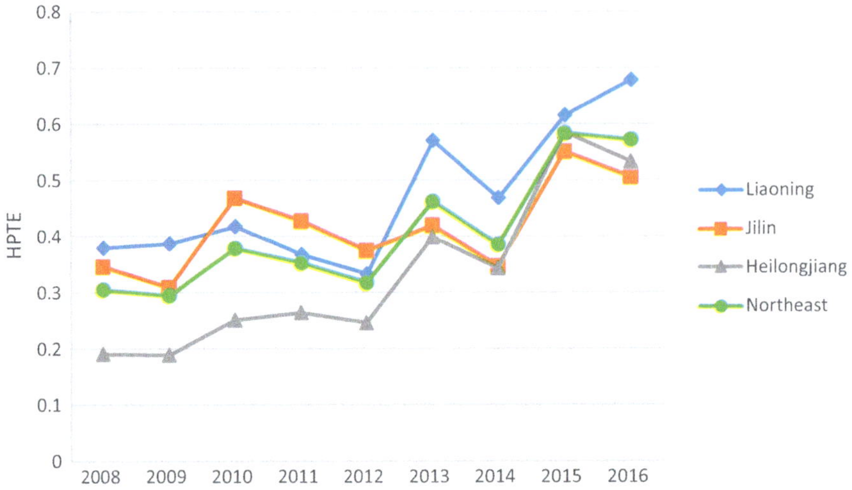

Fig. 5.19 The trend of HPTE in the Northeast from 2008 to 2016. *Data source* Edited by the authors

heavy industry made up of state-owned enterprises, and the critical industry of manufacturing has failed to innovate and transform since the beginning of the 1990s (Ren et al., 2020). In the 1990s, many state-owned enterprises encountered problems such as declining industrial output and incomes, and bankrupting. Many employees of state-owned enterprises lost their jobs and chose to leave the Northeast to earn their livings, which is the so-called "Northeast Phenomenon". The region has lost its leading position and begun to lag behind the coastal areas (Zhang, 2008). Since the 2000s the resident population growth in the Northeast has gradually slowed down, and the resident population even showed negative growth. In this context, the phenomenon of inefficient use of transport infrastructure appeared in many areas of the Northeast. The annual average values of HPTE and HFTE in the Northeast were maintained at a low level in the eight regions.

Under the guidance and support of the national strategy of revitalizing old industrial bases, the industrial development in Northeast China has been improved and upgraded in recent years, which has led to the development of highway logistics in the Northeast and promoted the improvement of HPTE and HFTE (Figs. 5.19, 5.20, 5.21 and 5.22).

5.2.3.5 The Middle Reaches of the Yellow River

During the study period, the annual average values of HPTE and HFTE of the middle reaches of the Yellow River were 0.493 and 0.509, respectively. The HPTE was lower than the national average, while the HFTE was higher than the national average level. The regional HPTE first declined and then rose, and it was at a low level from 2010

Fig. 5.20 The mean value of HPTE in the Northeast from 2008 to 2016. *Data source* Edited by the authors

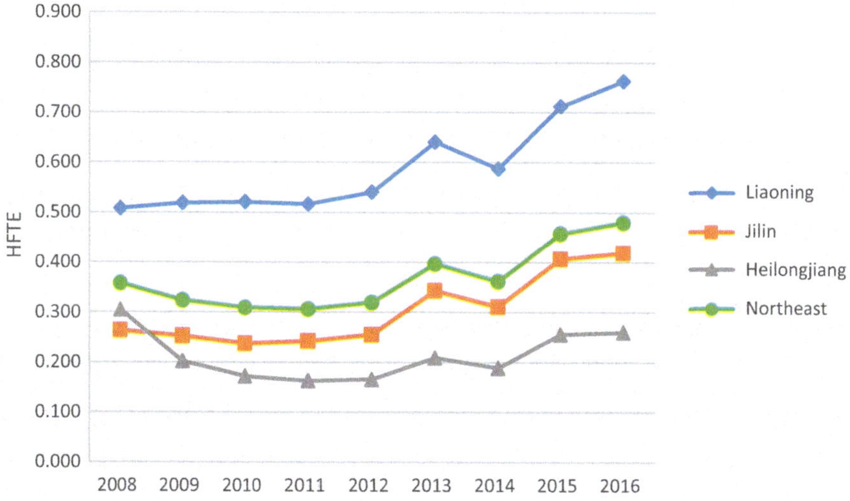

Fig. 5.21 The trend of HFTE in the Northeast from 2008 to 2016. *Data source* Edited by the authors

Fig. 5.22 The mean value of HFTE in the Northeast from 2008 to 2016. *Data source* Edited by the authors

to 2012. From 2008 to 2012, the regional HFTE was relatively stable, with little change. After 2012, it dropped before rising sharply, with the lowest level in 2014.

In terms of provincial differences, the HPTE and HFTE in Henan were relatively high in the region, especially the former, with an annual average value of more than 0.8. The HFTE in Inner Mongolia was at a high level. The HPTE in Henan, Shaanxi and Shanxi first declined and then rose, while that in Inner Mongolia showed a trend of fluctuation. The HFTE of each province showed obvious fluctuations (Figs. 5.23, 5.24, 5.25 and 5.26).

The middle reaches of the Yellow River are an important base for coal production in China, as well as an important base for power and chemical industries with coal as raw material. The secondary industry is in the leading position among the whole economy (Song et al., 2021), and the region has high demand of highway freight transport.

5.2.3.6 The Middle Reaches of the Yangtze River

The middle reaches of the Yangtze River had a high level of HPTE and HFTE. The regional annual average values of HPTE and HFTE were 0.6305 and 0.656, respectively, which were higher than the national average level. The regional HFTE was at the highest level in China.

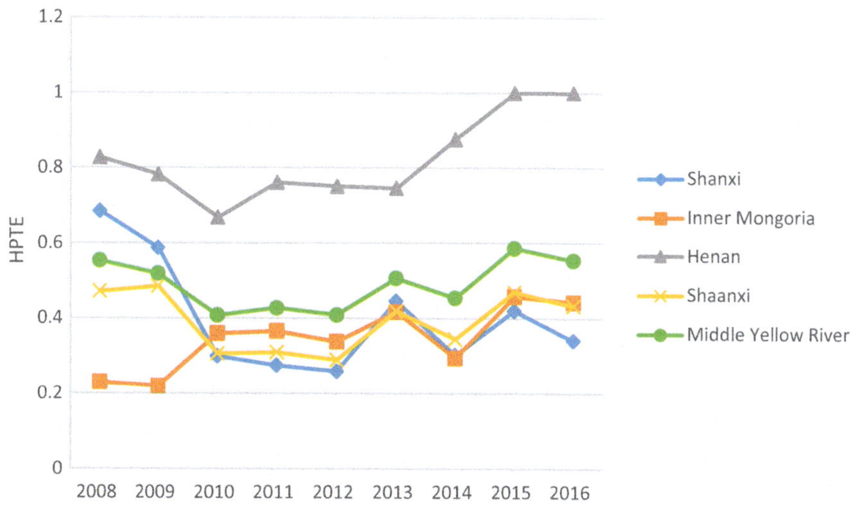

Fig. 5.23 The trend of HPTE in Middle reaches of the Yellow River from 2008 to 2016. *Data source* Edited by the authors

Fig. 5.24 The mean value of HPTE in Middle reaches of the Yellow River from 2008 to 2016. *Data source* Edited by the authors

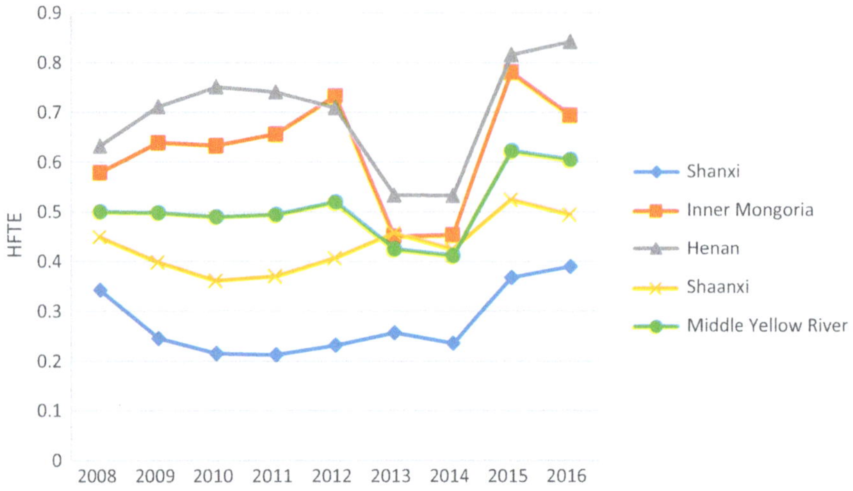

Fig. 5.25 The trend of HFTE in Middle reaches of the Yellow River from 2008 to 2016. *Data source* Edited by the authors

Fig. 5.26 The mean value of HFTE in Middle reaches of the Yellow River from 2008 to 2016. *Data source* Edited by the authors

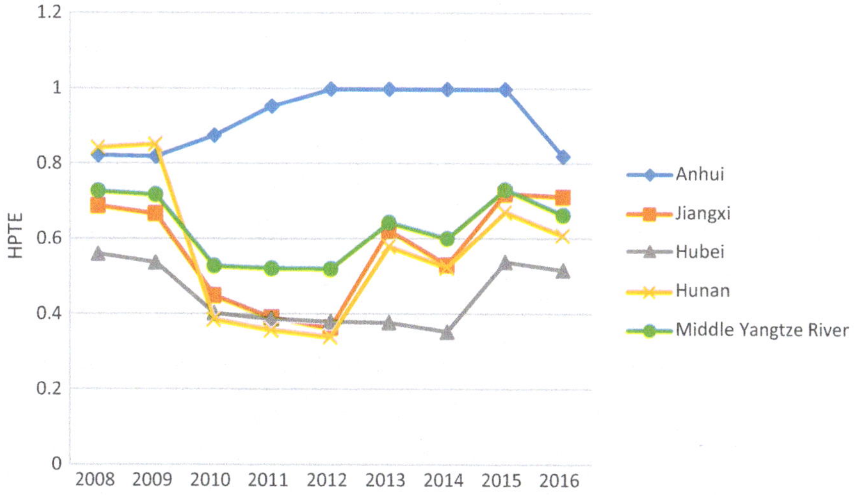

Fig. 5.27 The trend of HPTE in Middle reaches of the Yangtze River from 2008 to 2016 *Data source* Edited by the authors

The HPTE of the middle reaches of the Yangtze River was similar to that of the middle reaches of the Yellow River, showing a trend of first declining and then rising, which was at a low level from 2010 to 2012. The regional HFTE generally showed an upward trend, with a large increase in 2015, but a decline in 2016.

Although the middle reaches of the Yangtze River lag behind the eastern region in many aspects, they have advantages in water resources, shipping, mineral resources and traditional industrial foundation (Sun, 2021). Driven by the Rise of Central China Plan, after 2010, four national demonstration zones for industrial relocation have been established in the region, and many traditional manufacturing industries have been relocated from the eastern region to the middle reaches of the Yangtze River. Heavy chemical industries in the middle reaches of the Yangtze River, such as steel, petrochemical, energy and building materials industries, are developing rapidly (Sun, 2021), which have stimulated local transport demand and promoted the development of the logistics industry.

In terms of provincial differences, the HPTE and HFTE in Anhui were at the highest level not only in the region, but also across China. The HFTE in Anhui was always in the production frontier surface, and its highway freight turnover was the highest in China. During the study period, the HFTE in Jiangxi was also at a high level (above 0.75) (Figs. 5.27, 5.28, 5.29 and 5.30).

5.2.3.7 The Southwest

During the study period, the annual average values of HPTE and HFTE in Southwest China were 0.482 and 0.251, respectively, which were lower than the national average

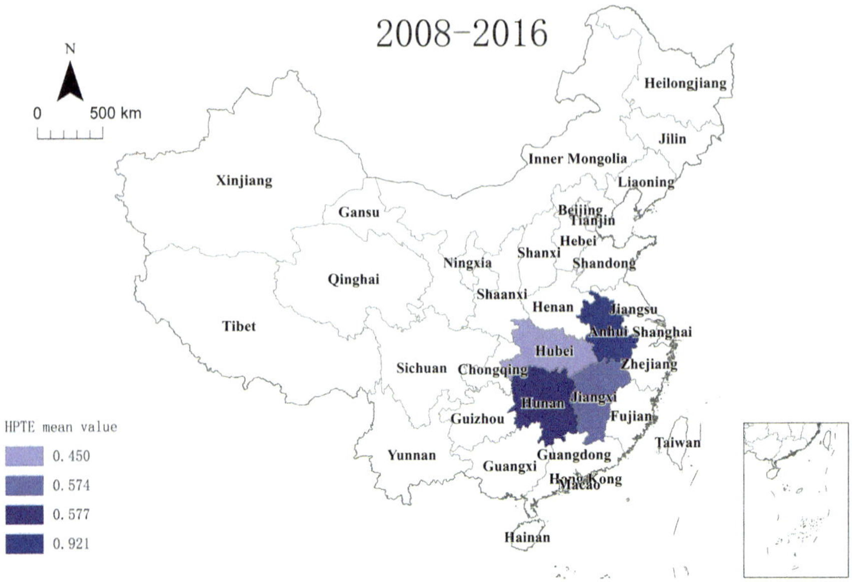

Fig. 5.28 The mean value of HPTE in Middle reaches of the Yangtze River from 2008 to 2016. *Data source* Edited by the authors

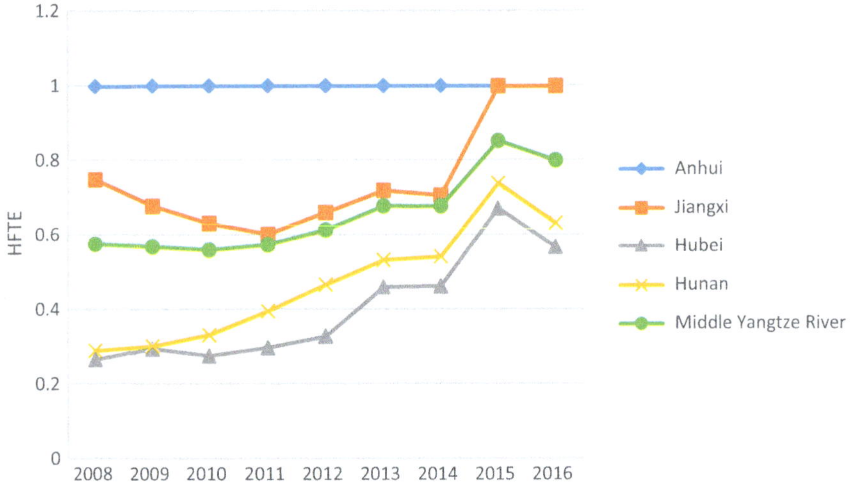

Fig. 5.29 The trend of HFTE in Middle reaches of the Yangtze River from 2008 to 2016. *Data source* Edited by the authors

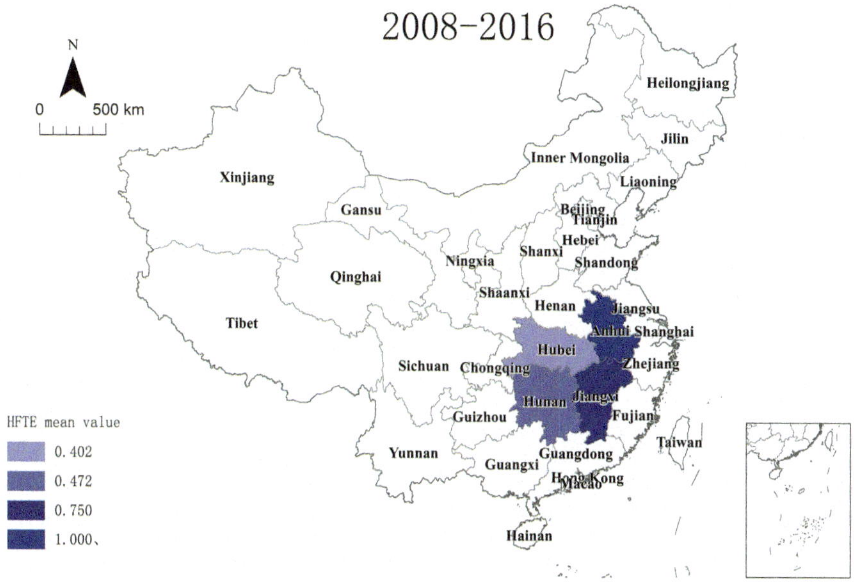

Fig. 5.30 The mean value of HFTE in Middle reaches of the Yangtze River from 2008 to 2016. *Data source* Edited by the authors

level. The regional HFTE was the lowest in China. The regional HPTE first declined and then rose, with the lowest level in 2012, while the HFTE was on the rise in the whole study time.

The Southwest is rich in natural resources, but there are many mountains in the region. The geological conditions are very complicated in some provinces of the Southwest, and the economic foundation is far weaker than coastal areas (Fu et al., 2020; Peng et al., 2021), which has a bad effect on the regional HPTE and HFTE.

In terms of provincial differences, the HPTE and HFTE in Guangxi were higher and those of Yunnan were the lowest in the region. Guangxi has the only estuary in the Southwest and is a famous tourist attraction in China, attracting a large number of tourists. The highway passenger and freight turnover is significantly at a high level. Yunnan also is a famous tourist attraction in China (Hipsher, 2017), but karst landscapes are widely distributed in Yunnan, which inhibits the highway transport demand (Figs. 5.31, 5.32, 5.33 and 5.34).

5.2.3.8 The Northwest

During the study period, the annual average values of HPTE and HFTE in Northwest China were 0.354 and 0.325, respectively, which were lower than the national average level, and the regional HPTE was the lowest in China. The HPTE and HFTE of

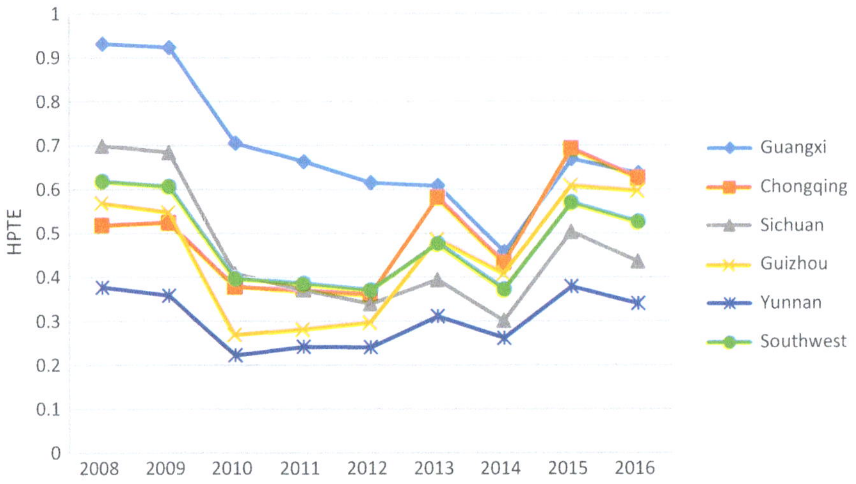

Fig. 5.31 The trend of HPTE in Southwest from 2008 to 2016. *Data source* Edited by the authors

Fig. 5.32 The mean value of HPTE in Southwest from 2008 to 2016. *Data source* Edited by the authors

the region first declined and then rose, with the lowest levels in 2010 and 2011, respectively.

The Northwest, sparsely populated, is limited by its geographical location and poor transport conditions. Its economic foundation is weak, resulting in a low level

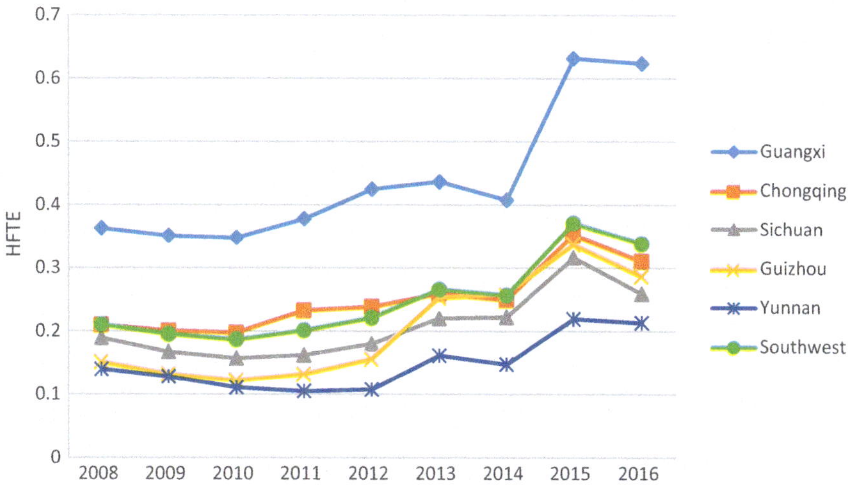

Fig. 5.33 The trend of HFTE in Southwest from 2008 to 2016. *Data source* Edited by the authors

Fig. 5.34 The mean value of HFTE in Southwest from 2008 to 2016. *Data source* Edited by the authors

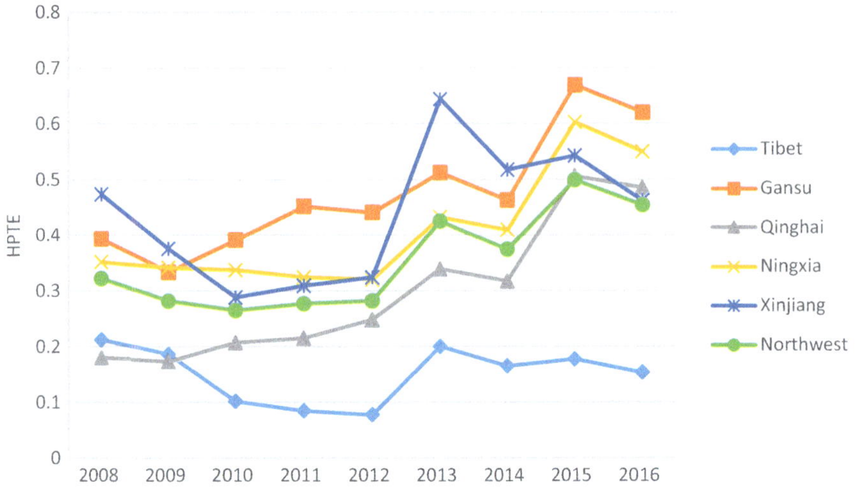

Fig. 5.35 The trend of HPTE in the Northwest from 2008 to 2016. *Data source* Edited by the authors

of transport infrastructure (Zeng, 2021; Zhao et al., 2020). With the implementation of the Western Development Program, the passenger and freight volume has increased significantly, which promotes HPTE and HFTE.

The HPTE in Gansu, Ningxia and Xinjiang was higher, with an average annual value of more than 0.4, while that in Tibet and Qinghai was lower, with an average annual value of less than 0.3. The average annual value of HFTE in Ningxia was more than 0.6, significantly higher than that of other northwest provinces. The HPTE and HFTE in Tibet were not only the lowest in the region, but also one of the lowest in China, with an average annual value of less than 0.2. This is mainly because Tibet is located in the Tibet Plateau, and its geographical location and population size determine its low demand for passenger and freight transport. The average annual value of HFTE in other northwest provinces was between 0.2 and 0.4 during the study time (Figs. 5.35, 5.36, 5.37 and 5.38).

5.3 Spatial Autocorrelation Analysis

Before using the regression analysis, a spatial autocorrelation analysis of TLEE should be conducted. If there is significant positive spatial autocorrelation in the dependent variables, spatial regression models are more appropriate than general regression models (Zeng et al., 2022).

When calculating and testing the existence of correlation of regional economic spaces, spatial statistics are frequently used in two statistics similar to the correlation coefficient. The Moran's I index and Geary index are the two spatial correlation index

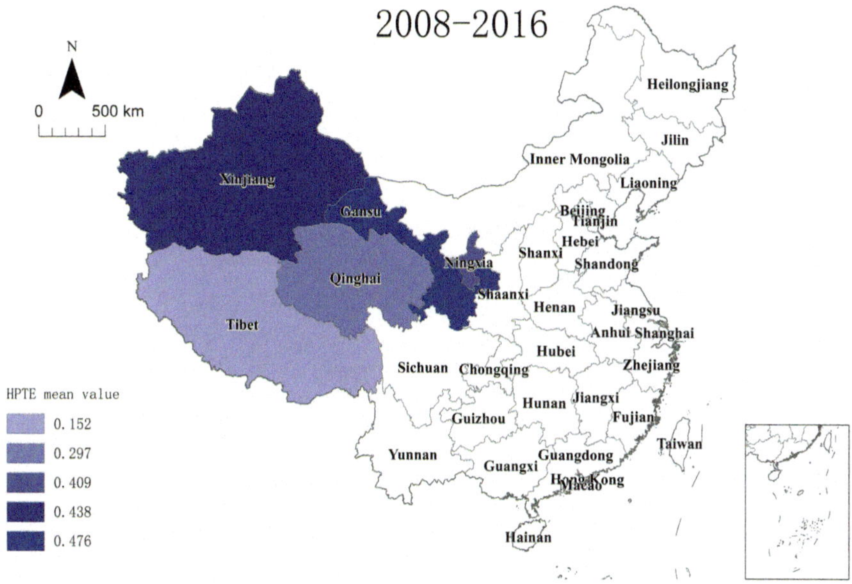

Fig. 5.36 The mean value of HPTE in Northwest from 2008 to 2016. *Data source* Edited by the authors

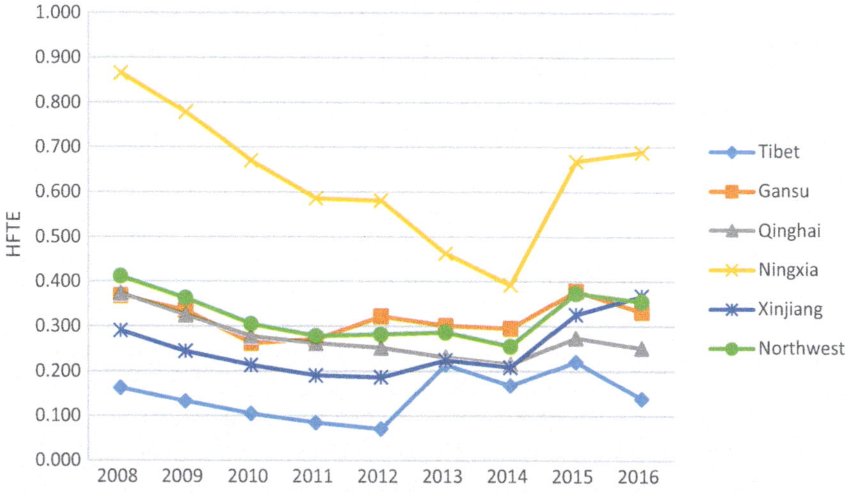

Fig. 5.37 The trend of HFTE in Northwest from 2008 to 2016. *Data source* Edited by the authors

Fig. 5.38 The mean value of HFTE in the Northwest from 2008 to 2016. *Data source* Edited by the authors

statistics. Comparing with the Geary index, the Moran's I index is stronger and less are less likely to be impacted by the deviation from the normal distribution (Wang et al., 2019b). Therefore, the application of this statistic is more extensive. At the functional level, Moran's I index can be further divided into local and global spatial autocorrelation Moran's I.

5.3.1 The Global Moran's I Analysis

This study uses Stata 13 software to analyze the spatial autocorrelation of HPTE and HFTE in provinces of China from 2008 to 2016. Tables 5.7 and 5.8 shows the results of the Global Moran's I analysis.

The positive value of the global Moran's I index of HPTE demonstrated that there existed evident spatial clustering among all regions. Obviously, the spatial correlations of regional HPTE connectivity were statistically significant during the whole research time, which indicated that provinces with high HPTE were inclined to cluster while those with relatively low HPTE crowded together.

More specifically, the Global Moran's I index of HPTE showed a fluctuating downward trend, which decreased from 0.482 in 2008 to 0.354 in 2016, indicating that the provinces with similar HPTE became more dispersed geographically, and

Table 5.7 The Global Moran's I of HPTEs in China from 2008 to 2016

Year	Moran's I	E(I)	Sd(I)	Z
2008	0.482***	– 0.033	0.121	4.259
2009	0.535***	– 0.033	0.121	4.699
2010	0.504***	– 0.033	0.120	4.464
2011	0.468***	– 0.033	0.120	4.181
2012	0.445***	– 0.033	0.119	4.011
2013	0.497***	– 0.033	0.120	4.428
2014	0.343***	– 0.033	0.118	3.186
2015	0.429***	– 0.033	0.119	3.890
2016	0.354***	– 0.033	0.119	3.255

Note: ***, **, and * indicate significance at the 1%, 5%, and 10% levels, respectively
Data source Edited by the authors

the spatial differences of HPTE between provinces became larger from 2008 to 2016 (Table 5.8).

The global Moran's I index of HFTE was significantly positive at 5% from 2008 to 2012 and from 2014 to 2016, indicating that there existed a positive spatial autocorrelation for HFTE during this period. However, the global Moran's I index was not significant in 2013. According to the research results of Yang et al. (2019) and Wang et al. (2019a), the HFTE in China generally exhibited a positive spatial correlation during the study period.

Table 5.8 The Global Moran's I of HFTEs in China from 2008 to 2016

Year	Moran's I	E(I)	Sd(I)	Z
2008	0.202**	– 0.033	0.118	1.989
2009	0.253**	– 0.033	0.118	2.426
2010	0.265**	– 0.033	0.118	2.529
2011	0.252**	– 0.033	0.118	2.426
2012	0.226**	– 0.033	0.118	2.192
2013	0.151	– 0.033	0.116	1.588
2014	0.196***	– 0.033	0.115	1.989
2015	0.281***	– 0.033	0.120	2.622
2016	0.304***	– 0.033	0.120	2.823

Note: ***, **, and * indicate significance at the 1%, 5%, and 10% levels, respectively
Data source Edited by the authors

5.3.2 *Local Spatial Autocorrelation Analysis*

5.3.2.1 The Local Spatial Autocorrelation Analysis of HPTE

The global spatial autocorrelation index Global Moran's *I* only verifies whether there exists a spatial effect problem from the overall perspective of the investigated object; however, it cannot be determined from the internal specific spatial characteristics. In order to verify the spatial correlation between local regions, the local spatial correlation index is often selected when analyzing the local characteristics of spatial association (Wang et al., 2019b). The local spatial autocorrelation index usually can be measured by the Moran scatterplot and local indicators of spatial association (LISA).

Figures 5.39, 5.40, and 5.41 present the Moran scatterplot of HPTE in China in 2008, 2012 and 2016, respectively. The scatterplot of Moran's I displays the value of a given position (X-axis) relative to its neighbor value (Y-axis). There are four quadrants in the scatterplot. The dots in the first quadrant (in which high HPTEs are surrounded by high HPTE areas, that is, a High-High (HH) agglomeration area) and the third quadrant (in which low HPTEs are surrounded by low HPTE areas, that is, a Low-Low (LL) agglomeration area) represent positive spatial correlations above or below the average value, respectively. The dots in the second quadrant (in which high HPTEs are surrounded by low HPTE areas, that is, a High-Low (HL) agglomeration area) and the fourth quadrant (in which low HPTEs are surrounded by high HPTE areas, that is, a Low-High (LH) agglomeration area) represent negative spatial correlations.

As shown in these figures, in 2008, there were 13 provinces in the first quadrant and 10 provinces in the third quadrant; specifically, provinces located in the first and third quadrants made up approximately 42% and 32.2% of the total sample provinces. In 2012, there were 11 provinces in the first quadrant and 15 provinces in the third quadrant; specifically, provinces located in the first and third quadrants made up approximately 35.5% and 48.4% of the total sample provinces. In 2016, there were nine provinces in the first quadrant and 12 provinces in the third quadrant; specifically, provinces located in the first and third quadrants made up approximately 29% and 38.7% of the total sample provinces.

As for the second and fourth quadrants, in 2008, there were six provinces in the second quadrant and two provinces in the fourth quadrant; in 2012, there were five provinces in the second quadrant, and there was no province in the fourth quadrant; in 2016, there were four provinces in the second quadrant and six provinces in the fourth quadrant.

As the above statistics indicate, most of the provinces were distributed in the first and third quadrants, while a few provinces were located in the second and fourth quadrants, indicating that the positive spatial correlation characteristics of HPTE were very obvious. The acronyms of 30 provinces in China are shown in Table 5.9.

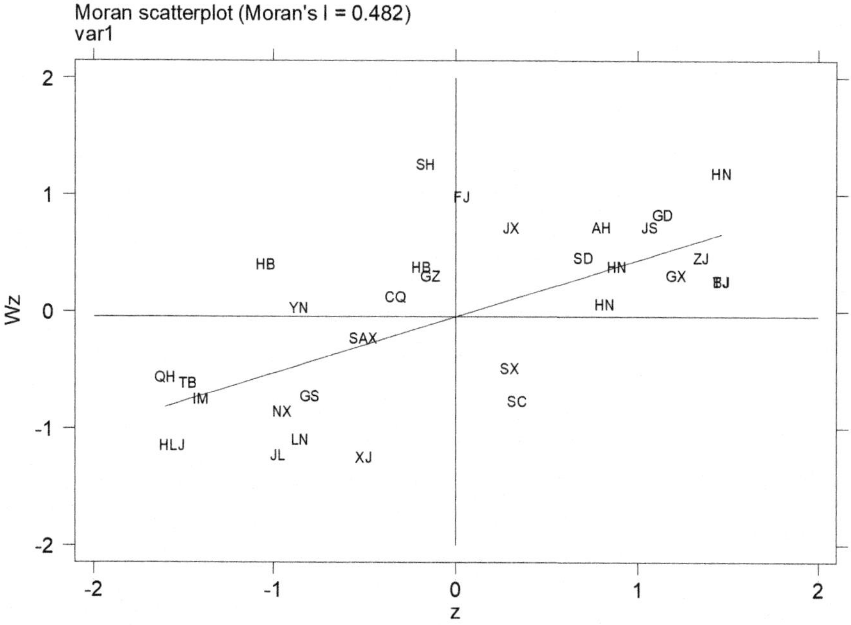

Fig. 5.39 The Moran's I scatterplot of HPTE in 31 provinces in 2008. *Data source* Edited by the authors

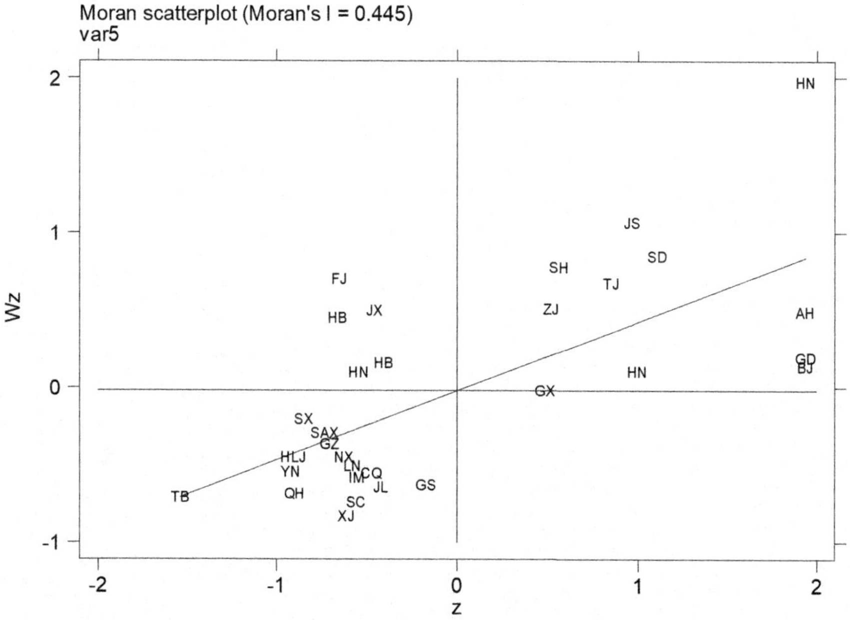

Fig. 5.40 The Moran's I scatterplot of HPTE in 31 provinces in 2012 *Data source* Edited by the authors

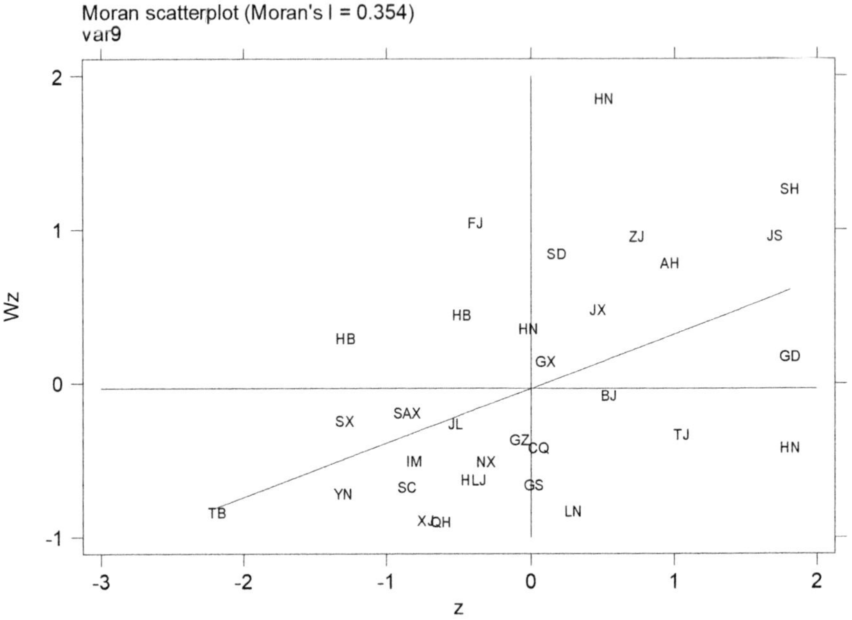

Fig. 5.41 The Moran's I scatterplot of HPTE in 31 provinces in 2016. *Data source* Edited by the authors

Table 5.9 The acronyms of provinces in China

Provinces	Acronyms	Provinces	Acronyms
Beijing	BJ	Hubei	HB
Tianjin	TJ	Hunan	HUN
Hebei	HB	Guangdong	GD
Shanxi	SX	Guangxi	GX
Inner Mongolia	IM	Hainan	HN
Liaoning	LN	Chongqing	CQ
Jilin	JL	Sichuan	SC
Heilongjiang	HLJ	Guizhou	GZ
Shanghai	SH	Yunnan	YN
Jiangsu	JS	Tibet	TB
Zhejiang	ZJ	Shaanxi	SAX
Anhui	AH	Gansu	GS
Fujian	FJ	Qinghai	QH
Jiangxi	JX	Ningxia	NX
Shandong	SD	Xinjiang	XJ
Henan	HN		

Data source Edited by the authors

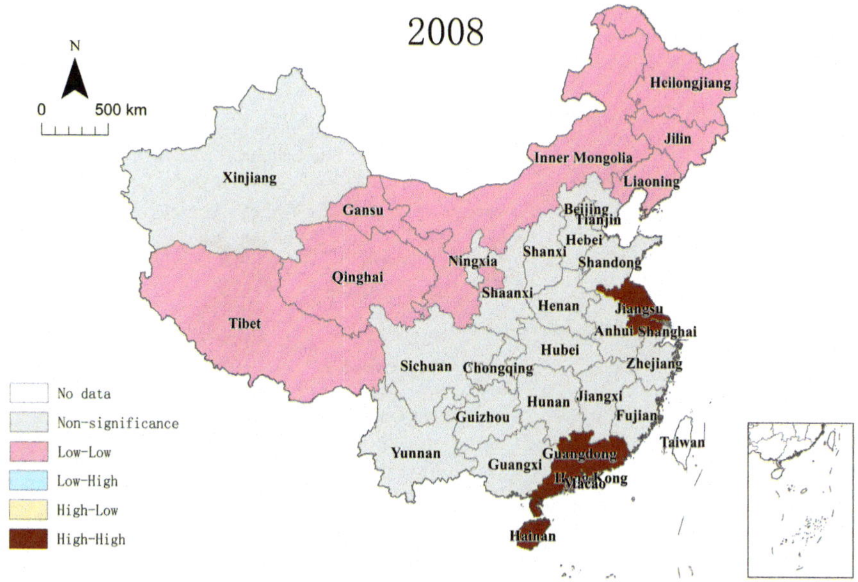

Fig. 5.42 LISA diagram of HPTE in 2008. *Data source* Edited by the authors

The LISA cluster maps in four groups—High-High cluster (red color), Low-Low cluster (pink color), High-Low cluster (beige color), and Low-High cluster (light blue color)—are provided, describing provinces at the significance level of 5%.

As shown in Fig. 5.42, in 2008, there were 10 provinces with significant locations color-coded by different types of LISA coefficients of HPTE consumption in China. More specifically, Jiangsu, Guangdong, and Hainan showed the HH-cluster of local spatial autocorrelation of HPTE. Inner Mongolia, Liaoning, Qinghai, Gansu, Jilin, Heilongjiang and Tibet revealed LL local spatial autocorrelation of HPTE; these provinces are mainly located in Northeast and Western China.

In 2012 and 2016, there were eight provinces with significant locations color-coded. More specifically, in 2012, Jiangsu, Shangdong, Anhui and Hainan revealed HH local spatial autocorrelation of HPTE, which are mainly distributed in Eastern China; Tibet and Qinghai revealed LL local spatial autocorrelation of HPTE. In 2016, Shanghai, Jiangsu, Zhejiang, Anhui and Henan revealed HH local spatial autocorrelation of HPTE, which are also mainly located in Eastern China. Sichuan, Yunnan and Tibet revealed LL local spatial autocorrelation of HPTE (Figs. 5.42, 5.43 and 5.44).

Fig. 5.43 LISA diagram of HPTE in 2012. *Data source* Edited by the authors

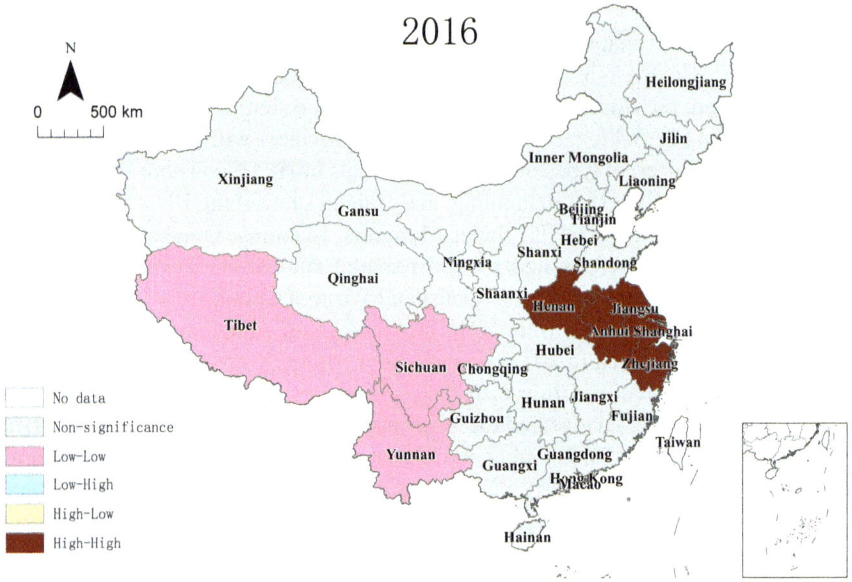

Fig. 5.44 LISA diagram of HPTE in 2016. *Data source* Edited by the authors

5.3.2.2 The Local Spatial Autocorrelation Analysis of HFTE

The Moran's I scatterplots of HFTE in 31 provinces in 2008, 2012 and 2016 are shown in Figs. 5.45, 5.46 and 5.47.

As shown in these figures, in 2008, 2012, and 2016, most provinces were located in the first, second and third quadrants. Beijing, Shanxi, Jilin, Heilongjiang and Jiangsu were stably located in the third quadrant (HH agglomeration area), which indicates that the HFTE of these five provinces was very high, and the HFTE in these five provinces had a diffusion effect on the surrounding areas. These five provinces are mainly distributed in the north and coastal areas of China.

Hebei, Inner Mongolia, Liaoning, Shandong and Henan were always in the second quadrant (LH agglomeration area) in 2008, 2012 and 2016; these five provinces are surrounded by provinces with higher HFTE. In 2008, 2012 and 2016, Sichuan, Chongqing, Yunnan, Guizhou, Tibet, Qinghai and Xinjiang were all located in the third quadrant (LL agglomeration area), indicating that the HFTE in these seven provinces was low, and these seven provinces are surrounded by LL agglomeration areas. The LL agglomeration areas are mainly distributed in Western China (Figs. 5.48, 5.49 and 5.50).

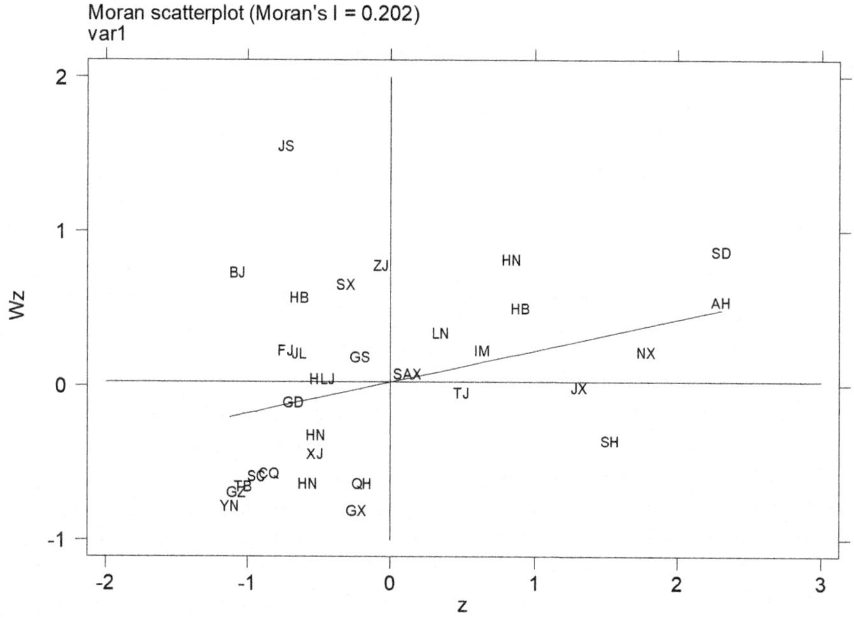

Fig. 5.45 The Moran's I scatterplot of HFTE in 31 provinces in 2008. *Data source* Edited by the authors

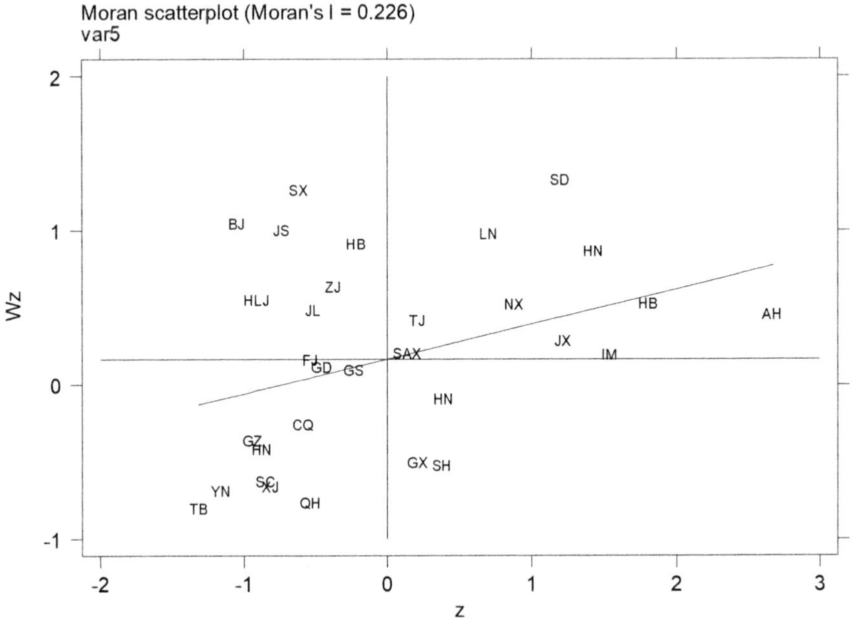

Fig. 5.46 The Moran's I scatterplot of HFTE in 31 provinces in 2012. *Data source* Edited by the author

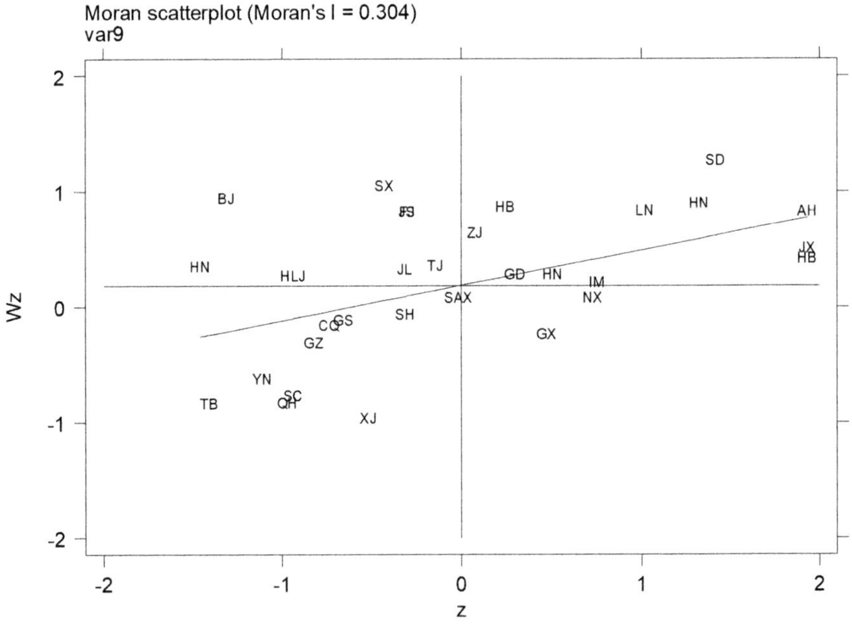

Fig. 5.47 The Moran's I scatterplot of HFTE in 31 provinces in 2016. *Data source* Edited by the authors

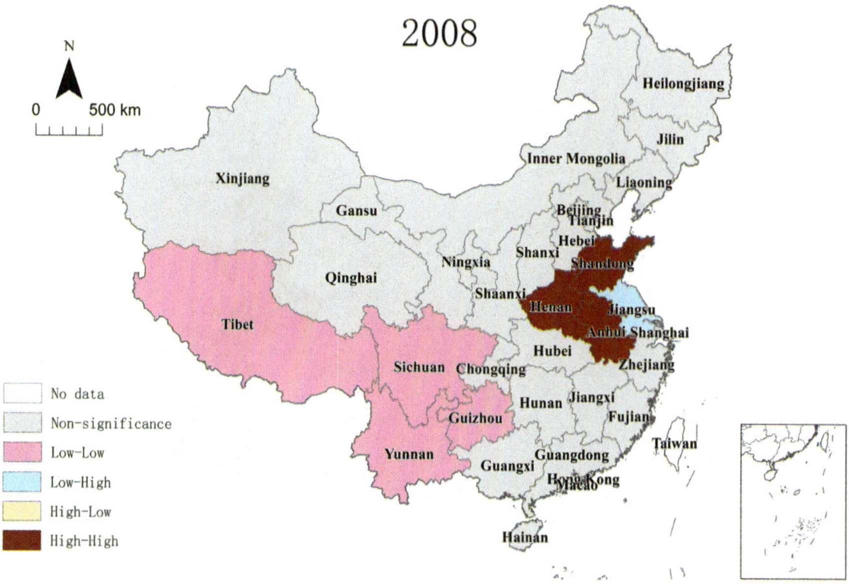

Fig. 5.48 LISA diagram of HFTE in 2008. *Data source* Edited by the authors

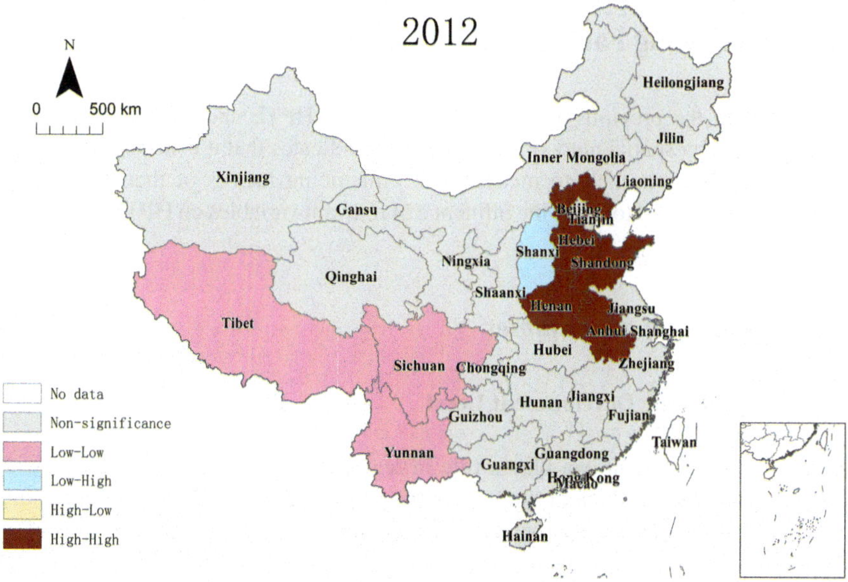

Fig. 5.49 LISA diagram of HFTE in 2012. *Data source* Edited by the authors

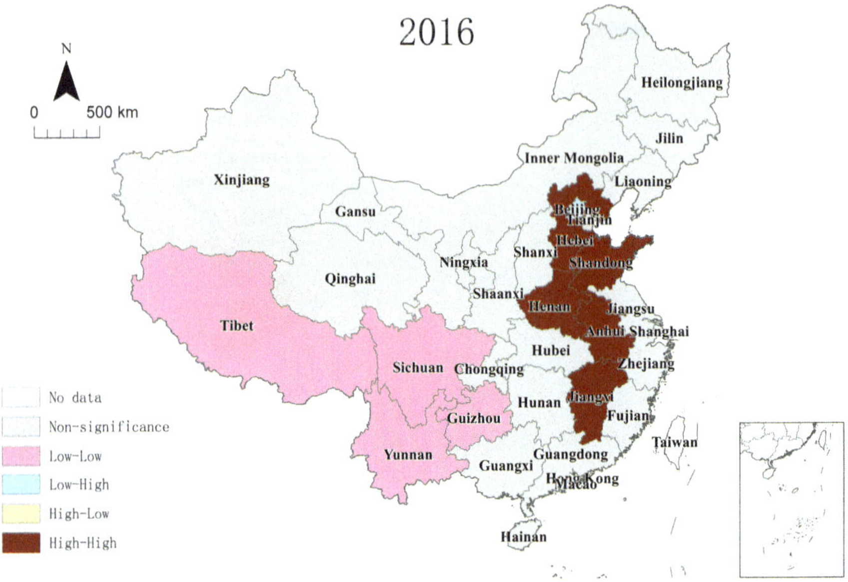

Fig. 5.50 LISA diagram of HFTE in 2016. *Data source* Edited by the authors

5.4 Influencing Factors of HPTE and HFTE

According to the previous analysis, the provincial HPTE and HFTE in China are characterized by spatial autocorrelation, which indicates that it is necessary to build spatial panel data models to measure the influencing factors of them. The SDM regression is adopted to study the influence of relevant variables on HPTE and HFTE.

5.4.1 Selection of Variables

5.4.1.1 Economic Development Level

Economic development can play a role in regional highway transport and provide capital, technology and talent support for the construction of regional infrastructure. The demand for highway passenger transport ultimately comes from the development of economy, the improvement of residents' income level and urbanization (Ren, 2009). The rapid development of the regional economy can also produce a large number of business activities and logistics demands (Zhao, 2014). China will have a large number of goods to be transported by road for a long time in the future. Therefore, this study chooses the economic development level as an important dependent variable in the regression analysis of the driving factors of HPTE and HFTE.

5.4.1.2 Passenger Transport Structure

With the improvement of the railway network and the development of high-speed railways, railway transport has had a fierce impact on the highway passenger transport market, which has affected HPTE to a certain extent. For example, the opening of Cheng-Chong high-speed rail line has led to a sharp decrease in highway passenger volume. Zhang and Liu (2011) believe that the railway competition significantly reduces highway transport efficiency. Therefore, this study chooses the passenger transport structure as an important dependent variable in the regression analysis of the driving factors of HPTE, referring to the proportion of the railway passenger volume in the total passenger volume.

5.4.1.3 Highway Structure

Highway structure affects the accessibility of highway transport. Expressways have the advantages of high driving speed, high transport capacity, convenience and comfort, and the reduction of energy consumption (Dong et al., 2016; Sun et al., 2018). Expressways are the primary choice for long-distance rapid transport, and are the skeleton of the national trunk road network, which play an irreplaceable important role in economic growth. This study uses the highway structure as a dependent variable to analyze the influencing factors of HPTE and HFTE, and selects the proportion of expressways to the total highways to represent highway structure.

5.4.1.4 Population Density

Population density affects the utilization frequency of transport infrastructure. The low population density will lead to low benefits of highway transport infrastructure. Yan and Wang (2011) believe that population density is positively related to infrastructure use efficiency. Ding and Liao (2016) believe that the moderate increase in population density plays a role in reducing marginal cost and improving the economic efficiency of infrastructure investment. Therefore, the study chooses the population density as an important variable affecting HFTE, which is equal to the ratio of regional permanent resident population to the regional area.

5.4.1.5 Urbanization Level

Urbanization can bring a lot of investment in and construction of transport infrastructure and heavy passenger and cargo demand (Xie et al., 2014), which will affect HPTE and HFTE. Therefore, urbanization level is chosen as an important dependent variable in the regression analysis of the driving factors of HPTE and HFTE, which is equal to the ratio of urban permanent resident population to the total permanent resident population.

5.4.1.6 Industrial Structure

The development of the second and tertiary industries can provide a large number of employment opportunities in cities, attracting more and more agricultural surplus labor to cities for better job opportunities, which will affect HPTE. The passenger transport sector belongs to the tertiary industry. Therefore, this book chooses the proportion of the tertiary industry in GDP to measure the impact of industrial structure on HPTE.

The primary and secondary industries provide mostly physical products, which have a large demand for transport and storage. Therefore, they have a greater impact on HFTE. Based on the research results of Liu et al. (2018) and Xu (2021), this book selects the proportion of the secondary industry to GDP to reflect the impact of industrial structure on HFTE (Tables 5.10 and 5.11).

Table 5.10 Influencing factors of HPTE

Explanatory variables	Definitions of variables
Economic development level (EDL)	GDP per capita (10^4 RMB)
Passenger transport structure (PTS)	Proportion of the railway passenger volume to the total passenger volume (%)
Highway structure (HS)	Proportion of expressways to the total highways (%)
Population density (PD)	Ratio of regional permanent resident population to the regional area (person/km^2)
Urbanization level (UL)	Proportion of urban permanent resident population in the total permanent resident population (%)

Data source The authors, edited from China Statistical Yearbook (2017)

Table 5.11 Influencing factors of HFTE

Explanatory variables	Definitions of variables
Economic development level (EDL)	GDP per capita (10^4 RMB)
Highway structure (HS)	Proportion of expressways to the total highways (%)
Population density (PD)	Ratio of regional permanent resident population to the regional area (person/km^2)
Urbanization level (UL)	Proportion of urban permanent resident population in the total permanent resident population (%)
Industrial structure (IS)	Proportion of the secondary industry to GDP (%)

Data source The authors, edited from China Statistical Yearbook (2017)

5.4.2 *Regression Analysis of HPTE*

5.4.2.1 Construction of the Regression Model

Before the regression analysis of the driving factors of HPTE, the LR test and Wald test are applied to select the suitable model. The LR and Wald test results reject the null hypothesis that the SDM can be transformed into the SLM or SEM. Additionally, the Hausman results suggest that fixed effects should also be selected. Based on the above test results, the study finally selects the spatial Durbin model under fixed effects to empirically investigate the driving factors of HPTE.

We consider the efficiency of each province as the dependent variable and influencing factors as explanatory variables to analyze their impacts on HPTE, and establish the SDM regression model based on Eq. (2.19) based on the Zhao et al. (2020), which can be expressed as follows:

$$
\begin{aligned}
HPTEi,t = {} & \rho \sum_{j=1}^{N} Wi,jHPTEi,t + \beta 1DELi,t + \beta 2HSi,t + \beta 3RTSi,t \\
& + \beta 4PDi,t + \beta 5ULi,t + \beta 6ISi,t + \theta 1 \sum_{j=1}^{N} Wi,jDELi,t \\
& + \theta 2 \sum_{j=1}^{N} Wi,jHSi,t + \theta 3 \sum_{j=1}^{N} RTSi,t + \theta 4 \sum_{j=1}^{N} PDi,t \\
& + \theta 5 \sum_{j=1}^{N} Wi,jULi,t + \theta 6 \sum_{j=1}^{N} Wi,jISi,t + \mu + \varepsilon i,t
\end{aligned} \tag{5.1}
$$

The SDM regression model based on fixed effects contains three types: spatial fixed-effects, time fixed-effects, and spatial and time fixed-effects. Based on the LR test, the spatial fixed-effects SDM model is chosen as the best-fit model (Table 5.12).

5.4.2.2 The Regression Results

The results of SDM with spatial and time fixed-effects show that the economic development level, passenger transport structure, highway structure and industrial structure show a significant positive impact on HPTE, while the impact of population density and urbanization level on HPTE is not significant.

Table 5.12 Durbin model regression results of HPTE

	Spatial fixed-effects	Time fixed-effects	Spatial and time fixed-effects
EDL	0.0157261	– 0.02505555**	0.0411086 ***
HS	– 1.217424***	– 0.5944091***	– 1.276832***
RTS	2.047809*	2.592125**	2.447026**
PD	0.0005886***	0.0000786***	0.0001626
UL	– 0.7357372	0.4037739	– 0.5158126
IS	0.3778031	– 0.513978	0.6637999***
W*EDL	– 0.017414	– 0.040028**	0.0484112*
W*HS	1.540297***	0.6808313***	0.9160301***
W*RD	– 0.820538	6.530794**	– 0.0048367
W*PD	– 0.0021446***	0.000267***	– 0.0021191***
W*UL	– 0.861292	0.2316134	– 0.1751384*
W*IS	1.398836***	0.5033556	2.235505
ρ	0.5752435***	0.3710904***	0.3784417***
Variance sigma2_e	0.0053695***	0.0250187***	0.0051901 ***
R-squared	0.4196	0.0671	0.1188
Log-likelihood	332.3546	122.748	345.3449

Note: ***, **, and * indicate significance at the 1%, 5%, and 10% levels, respectively
Data source Edited by the authors

Economic Development Level

There is a significant positive correlation between economic development level and HPTE as expected. It can be seen from the previous analysis that the overall HPTE of China presents a spatial decreasing trend from coastal areas to the middle reaches to inland areas. The overall level of HPTE in coastal areas is the highest, but the rapid economic development has led to the gradual increase of traffic pressure and congestion in some urban districts, resulting in the inconvenience of highway transport. In everyday management, traffic management departments should pay attention to the introduction of talent and science and technology, and carry out comprehensive treatment of roads, in a bid to reduce traffic congestion and to ensure convenient and fast transport. Meanwhile, the southwest and northwest regions belong to inland areas, where the economic development and highway transport infrastructure lag behind. Therefore, the local governments should increase investment in highway construction in these inland areas.

Passenger Transport Structure

Passenger transport structure has a significant negative impact on HPTE at the level of 1%. It indicates that high-speed rail has an inhibitory effect on HPTE. It is necessary to carefully analyze the impact of high-speed rail on highway passenger transport in various regions, then adjust the number of passenger trains and routes, and try to improve the passenger carrying rate. It is advisable to adjust the structure of highway passenger transport tools; for example, national highway and expressway passenger transport needs to be dominated by high-grade buses, while rural highway passenger transport should be dominated by ordinary buses and intermediate buses. Therefore, the government departments should roll out some preferential policies for the production and sales of high-grade passenger cars, and increase the proportion of high-grade passenger cars in national highway and expressway passenger transport.

Highway Structure

The regression coefficient of highway structure is significantly positive as expected. From 2008 to 2016, the expressway construction made remarkable achievements, and the length of expressways in operation increased from 60,300 km to 131,000 km. However, the proportion of expressways to the total length of highways in operation only increased from 1.6 to 2.8%, which was still at a low level. The local governments need to increase investment in expressway construction. Efforts are necessary to establish a complete system of laws and regulations to regularly maintain highway safety, and renovate problematic highway sections.

Industrial Structure

The regression coefficient of industrial structure is significantly positive, indicating that the greater the proportion of the tertiary industry in GDP, the higher the HPTE. The highway passenger transport sector can integrate with other tertiary industries to realize the transformation and upgrading of the transport industry; for example, highway passenger transport enterprises can develop the internet technology for the network information management platform, providing passenger services online.

Population Density and Urbanization Level

The regression coefficients of population density and urbanization level are not significant, which indicates that they are not the factors that affect HPTE. In the twenty-first century, there are some great changes in passenger transport modes in China. The proportion of highway passenger turnover in the total passenger turnover presents a downward trend, which decreased from 54.3% in 2000 to 32.7% in 2016. In the same period, the proportion of railway and aviation passenger turnover increased from 36.7% and 7.9% to 40.3% and 26.8%, respectively. Highway transport, railway transport and air transport show parallel development. Therefore, the passenger transport demand caused by population growth and urbanization is no longer dominated by highway passenger transport. That is why the population density and urbanization level are not the driving factors of HPTE.

5.4.3 *Regression Analysis of HFTE*

5.4.3.1 Construction of the Regression Model

According to the test results of Wald statistics and LR statistics, both Wald spatial error and Wald spatial lag reject the null hypothesis at a significance level of 1% regarding the question of whether SDM can be simplified to SLM or SEM, which further verifies that the SDM is more suitable for the application of regression analysis. Meanwhile, Hausman's test results show that the fixed effects model is the most suitable one.

With HFTE as the dependent variable and the economic development level, highway structure, population density, urbanization level and industrial structure as the analytic variables, the SDM based on fixed effects is constructed:

$$\begin{aligned} RFTEi, t = {} & \rho \sum_{j=1}^{N} Wi, jRFTEi, t + \beta 1DELi, t + \beta 2HSi, t + \beta 3PDi, t \\ & + \beta 4ULi, t + \beta 5ISi, t + \theta 1 \sum_{j=1}^{N} Wi, jDELi, t + \theta 2 \sum_{j=1}^{N} Wi, HSji, t \\ & + \theta 3 \sum_{j=1}^{N} Wi, jPDi, t + \theta 4 \sum_{j=1}^{N} Wi, jULi, t + \theta 5 \sum_{j=1}^{N} Wi, jISji, t \\ & + \mu + \varepsilon i, t \end{aligned} \tag{5.2}$$

The LR test found that the SDM based on spatial and time fixed effects is fitter than the other fixed effects models. Therefore, the results of SDM based on spatial and time fixed effects are selected to conduct regression analysis (Table 5.13).

5.4.3.2 The Regression Results

The results of the spatial and time fixed effects of SDM indicate that highway structure and industrial structure show a significant positive impact on HFTE, and that population density shows a significant negative impact on HFTE. The economic development level and urbanization level have no significant impact on HFTE.

Table 5.13 Durbin model regression results of HFTE

	Spatial fixed-effects	Time fixed-effects	Spatial and time fixed-effects
EDL	– 0.0134612*	– 0.0232198*	– 0.0207185
HS	3.949959**	3.2427833**	4.413891***
PD	– 0.0005283***	0.0001657***	– 0.0003519**
UL	0.3221291***	– 0.8717211***	0.2902116
IS	0.3652746***	1.770373***	0.8769116***
W*EDL	– 0.0235264	0.0307543	0.0075875
W*HS	2.508061	12.72758***	0.1009031
W*PD	0.0011866***	– 0.001388**	0.0014324***
W*UL	0.018161	– 0.0555263	0.0540423
W*IS	– 1.212304***	0.8744742**	0.7267649*
ρ	0.3769752***	0.1387432	0.046186***
Variance sigma2_e	0.0049505 ***	0.0267261***	0.0042774***
R-squared	0.0220	0.3859	0.0384
Log-likelihood	339.5938	109.3471	364.8875

Note: ***, **, and * indicate significance at the 1%, 5%, and 10% levels, respectively
Data source Edited by the authors

Highway Structure

The regression coefficient of highway structure is significantly positive as expected, which indicates that the higher the ratio of expressways to the total length of highways in operation, the higher the HFTE. The construction of expressways can not only better promote the rapid development of the logistics industry, but also reduce the service cost of the logistics industry and ultimately benefit the public. At present, expressway transport tools are gradually developing towards the direction of high speed, large capacity and intelligence. Large vehicles can transport goods and passengers with higher efficiency, lower energy consumption, and less pollution to the environment; therefore, the development of vans, container vehicles, and special vehicles is necessary. Government departments should formulate certain preferential policies for the production and sales of large freight vehicles to increase the proportion of large freight vehicles to the total freight vehicles.

Population Density

Population density has a significant negative effect on HFTE, which is contrary to expectations. In China, many workforces flow to the regions with a relatively high level of economic development, such coastal provinces, and the population density of these regions has surged, producing population pressure. These regions should improve the proportion of education and scientific research in GDP so as to improve and optimize the quality of the population, which will promote the coordinated, stable and sustainable development of the national economy.

Industrial Structure

Industrial structure has a significant positive impact on HFTE at the level of 1%. It indicates that the secondary industry has a positive effect on the regional HFTE, which is consistent with the research results of Xu (2021). For the freight transport enterprises, there exist some problems in the traditional highway freight transport mode, such as dispersed scale and lack of intensive management, all of which lead to low utilization efficiency and serious waste of transport resources. Therefore, transport managers should establish a scientific and reasonable system of freight transport, and set up an open information platform to realize transport resource sharing. Small highway freight transport enterprises should be consolidated to form a large powerful company and a large-scale highway operation system.

Economic Development Level and Urbanization Level

The regression coefficients of economic development level and urbanization level are not significant, which shows that they cannot have a significant impact on HFTE. From 2008 to 2016, the proportion of highway freight turnover in the total freight turnover varied between 29.8 and 34.3% in China, which means that the new freight demand brought about by economic development and urbanization has a little dependence on highways; this may be one of the important reasons that the level of economic development and urbanization does not have a significant impact on HFTE.

5.5 Conclusions

5.5.1 The Spatial Distribution Differences of Both HPTE and HFTE in China Are Pretty Evident

Generally speaking, the HPTE in coastal areas (the eastern coast, the northern coast and the southern coast) was higher than that in the middle reaches (the middle reaches of the Yangtze River, the middle reaches of the Yellow River and the Northeast), while the HPTE in the middle reaches was significantly higher than that in inland areas (the Southwest and the Northwest). (For the convenience of description, the Northeast is temporarily included in the middle reaches.) The HFTE in Middle reaches of the Yangtze River, Northern coast, and Middle reaches of the Yellow River was higher than the national level during the study period, followed by the northern coast, the

Northeast and the Southwest. The southern coast and the eastern coast had the lowest HFTE.

5.5.2 *The National HPTE and HFTE Generally Showed a Trend of First Decreasing and then Increasing*

The HPTE and HFTE of all provinces in China showed a U-shaped trend, and the lowest level was observed in 2012 and 2011, respectively. In response to the global financial crisis in 2008, China implemented a loose fiscal and monetary policy (Li et al., 2015), and many regions began to increase investment in highway infrastructure construction. The total investment in the highway transport sector increased from RMB 741.151 billion in 2008 to RMB 1385.635 billion in 2011 (Statistical Yearbook of the Chinese Investment in Fixed Assets, 2009, 2012). The continuous large-scale investment in highway construction brought about repeated construction, incompatibility management competition and other problems of overcapacity, inhibiting highway transport efficiency. The Chinese government stopped the loose fiscal policy and adopted a prudent fiscal policy from 2011. As a result, the investment increment and stock of the highway transport sector were optimized, and then the overall HPTE and HFTE of China showed a fluctuating upward trend rather than falling after 2011.

5.5.3 *The Impact Mechanism of HPTE and HFTE*

We found that the economic development level, passenger transport structure, highway structure and industrial structure show a significant positive impact on HPTE, while the impact of population density and urbanization level on HPTE is not significant.

In addition, highway structure and industrial structure show a significant positive impact on HFTE, while population density shows a significant negative impact on HFTE. The economic development level and urbanization level have no significant impact on HFTE.

References

China Statistical Yearbooks (2017). China Statistical Publishing House. https://data.cnki.net/yearbook/Single/N2020100004

China Statistical Yearbook (2009–2017). China Statistical Publishing House. https://data.cnki.net/yearbook/Single/N2020100004

China Transport Statistical Yearbook (2009–2017). China Statistical Publishing House. https://data.cnki.net/yearBook/single?id=N2022010162. Accessed April 24, 2021)

Ding, H. Y., & Liao, Y. H. (2016). Infrastructure investment efficiency and its spatial and time spillover effect in the three gorges region. *Journal of Guizhou University of Finance and Economics, 35*(S1), 21–25 (In Chinese). https://d.wanfangdata.com.cn/periodical/gzcjxyxb201606011

Dong, W. H., Li, X. Q., & Song, Y. (2016). Seasonal variation in diversity of soil macro-fauna on both sides of Beijing-Harbin expressway. *Journal of Nanjing Forestry University (Natural Sciences Edition), 40*(4), 10–16 (In Chinese). https://wenku.baidu.com/view/de9f85f86194dd88d0d233d4b14e852458fb390e?fr=xueshu_top

Fu, W., Luo, M., Chen, J., & Udimal, T. B. (2020). Carbon footprint and carbon carrying capacity of vegetation in ecologically fragile areas: A case study of yunnan. *Physics and Chemistry of the Earth Parts A/B/C, 120*(12), 102904. https://doi.org/10.1016/j.pce.2020.102904

Goldsmith, R. M. (1951). A perpetual inventory of national wealth. *Studies in Income and Wealth 14*, 5–73. http://www.nber.org/books/unkn51-2

Hipsher, S. (2017) Yunnan, China: Background, economic conditions, and tourism. In: Poverty reduction, the private sector, and tourism in Mainland Southeast Asia. Palgrave Macmillan. https://doi.org/10.1007/978-981-10-5948-3_10

Li, G. P., & Luo, X. R. (2017). Coordinated development between population and economy in the Beijing-Tianjin-Hebei region. *Progress in Geography, 36*(1), 25–33 (In Chinese). http://www.cnki.com.cn/Article/CJFDTotal-DLKJ201701004.htm

Li, T., Yang, W. Y., Zhang, H. R., & Cao, X. S. (2015). Evaluating the impact of transport investment on the efficiency of regional integrated transport systems in China. *Transport Policy, 45*, 66–76. https://doi.org/10.1016/j.tranpol.2015.09.005

Liu, B. L., & Liu, Y. (2007). Estimation in China cumulative amount of highway and waterway capital. *Journal of Beijing Jiaotong University (Social Sciences Edition), 6*(3), 44–48 (In Chinese). https://doi.org/10.16797/j.cnki.11-5224/c.2007.03.010

Liu, H. W., Wu, J., & Liang, W. (2018). Efficiency of highway cargo transportation and its effect factors based on perspective of Wanjiang demonstration zone. *Journal of Transportation Systems Engineering and Information Technology, 18*(2), 27–32. (In Chinese). https://d.wanfangdata.com.cn/periodical/ChlQZXJpb2RpY2FsQ0hJTmV3UzIwMjIwMzAyEhRqdHlzeHRnY3l4eDIwMTgwMjAwNRoIeHp5cXA3djk%3D

National Bureau of Statistics of China (NBSC). (2021). http://data.stats.gov.cn/easyquery.htm?cn=E0103. Accessed April 29, 2021.

Peng, J., Jiang, H., Liu, Q. H., Green, S. M., Quine, T. A., Liu, H. Y., Qiu, S. J., Liu, Y. X., & Meersmans, J. (2021). Human activity vs. climate change: Distinguishing dominant drivers on LAI dynamics in karst region of southwest China. *Science of the Total Environment*, 144297. https://doi.org/10.1016/j.scitotenv.2020.144297

Ren, Q. L. (2009). Influence factor and mechanism of highway transportation demand. *Transport Research, 9*, 127–131. (In Chinese). https://max.book118.com/html/2017/1223/145601904.shtm

Ren, W. X., Xue, B., Yang, J., & Lu, C. P. (2020). Effects of the northeast China revitalization strategy on regional economic growth and social development. *Chinese Geographical Science, 30*(5), 791–809. https://doi.org/10.1007/s11769-020-1149-5

Song, Yin, G., & Lu, Z., et al. (2021). Industrial ecological efficiency of cities in the Yellow River basin in the background of China's economic transformation: spatial-temporal characteristics and influencing factors. *Environmental Science and Pollution Research.* https://doi.org/10.1007/s11356-021-15964-2-008-0109-2

Statistical Yearbook of the Chinese Investment in Fixed Assets (2009–2013, 2015–2017). China Statistical Publishing House. https://data.cnki.net/yearBook/single?id=N2019030174

Sun, L. (2021). Coordinated development of ecological environment in Yangtze River economic belt. In: X. Jing (Eds.), *Development of the Yangtze river economic belt in China. Urban governance practices in China.* Springer. https://doi.org/10.1007/978-981-16-1578-8_4

Sun, Q. Y., Mo, R. J., & Liu, Y. H. (2018). Study on short-term traffic state forecasting model of SVM considering proportion of large vehicles. *Journal of Highway and Transportation Research and Development, 35*(10), 126–132. (In Chinese). https://d.wanfangdata.com.cn/periodical/gljtkj201810017

Sun, Z. l., & Hou, Y. L. (2019). How does industrial intelligence reshape the employment structure of Chinese labor force. *China Industrial Economics, 05*, 004. (In Chinese) http://en.cnki.com.cn/Article_en/CJFDTotal-GGYY201905005.htm

Wang, G. F., Deng, X. Z., Wang, J. G., Zhang, F., & Liang, S. Q. (2019a). Carbon emission efficiency in China: A spatial panel data analysis. *China Economic Review, 56*, 101313. https://doi.org/10.1016/j.chieco.2019.101313

Wang, R., Zameer, H., Feng, Y., Jiao, Z., Xu, L., & Gedikli, A. (2019b). Revisiting Chinese resource curse hypothesis based on spatial spillover effect: A fresh evidence. *Resource Policy, 64*, 101521. https://doi.org/10.1016/j.resourpol.2019.101521

Xu, L. (2021). Evaluation of my country's highway transportation efficiency and analysis of its temporal and spatial differences. *Anhui University of Science and Technology*. https://d.wanfangdata.com.cn/thesis/Y3804447

Xie, S. H., Wang, L. X., & Shao, Z. L. (2014). Review on carbon emissions researches at home and abroad. *Arid Land Geography, 37*(04):720–730. (In Chinese). https://d.wanfangdata.com.cn/periodical/ChlQZXJpb2RpY2FsQ0hJTmV3UzIwMjIwMzAyEg5naHFkbDIwMTQwNDAxMBoIeDE0aGhreng%3D

Yan, S. Y., & Wang, Z. (2011). Compact city spatial planning of Yuechi county town, Sichuang. *City Planning Review, 35*(S1), 21–25 (In Chinese). https://wenku.baidu.com/view/1d172d7e02d8ce2f0066f5335a8102d277a2614b?fr=xueshu_top

Yang, L., Wang, K. L., & Geng, J. C. (2018). China's regional ecological energy efficiency and energy saving and pollution abatement potentials: An empirical analysis using epsilon-based measure model. *Journal of Cleaner Production, 194*, 300–308. https://doi.org/10.1016/j.jclepro.2018.05.129

Yang, W. Y., Wang, W. L., & Ouyang, S. S. (2019). The influencing factorsand spatial spillover effects of CO_2 emissionsfrom transportation in China. *Science of the Total Environment, 696*, 133900. https://doi.org/10.1016/j.scitotenv.2019.133900

Zeng, L. (2021). China's eco-efficiency: Regional differences and influencing factors based on a spatial panel data approach. *Sustainability., 13*(6), 3143. https://doi.org/10.3390/su13063143

Zeng, L., Li, H., Wang, X., Yu, Z., Hu, H., Yuan, X., Zhao, X., Li, C., Yuan, D., Gao, Y., Nie, Y., & Huang, L. (2022). China's transport land: Spatiotemporal expansion characteristics and driving mechanism. *Land, 11*, 1147. https://doi.org/10.3390/land11081147

Zhang, J., Wu, G. Y., & Zhang, J. P. (2004). The estimation of China's provincial capital stock: 1952–2000. *Economic Research Journal, 10*, 35–44. http://en.cnki.com.cn/Article_en/CJFDTOTAL-JJYJ200410004.htm

Zhang, P. Y. (2008). Revitalizing old industrial base of Northeast China: Process, policy and challenge. *Chinese Geographical Science, 18*(2), 109–118. https://doi.org/10.1007/s11769

Zhang, Z. J., & Liu, Y. H. (2011). Induced demand, network density and performance measurement of the motor carrier industry in China: 2000–2009. *Shanghai Journal of Economics, 10*, 58–66+76. http://www.cnki.com.cn/Article/CJFDTotal-HSYJ201110008.htm

Zhao, L. (2014). *The research on efficiency of China's provincial logistics industry* [Master Dissertation of Beijing, Jiaotong University, Beijing] (In Chinese). https://d.wanfangdata.com.cn/thesis/Y586044

Zhao, P. J., Zeng, L. E., Lu, H. Y., Zhou, Y., Hu, H. Y., & Wei, X. Y. (2020). Green economic efficiency and its influencing factors in China from 2008 to 2017: Based on the super-SBM model with undesirable outputs and spatial Dubin model. *Science of the Total Environment*, 140026. https://doi.org/10.1016/j.scitotenv.2020.140026

Chapter 6
Railway Transport Efficiency

6.1 Background and Methods

6.1.1 Background

Due to its advantages such as big transport capacity, low energy consumption, low pollution, safety and comfort, railway transport is one of the key modes of transport in many countries. With 6% of the world's total length of railways in operation, China has completed 25% of global railway transport, and become the country with the highest railway transport efficiency in the world. By the end of 2016, China's length of railways in operation totaled about 124,000 km; the length of high-speed railways that were completed and open to traffic in China reached 16,000 km (Fig 6.1). At present, China has one of the world's largest railway networks, second only to the United States, and the world's largest high-speed rail network. While China's railway transport has made great achievements in the world, it still faces some dilemma. More specifically, the railway development has lagged behind China's social and economic development for a long time; with the rapid economic growth and the continuous upgrading of consumption structure, China is facing increasingly serious pressure on resources and environmental protection. As a transport mode with a large traffic volume and low energy consumption, railways should provide more effective support for economic growth. In this context, the shortage of railway transport capacity has become a serious bottleneck in social and economic development. The causes of China's railway capacity shortage are very complex, which may be related to the insufficient output caused by insufficient railway investment. Based on this, the railway performance evaluation is very important, which is not only conducive to a comprehensive understanding of China's railway efficiency and the in-depth exploration of the causes of low efficiency of railway production, but also provides targeted policy suggestions for relevant departments.

The research on railway passenger transport efficiency (RPTE) and railway freight transport efficiency (RFTE) covers 31 provinces from 2008 to 2016 in China, while Taiwan, Hong Kong and Macao are not included in the study region (Fig. 6.1).

P. Zhao and L. Zeng, *Transport Efficiency and Safety in China*, Population, Regional Development and Transport, https://doi.org/10.1007/978-981-99-1055-7_6

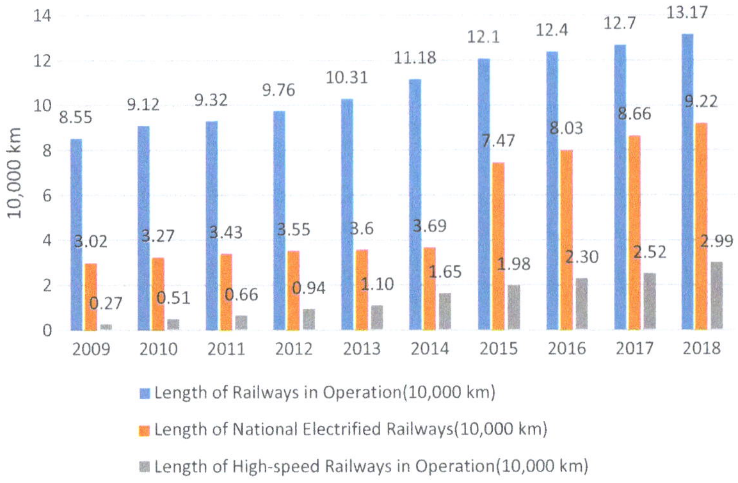

Fig. 6.1 The length of railways in operation of China (2009–2018). *Data source* The authors, edited from National Bureau of Statistics of China (NBSC) (2021)

6.1.2 Methods

The research on RPTE and RFTE covers 31 provinces, municipalities and autonomous regions in China from 2008 to 2016, while Taiwan, Hong Kong and Macao are not included in the study scope.

The input indexes of RPTE include the capital stock of the railway transport sector and the length of railways. The output index is the railway passenger turnover. The input indexes of RFTE include the capital stock of the railway transport sector and the length of railways. The output index is the railway freight turnover (Table 6.1).

The capital stock of the railway transport sector is also calculated using the perpetual inventory method. The basic formula of the perpetual inventory method is: $Ki, t = Ii, t + (1 - \delta i, t)Ki, t - 1$. K represents the capital stock of the railway transport sector, I represents the total fixed capital investment in the railway transport sector, and δ represents the capital depreciation rate of the railway transport sector. Because the total fixed capital investment in the railway transport sector in some provinces fluctuates greatly, in order to prevent the occurrence of special deviation values, this study takes the average value of the total fixed asset investment from 2007 to 2009, and then divides it by 10% as the capital stock of the railway transport sector in the base period. The annual capital depreciation rate of the railway transport sector δ is 8.76% based on the research results of Li and Zhang (2016).

Table 6.1 The measurement index system of railway transport efficiency

	Primary indexes	Secondary indexes
Railway passenger transport efficiency	Inputs	The capital stock of the railway transport sector (billion RMB)
		The length of railways (kilometers)
	Output	Railway passenger turnover (passenger-kilometer)
Railway freight transport efficiency	Inputs	The capital stock of the railway transport sector (billion RMB)
		The length of railways (kilometers)
	Output	Railway freight turnover (ton-kilometer)

Data source China Statistical Yearbook (2009–2017), Statistical Yearbook of the Chinese Investment in Fixed Assets (2009–2013, 2015–2017), China Transport Statistical Yearbook (2009–2017) and statistical yearbooks of provinces over the years

6.2 Measurement Results

According to the EBM method, using Maxdea 7.9 ultra software, we calculate the RPTEs and RFTEs of 31 provinces in China from 2008 to 2016. The results are listed in Tables 6.2 and 6.3 This paper applies ArcGIS 10.0 software to draw spatial distribution maps of RPTE and RFTE (Figs. 6.2 and 6.3).

6.2.1 The Overall Characteristics

6.2.1.1 The Overall National RPTE and RFTE Were Low

The national average values of RPTE and RFTE from 2008 to 2016 were 0.369 and 0.158, respectively. The number of provinces whose RPTE and RFTE were above the national level was 14 and 9, respectively. Therefore, on the whole, the RPTE and RFTE in China were very low. The overall spatial distribution characteristics of RPTE and RFTE were similar: the RPTE and RFTE of the three coastal areas were at a high level, the RPTE and RFTE of Middle reaches of the Yellow River and the Yangtze River, and the Northeast were at an intermediate level, and the RPTE and RFTE of inland areas were at a low level (Fig. 6.4).

Table 6.2 RPTE of 31 provinces in China from 2008 to 2016

Provinces	2008	2009	2010	2011	2012	2013	2014	2015	2016	Mean
Beijing	0.329	0.337	0.377	0.366	0.307	0.333	0.362	0.348	0.355	0.346
Tianjin	0.356	0.350	0.390	0.338	0.326	0.455	0.420	0.404	0.408	0.383
Hebei	0.267	0.263	0.309	0.348	0.306	0.333	0.364	0.299	0.325	0.313
Shanxi	0.143	0.116	0.115	0.138	0.124	0.175	0.135	0.122	0.119	0.132
Inner Mongolia	0.061	0.051	0.051	0.055	0.051	0.061	0.063	0.051	0.054	0.055
Liaoning	0.236	0.229	0.248	0.277	0.226	0.323	0.344	0.263	0.295	0.271
Jilin	0.137	0.130	0.143	0.171	0.147	0.166	0.168	0.142	0.150	0.150
Heilongjiang	0.098	0.097	0.104	0.124	0.114	0.141	0.147	0.131	0.134	0.121
Shanghai	0.558	0.582	0.524	0.361	0.335	0.662	0.709	0.690	0.730	0.572
Jiangsu	1.000	1.000	1.000	1.000	1.000	1.000	1.000	1.000	1.000	1.000
Zhejiang	1.000	0.726	0.806	0.761	0.675	0.871	0.797	0.675	0.705	0.780
Anhui	0.514	0.519	0.604	0.691	0.670	0.651	0.721	0.493	0.511	0.597
Fujian	0.335	0.249	0.269	0.273	0.239	0.345	0.389	0.295	0.326	0.302
Jiangxi	0.370	0.341	0.382	0.452	0.421	0.549	0.480	0.439	0.457	0.432
Shandong	0.507	0.514	0.523	0.553	0.516	0.445	0.408	0.356	0.382	0.467
Henan	0.488	0.501	0.510	0.639	0.538	0.592	0.611	0.543	0.572	0.555
Hubei	0.418	0.361	0.372	0.476	0.431	0.479	0.527	0.501	0.532	0.455
Hunan	0.547	0.413	0.478	0.613	0.578	0.716	0.653	0.578	0.579	0.573
Guangdong	0.951	0.848	1.000	1.000	1.000	0.870	1.000	0.725	0.769	0.907
Guangxi	0.349	0.307	0.331	0.405	0.383	0.271	0.249	0.247	0.264	0.312
Hainan	0.347	0.386	0.256	0.255	0.226	0.272	0.290	0.183	0.208	0.269
Chongqing	0.378	0.371	0.396	0.523	0.493	0.496	0.508	0.481	0.444	0.455
Sichuan	0.405	0.342	0.354	0.367	0.351	0.444	0.406	0.361	0.351	0.376
Guizhou	0.248	0.235	0.281	0.341	0.349	0.485	0.462	0.384	0.358	0.349
Yunnan	0.187	0.171	0.206	0.248	0.235	0.286	0.265	0.256	0.210	0.229
Tibet	0.077	0.070	0.077	0.084	0.073	0.146	0.100	0.079	0.086	0.088
Shaanxi	0.259	0.244	0.221	0.293	0.274	0.301	0.320	0.287	0.298	0.277
Gansu	0.252	0.248	0.271	0.359	0.335	0.417	0.324	0.287	0.298	0.310
Qinghai	0.056	0.058	0.059	0.074	0.069	0.089	0.087	0.088	0.099	0.076
Ningxia	0.143	0.127	0.118	0.142	0.138	0.160	0.167	0.175	0.159	0.148
Xinjiang	0.178	0.119	0.118	0.140	0.128	0.202	0.161	0.134	0.138	0.146
China	0.361	0.332	0.351	0.383	0.357	0.411	0.408	0.355	0.365	0.369

Data source Edited by the authors

Table 6.3 RFTE of 31 provinces in China from 2008 to 2016

Provinces	2008	2009	2010	2011	2012	2013	2014	2015	2016	Mean
Beijing	0.019	0.021	0.021	0.023	0.025	0.037	0.029	0.025	0.023	0.025
Tianjin	0.123	0.445	0.359	0.336	0.257	0.138	0.111	0.066	0.065	0.211
Hebei	0.304	0.350	0.320	0.260	0.248	0.301	0.219	0.186	0.182	0.263
Shanxi	0.068	0.068	0.055	0.052	0.052	0.055	0.037	0.031	0.034	0.050
Inner Mongolia	0.038	0.045	0.035	0.034	0.036	0.038	0.025	0.021	0.023	0.033
Liaoning	0.118	0.152	0.136	0.145	0.165	0.238	0.151	0.120	0.131	0.151
Jilin	0.051	0.054	0.040	0.034	0.034	0.044	0.029	0.022	0.020	0.036
Heilongjiang	0.053	0.057	0.039	0.035	0.035	0.046	0.026	0.018	0.015	0.036
Shanghai	1.000	1.000	1.000	1.000	1.000	1.000	1.000	1.000	1.000	1.000
Jiangsu	0.741	0.761	0.542	0.494	0.585	1.000	0.827	0.605	0.385	0.660
Zhejiang	0.096	0.114	0.110	0.120	0.127	0.172	0.129	0.111	0.106	0.121
Anhui	0.205	0.235	0.212	0.236	0.294	0.444	0.302	0.225	0.220	0.264
Fujian	0.046	0.050	0.047	0.051	0.062	0.084	0.078	0.088	0.108	0.068
Jiangxi	0.098	0.109	0.103	0.109	0.139	0.219	0.170	0.174	0.125	0.138
Shandong	0.418	0.510	0.361	0.294	0.250	0.224	0.134	0.115	0.100	0.267
Henan	0.274	0.342	0.283	0.272	0.290	0.298	0.212	0.179	0.181	0.259
Hubei	0.078	0.089	0.083	0.093	0.109	0.159	0.122	0.113	0.092	0.104
Hunan	0.173	0.188	0.150	0.161	0.195	0.181	0.117	0.081	0.075	0.147
Guangdong	0.078	0.087	0.079	0.082	0.105	0.134	0.146	0.123	0.161	0.111
Guangxi	0.135	0.160	0.124	0.103	0.098	0.121	0.091	0.085	0.086	0.111
Hainan	0.027	0.040	0.030	0.047	0.053	0.031	0.057	0.042	0.042	0.041
Chongqing	0.166	0.191	0.182	0.221	0.211	0.230	0.169	0.144	0.147	0.185
Sichuan	0.021	0.021	0.018	0.018	0.020	0.026	0.019	0.017	0.017	0.020
Guizhou	0.042	0.051	0.040	0.039	0.041	0.036	0.027	0.023	0.024	0.036
Yunnan	0.023	0.026	0.021	0.020	0.021	0.035	0.024	0.021	0.021	0.024
Tibet	0.009	0.010	0.008	0.005	0.005	0.014	0.010	0.010	0.006	0.009
Shaanxi	0.097	0.111	0.093	0.095	0.102	0.094	0.076	0.067	0.067	0.089
Gansu	0.123	0.135	0.102	0.107	0.130	0.163	0.102	0.078	0.078	0.113
Qinghai	0.021	0.024	0.022	0.024	0.025	0.029	0.020	0.017	0.018	0.022
Ningxia	0.221	0.231	0.148	0.142	0.170	0.188	0.142	0.133	0.082	0.162
Xinjiang	0.025	0.025	0.020	0.019	0.021	0.034	0.024	0.020	0.021	0.023
China	0.158	0.184	0.154	0.151	0.158	0.187	0.149	0.128	0.118	0.154

Data source Edited by the authors

Fig. 6.2 The mean value of RPTE intervals in 31 provinces from 2008 to 2016. *Data source* Edited by the authors

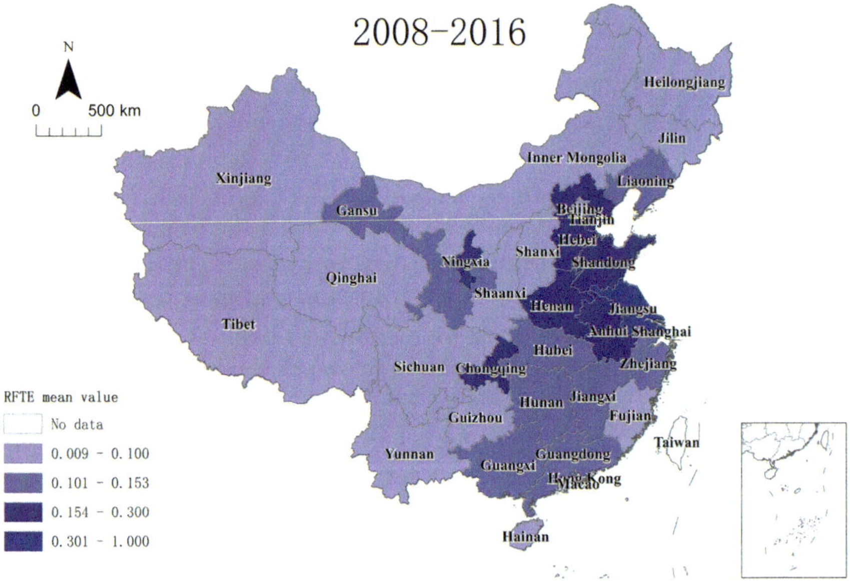

Fig. 6.3 The mean value of RFTE intervals in 31 provinces from 2008 to 2016. *Data source* Edited by the authors

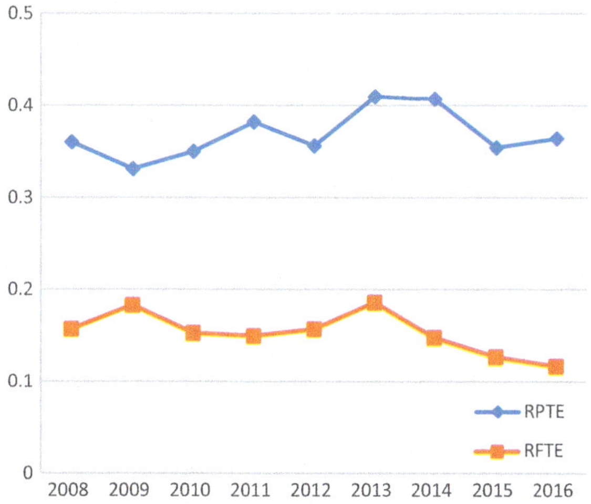

Fig. 6.4 The trend of national RPTE and RFTE from 2008 to 2016. *Data source* Edited by the authors

6.2.1.2 The National RPTE First Rose and then Declined, While RFTE Fluctuated in M Shape

The national RPTE fluctuated, with the highest point in 2013. From the 2010s, air transport developed very fast and captured a large share of the passenger transport market, which had a competitive relationship with railway passenger transport. RFTE showed a trend of first rising and then declining, with the highest point in 2013. For one thing, the railway transport sector is impacted by the air transport sector. For another, since 2013, China has increased investment in railway construction to build high-speed railways, which has a bad impact on RPTE and RFTE.

The national RFTE showed an M-shaped trend, which peaked in 2009 and 2013, respectively. The improvement of RFTE from 2011 to 2013 is related to the recovery of the macro-economy of China. In 2012, the Chinese Government proposed the establishment of ecological civilization and began to implement strict emission reduction policies (Zhao et al., 2022), resulting in the substantial reduction of the transport volume of many bulk commodities in 2014 and 2015. Through the freight data analysis, it is found that the national railway freight turnover had a negative growth after 2013, which decreased by 13.72% in 2015 compared with the previous year (NBSC, 2021), leading to the gradual decline of RFTE in recent years (Tables 6.4 and 6.5; Figs. 6.5 and 6.6).

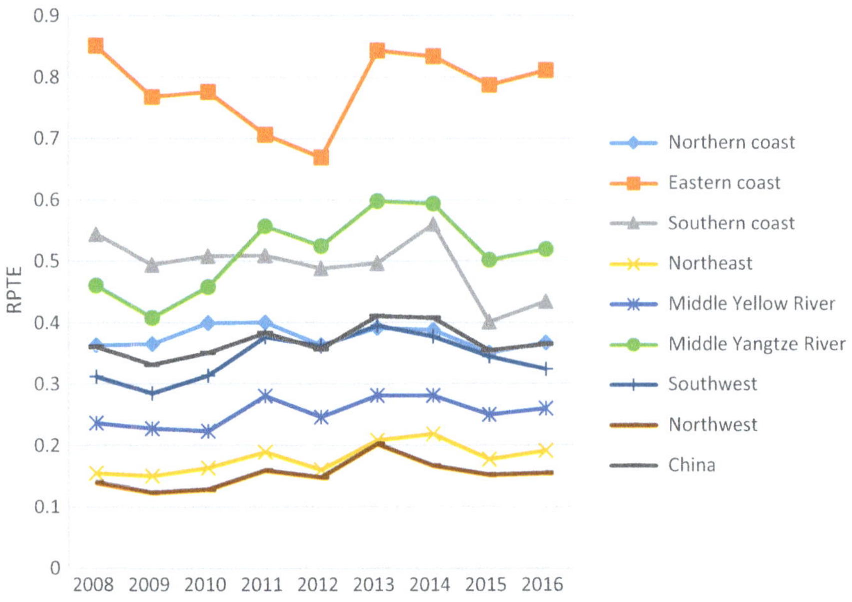

Fig. 6.5 The trend of RPTE nationwide and in eight economic zones from 2008 to 2016. *Data source* Edited by the authors

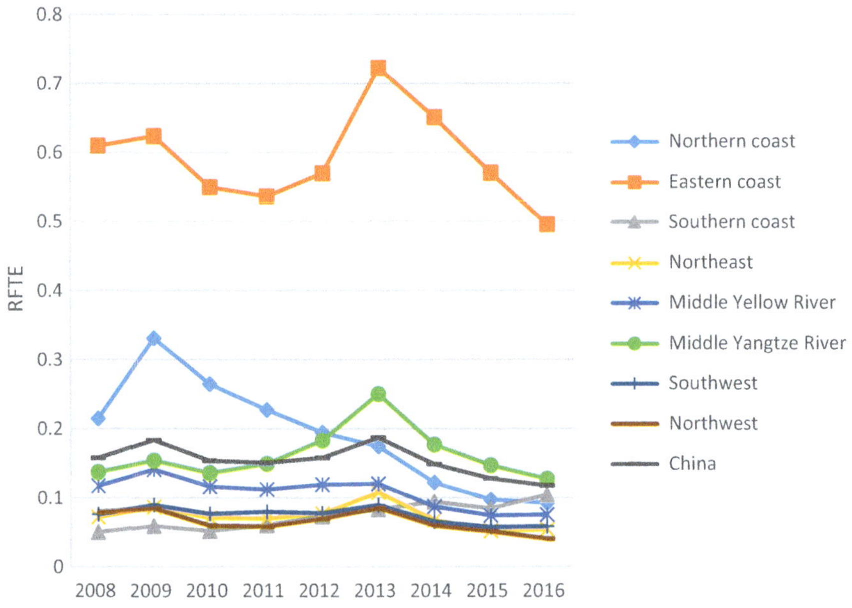

Fig. 6.6 The trend of RFTE nationwide and in eight economic zones from 2008 to 2016. *Data source* Edited by the authors

Table 6.4 RPTE of eight economic zones from 2008 to 2016

Economic zones	2008	2009	2010	2011	2012	2013	2014	2015	2016	Mean
Northern coast	0.365	0.366	0.400	0.401	0.364	0.392	0.389	0.352	0.368	0.377
Eastern coast	0.853	0.769	0.777	0.707	0.670	0.844	0.835	0.788	0.812	0.784
Southern coast	0.544	0.494	0.508	0.509	0.488	0.496	0.560	0.401	0.434	0.493
Northeast	0.157	0.152	0.165	0.191	0.162	0.210	0.220	0.179	0.193	0.181
Middle Yellow River	0.238	0.228	0.224	0.281	0.247	0.282	0.282	0.251	0.261	0.255
Middle Yangtze River	0.462	0.409	0.459	0.558	0.525	0.599	0.595	0.503	0.520	0.514
Southwest	0.313	0.285	0.314	0.377	0.362	0.396	0.378	0.346	0.325	0.344
Northwest	0.141	0.124	0.129	0.160	0.149	0.203	0.168	0.153	0.156	0.154
China	0.361	0.332	0.351	0.383	0.357	0.411	0.408	0.355	0.365	0.369

Data source Edited by the authors

Table 6.5 RFTE of eight economic zones from 2008 to 2016

Economic zones	2008	2009	2010	2011	2012	2013	2014	2015	2016	Mean
Northern coast	0.216	0.332	0.265	0.228	0.195	0.175	0.123	0.098	0.093	0.192
Eastern coast	0.612	0.625	0.551	0.538	0.571	0.724	0.652	0.572	0.497	0.594
Southern coast	0.050	0.059	0.052	0.060	0.073	0.083	0.094	0.084	0.104	0.073
Northeast	0.074	0.088	0.072	0.071	0.078	0.109	0.069	0.053	0.055	0.074
Middle Yellow River	0.119	0.142	0.117	0.113	0.120	0.121	0.088	0.075	0.076	0.108
Middle Yangtze River	0.139	0.155	0.137	0.150	0.184	0.251	0.178	0.148	0.128	0.163
Southwest	0.077	0.090	0.077	0.080	0.078	0.090	0.066	0.058	0.059	0.075
Northwest	0.080	0.085	0.060	0.059	0.070	0.086	0.060	0.052	0.041	0.066
China	0.158	0.184	0.154	0.151	0.158	0.187	0.149	0.128	0.118	0.154

Data source Edited by the authors

6.2.2 Spatial Variations

6.2.2.1 The Northern Coast

In the northern coast, the annual average values of RPTE and RFTE among provinces were 0.377 and 0.216, respectively, slightly higher than the national average. The regional RPTE and RFTE showed an M-shape tendency, which peaked in 2010 and 2013. The regional RFTE showed an obvious downward trend.

In terms of provincial differences, the RPTE in Shandong was relatively high compared with the rest of the country, with an average annual value of 0.467, while

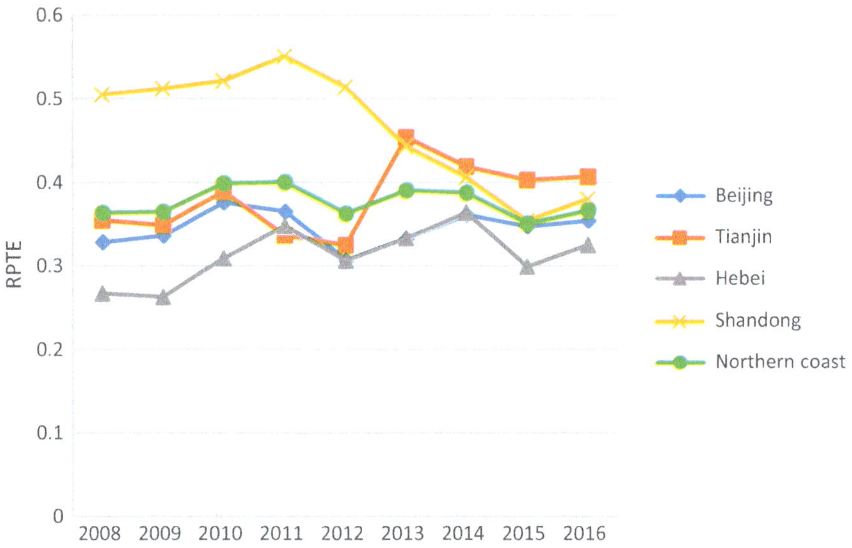

Fig. 6.7 The trend of RPTE in the northern coast from 2008 to 2016. *Data source* Edited by the authors

the RPTE in Hebei was relatively low in the region. Tianjin, Hebei and Shandong had similar levels of RFTE, and their average annual values were higher than the national level, while Beijing had a very low level of RFTE, which was the lowest not only in the region, but also across China.

Hebei is the channel between Northeast China and other provinces and is on the only route from Shanxi, Inner Mongolia and Northwest region to the Port of Tianjin; many important railway lines pass through Hebei; at the same time, Hebei undertakes the population and industrial relief of Beijing and Tianjin (Wang et al., 2019), which brings a huge freight volume. The freight volume of Shandong is second only to that of Hebei among the whole country (NBSC, 2021), which leads to a relatively high RFTE. Beijing is the capital of China, and there is heavy railway passenger traffic between Beijing and other parts of the country; however, due to the disassociation of non-capital functions and the relocation of the secondary industry (Mao, 2017), the railway freight volume is low, which inhibits its RFTE (Figs. 6.7, 6.8, 6.9 and 6.10).

6.2.2.2 The Eastern Coast

The eastern coast had the highest level of RPTE and RFTE among the eight regions. The regional RPTE and RFTE first declined, then rose, and then declined again. In the process of economic globalization, the eastern coast has become the area with the fastest growing economy in China, and economic and trade growth has produced heavy passenger and freight transport demand.

Fig. 6.8 The mean value of RPTE in Northern coast from 2008 to 2016. *Data source* Edited by the authors

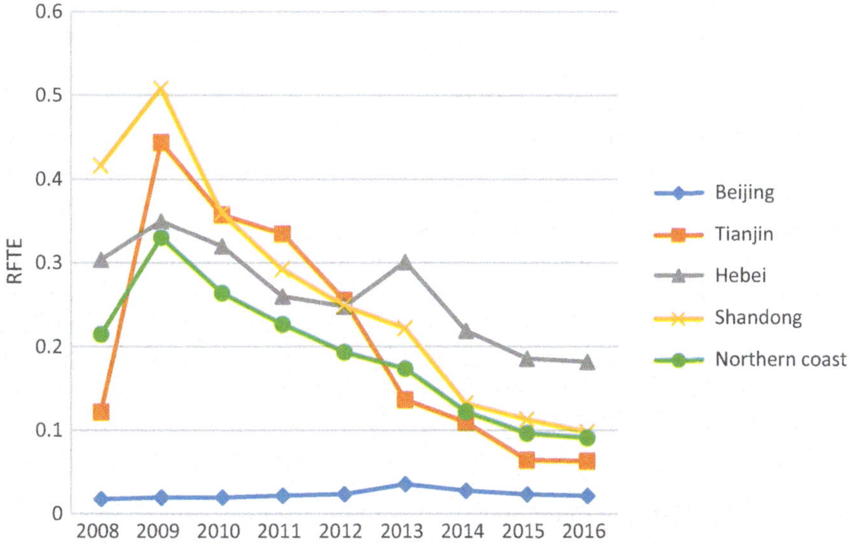

Fig. 6.9 The trend of RFTE in Northern coast from 2008 to 2016. *Data source* Edited by the authors

Fig. 6.10 The mean value of RFTE in Northern coast from 2008 to 2016. *Data source* Edited by the authors

In terms of provincial differences, the RPTE of Jiangsu and the RFTE of Shanghai were at the production frontier surface during the research period, which were the highest in China. However, the RFTE in Zhejiang was low. The railway passenger turnover in Jiangsu was always in the forefront of the country, which strongly promoted its RPTE at the production frontier surface. Backed by the Yangtze River Basin, Shanghai has a vast economic hinterland and a sufficient railway freight volume (Xu & Xu, 2018). During the study period, although the railway freight turnover in Zhejiang maintained a stable growth, a series of new high-speed railways were built in Zhejiang. The length of railways in Zhejiang increased rapidly from 1306 km in 2008 to 2540 km in 2016, an increase of nearly 94.8%, while the total length of railways in China only increased by 55% in the same period. The over-investment in railway lines inhibited the RFTE of Zhejiang (NBSC, 2021) (Figs. 6.11, 6.12, 6.13 and 6.14).

6.2.2.3 The Southern Coast

The annual average values of RPTE and RFTE in the southern coast during the study period were 0.493 and 0.073, respectively. The regional RPTE was higher than the national level, while the regional RFTE was lower than the national average. The regional RPTE first showed a stable trend and then decreased, while the regional RFTE increased rapidly, mainly due to the significant improvement of RFTEs in

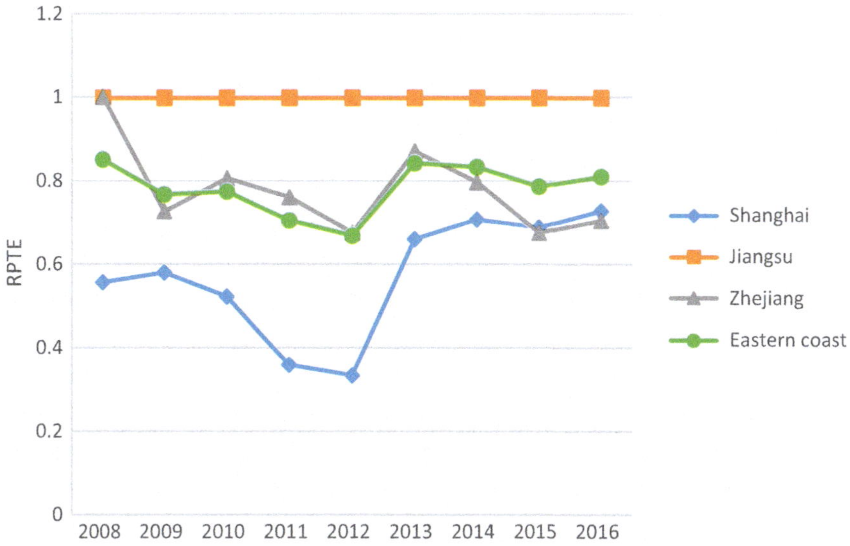

Fig. 6.11 The trend of RPTE in Eastern coast from 2008 to 2016. *Data source* Edited by the authors

Fig. 6.12 The mean value of RPTE in Eastern coast from 2008 to 2016. *Data source* Edited by the authors

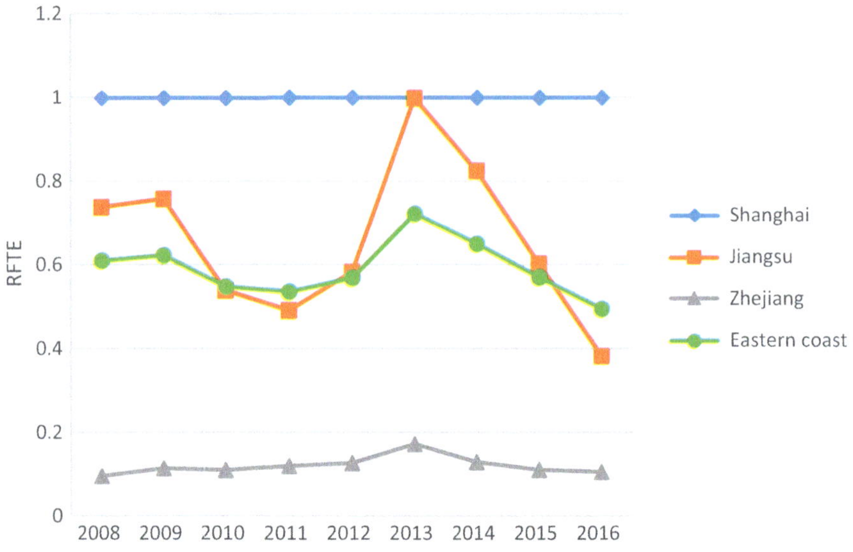

Fig. 6.13 The trend of RFTE in Eastern coast from 2008 to 2016. *Data source* Edited by the authors

Fig. 6.14 The mean value of RFTE in Eastern coast from 2008 to 2016. *Data source* Edited by the authors

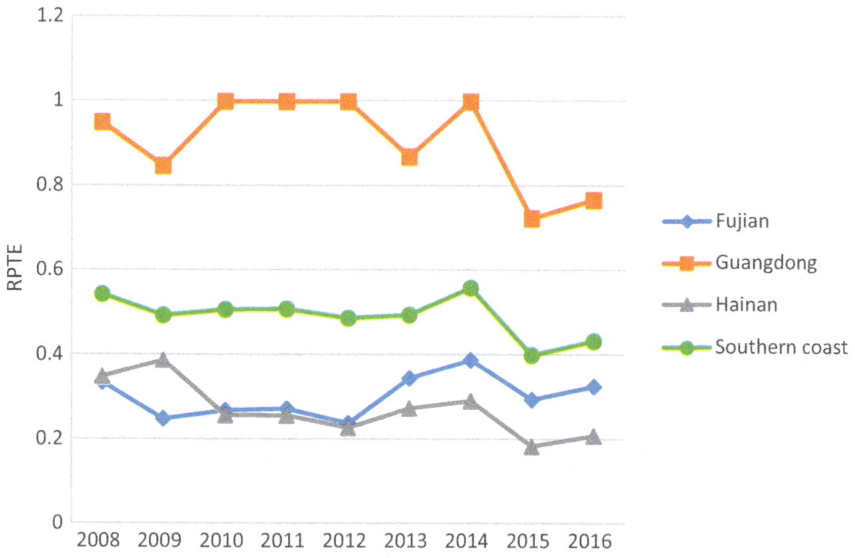

Fig. 6.15 The trend of RPTE in Southern coast from 2008 to 2016. *Data source* Edited by the authors

Guangdong and Fujian. During the study period, the railway freight turnover of Guangdong and Fujian increased significantly (NBSC, 2021), which promoted their RFTE.

The RPTE and RFTE in Guangdong were high, while the RPTE and RFTE in Hainan were low. Although Hainan is rich in natural and tourism resources, the railway transport sector is largely restricted by its geographical location. As Hainan is surrounded by the sea, people mainly travel to Hainan Island by plane. Highway transport is the major mode of transport in Hainan, and therefore there is little investment in railways. Furthermore, there is not much demand for transport, which has bad effects on the RPTE and RFTE in Hainan Island (Figs. 6.15, 6.16, 6.17 and 6.18).

6.2.2.4 The Northeast

The annual average values of RPTE and RFTE in Northeast China during the study period were 0.1807 and 0.074, respectively, which were lower than the national level, indicating that the regional railway transport efficiency was at a low level in the country. The regional RPTE showed a fluctuating upward trend, while the regional RFTE first rose and then declined, with the highest point in 2013.

As an old industrial base, the Northeast has a special geographical location and historical background. The region is one of the earliest regions to develop railways and high-speed railways in China; it has a relatively complete and dense passenger and freight railway network. However, there are too many railway branch lines in

Fig. 6.16 The mean value of RPTE in Southern coast from 2008 to 2016. *Data source* Edited by the authors

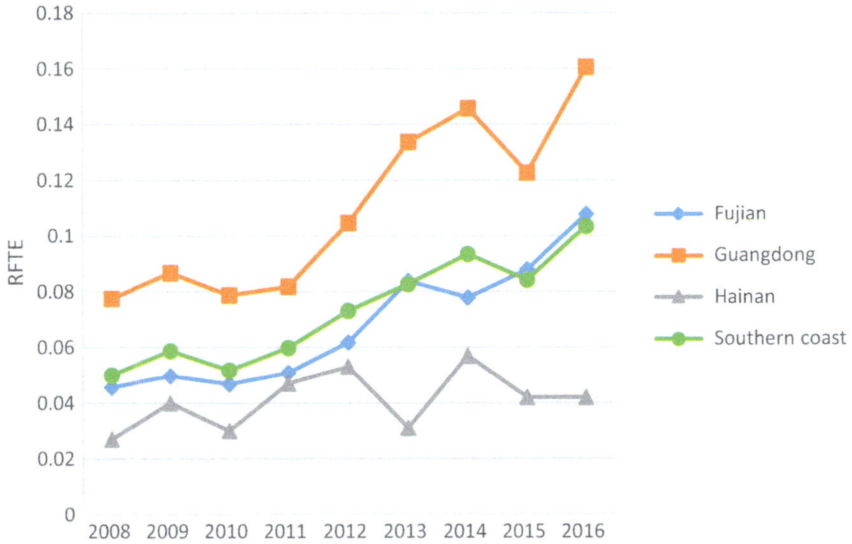

Fig. 6.17 The trend of RFTE in Southern coast from 2008 to 2016. *Data source* Edited by the authors

Fig. 6.18 The mean value of RFTE in the southern coast from 2008 to 2016. *Data source* Edited by the authors

Northeast China; the density of stations is large; the utilization rate of railway equipment is low; and there is a significant misallocation of resources. Despite high investment in railway infrastructure, the railway passenger turnover in the Northeast is characterized by slow growth; from 2008 to 2016, the railway passenger turnover in Northeast China increased by 11.92%, lower than the national growth level (61.7%) at the same period (NBSC, 2021). As a result, the RPTE in Northeast China was at a low level.

Under the guidance of the national strategy of revitalizing old industrial bases, the industrial development in Northeast China has been improved and upgraded in recent years, which has led to the development of logistics in the Northeast. From 2008 to 2011, the railway freight turnover in the Northeast increased year by year and reached the highest level in history in 2011. Since 2012, the freight turnover of heavy industries in the Northeast has decreased year by year; as a result, the RFTE showed a downward trend after 2013 (Figs. 6.19, 6.20, 6.21 and 6.22).

6.2.2.5 Middle Reaches of the Yellow River

The annual average values of RPTE and RFTE in Middle reaches of the Yellow River during the study period were 0.255 and 0.108, respectively, which were lower than the national average. The regional RPTE showed a slow downward trend, with fluctuations, after reaching the peak in 2010. During the study period, the regional

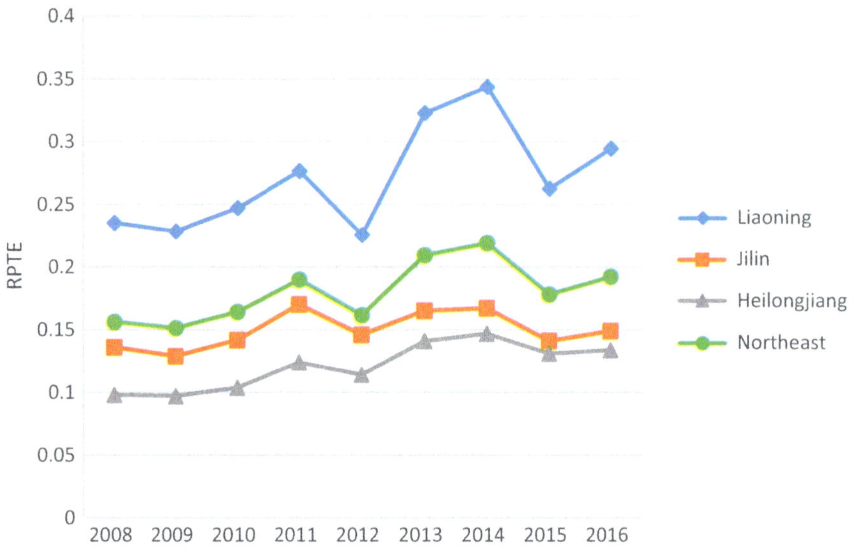

Fig. 6.19 The trend of RPTE in the Northeast from 2008 to 2016. *Data source* Edited by the authors

Fig. 6.20 The mean value of RPTE in the Northeast from 2008 to 2016. *Data source* Edited by the authors

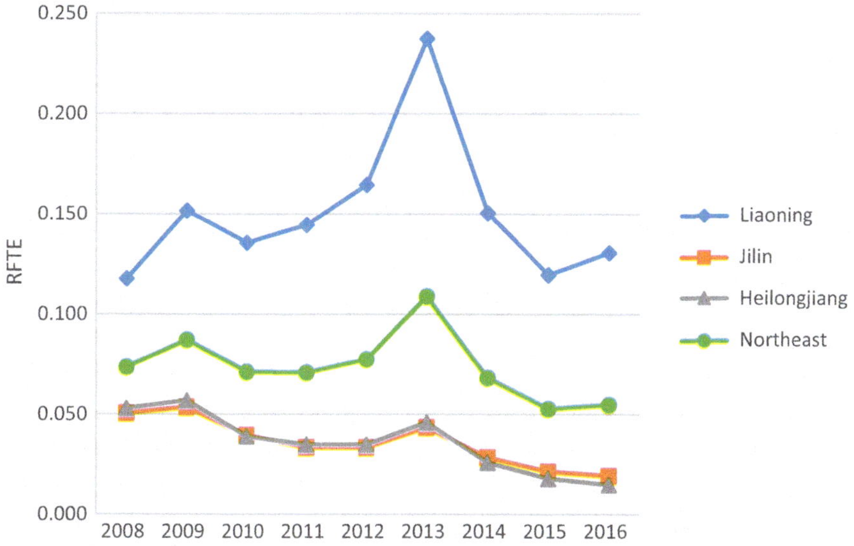

Fig. 6.21 The trend of RFTE in the Northeast from 2008 to 2016. *Data source* Edited by the authors

Fig. 6.22 The mean value of RFTE in the Northeast from 2008 to 2016. *Data source* Edited by the authors

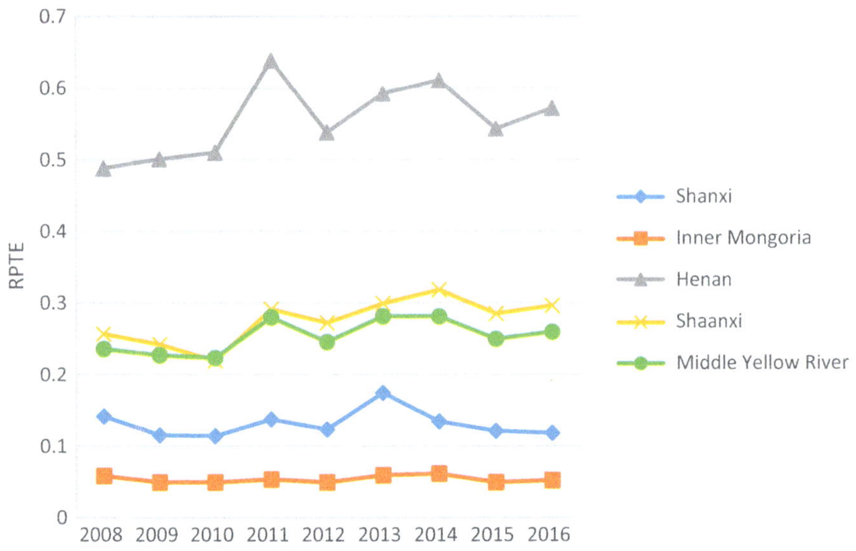

Fig. 6.23 The trend of RPTE in Middle reaches of the Yellow River from 2008 to 2016. *Data source* Edited by the authors

RFTE generally showed a slow downward trend. Middle reaches of the Yellow River is the energy and petrochemical base of China (Wang et al., 2021), and the regional railway freight sector is greatly affected by the structural adjustment policy.

In terms of provincial differences, the RPTE and RFTE in Henan were relatively high, taking Henan to the forefront of the field in China, while the RPTE and RFTE in Inner Mongolia were relatively low. Henan is located in the center of the national railway network, consisting of three rail corridors running north–south, called verticals, and five rail corridors running east–west, called horizontals (Zhang & Chen, 2015), which brings huge passenger and freight flows and high utilization efficiency of railway facilities. As for Inner Mongolia, it is vast in territory and has long railway lines in operation, with the population dispersed in a long spatial span from east to west and from north to south; all of these factors have a bad impact on the utilization efficiency of railway facilities (Figs. 6.23, 6.24, 6.25 and 6.26).

6.2.2.6 Middle Reaches of the Yangtze River

The annual average values of RPTE and RFTE in Middle reaches of the Yangtze River were 0.5143 and 0.163, respectively, which were higher than the national average. The RPTE and RFTE first increased and then decreased, with the highest level in 2013.

Located in the center of China, Middle reaches of the Yangtze River play a pivotal role in North–South and East–West communication of China. The transport of central

Fig. 6.24 The mean value of RPTE in Middle reaches of the Yellow River from 2008 to 2016. *Data source* Edited by the authors

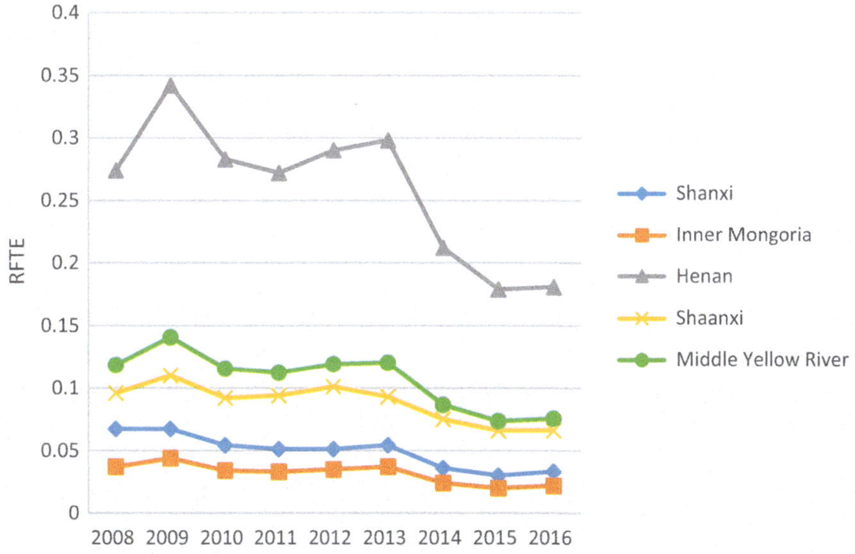

Fig. 6.25 The trend of RFTE in Middle reaches of the Yellow River from 2008 to 2016. *Data source* Edited by the authors

Fig. 6.26 The mean value of RFTE in Middle reaches of the Yellow River from 2008 to 2016. *Data source* Edited by the authors

China is mainly used for large-scale capital and material flows between East and West China, and the infrastructure exists mainly for this purpose (Liu, 2019). Driven by the Rise of Central China Plan, after 2010, four national demonstration zones for industrial relocation have been established in the region, namely the Wanjiang City Belt in Anhui, southern Hunan, Jingzhou in Hubei and Ganzhou in Jiangxi. It has facilitated the development of the local logistics industry, and the regional railway transport efficiency is at a high level.

Due to the influence of the policy of economic structural adjustment, the railway freight turnover of bulk commodities has been growing slowly since 2012, and the railway freight volume of bulk commodities in Hubei and Hunan has even been declining (NBSC, 2021). Therefore, at the end of the study period, the RFTEs in all provinces in the region showed a downward trend (Figs. 6.27, 6.28, 6.29 and 6.30).

With the proposal of the Rise of Central China Plan in 2004, the construction of large-scale coal bases in Anhui and Hubei has been strengthened. In addition, the cultural industry in Hunan, circular economy pilot zones in Jiangxi, the pilot zone for building a "resource-conserving, environment-friendly society" in Hubei, and the unique geographical location of Anhui will make the central region occupy an important position in the future railway transport market.

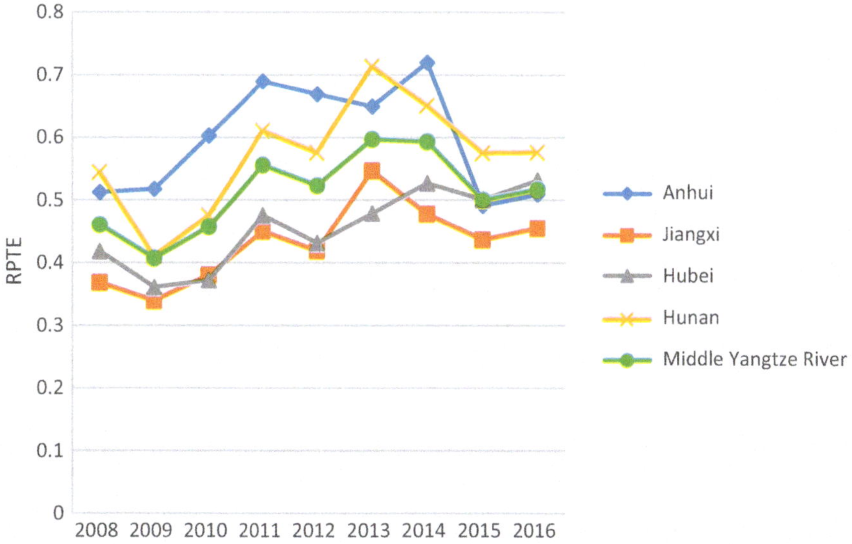

Fig. 6.27 The trend of RPTE in Middle reaches of the Yangtze River from 2008 to 2016. *Data source* Edited by the authors

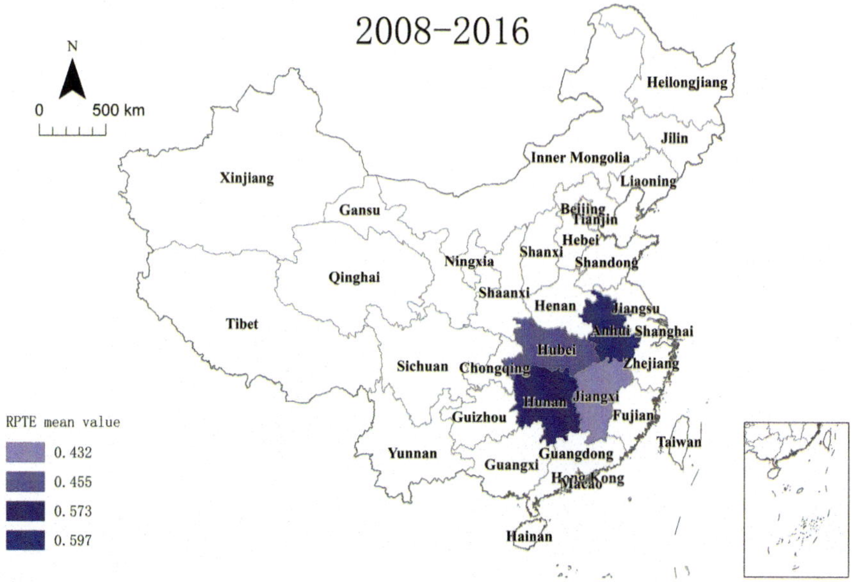

Fig. 6.28 The mean value of RPTE in Middle reaches of the Yangtze River from 2008 to 2016. *Data source* Edited by the authors

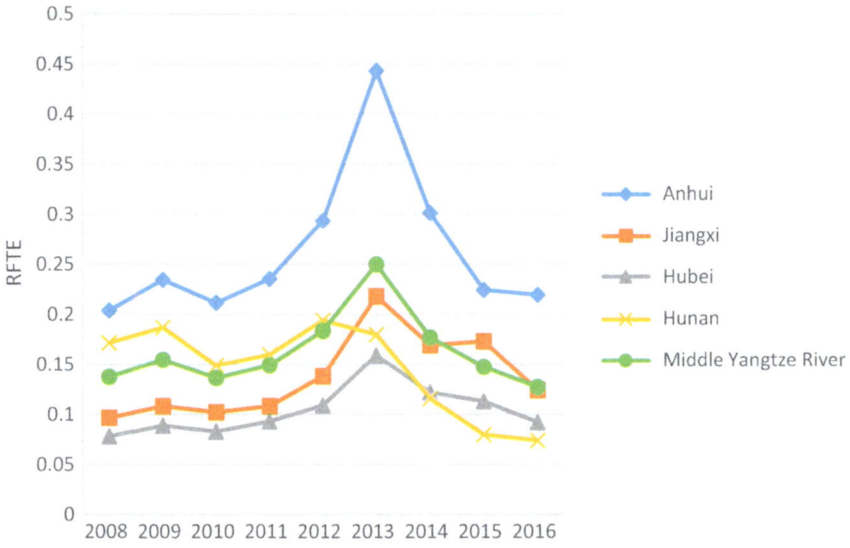

Fig. 6.29 The trend of RFTE in Middle reaches of the Yangtze River from 2008 to 2016. *Data source* Edited by the authors

Fig. 6.30 The mean value of RFTE in Middle reaches of the Yangtze River from 2008 to 2016. *Data source* Edited by the authors

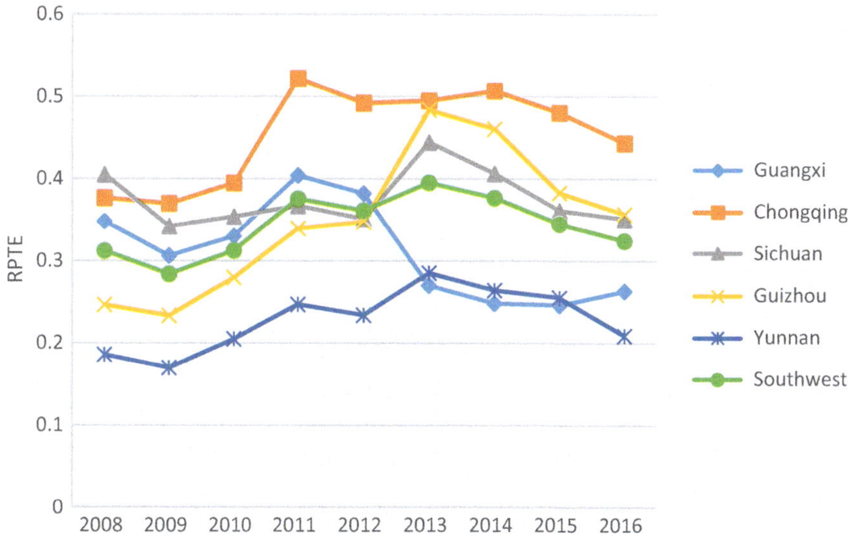

Fig. 6.31 The trend of RPTE in the Southwest from 2008 to 2016. *Data source* Edited by the authors

6.2.2.7 The Southwest

The annual average values of RPTE and RFTE in Southwest China were 0.344 and 0.075, respectively. The regional RPTE was lower than the national average, while the regional RFTE was higher than the national average. The regional RPTE first increased and then decreased, with the highest point in 2013. The regional RFTE was relatively stable from 2008 to 2013, and it showed a declining trend after 2013.

Southwest China is rich in resources, but the regional economic development is slow. Moreover, many areas in Southwest China are mountainous, and the geologic conditions for railway construction are poor, setting out higher technical requirements; all of these factors have a bad impact on the development of railway transport.

In terms of provincial differences, Chongqing had the highest level of RPTE and RFTE in the region, while the RPTE and RFTE in Yunnan were the lowest in the region. The topography in Yunnan is complex with longitudinal mountain ranges and deep canyons in the west and karst landscapes in the east (Li et al., 2015a, b), which has a bad influence on the development of the railway transport sector (Figs. 6.31, 6.32, 6.33 and 6.34).

6.2.2.8 Northwest

During the study period, the annual average values of RPTE and RFTE in Northwest China were 0.154 and 0.066, respectively, which were the lowest in China. The

Fig. 6.32 The mean value of RPTE in the Southwest from 2008 to 2016. *Data source* Edited by the authors

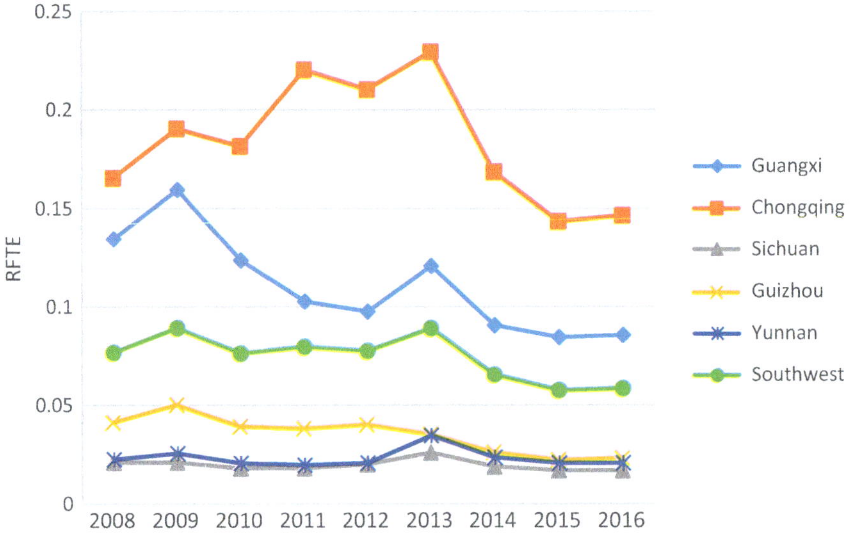

Fig. 6.33 The trend of RFTE in the Southwest from 2008 to 2016. *Data source* Edited by the authors

Fig. 6.34 The mean value of RFTE in the Southwest from 2008 to 2016. *Data source* Edited by the authors

regional RPTE first increased and then decreased, with the highest point in 2014. From 2008 to 2013, the regional RFTE showed a U-shaped trend, and then decreased after 2013.

Northwest China is a vast area with a sparse population, poor natural conditions, and an incomplete railway network. Railway construction in remote plateau areas of Northwest China is very difficult, which leads to the low level of railway infrastructure in some provinces of Western China and poor transport connectivity with other regions. Most of the provinces in Northwest China feature a poor ecological environment (Cao et al., 2021), a weak economic base, and a low level of productivity and literacy of the workforce (Zhao et al., 2020). It is difficult for these provinces to attract factors of production; instead, local production factors are found to flow to developed areas (Zeng et al., 2019).

With the implementation of the Western Development Program, the passenger and freight volume increased significantly, which promoted the RPTE and RFTE. However, after 2013, in order to construct a railway network consisting of eight rail corridors running north–south and eight rail corridors running east–west, the investment in railway infrastructure construction in Northwest China has increased sharply, which has inhibited the railway transport efficiency.

In terms of provincial differences, the RPTE and RFTE in Gansu, and the RFTE in Ningxia were relatively high. The RPTE and RFTE in Tibet and Xinjiang were very low. It is worth noting that the RPTE and RFTE in Tibet were the lowest not only in the region, but also across China. This is mainly because that Tibet is located

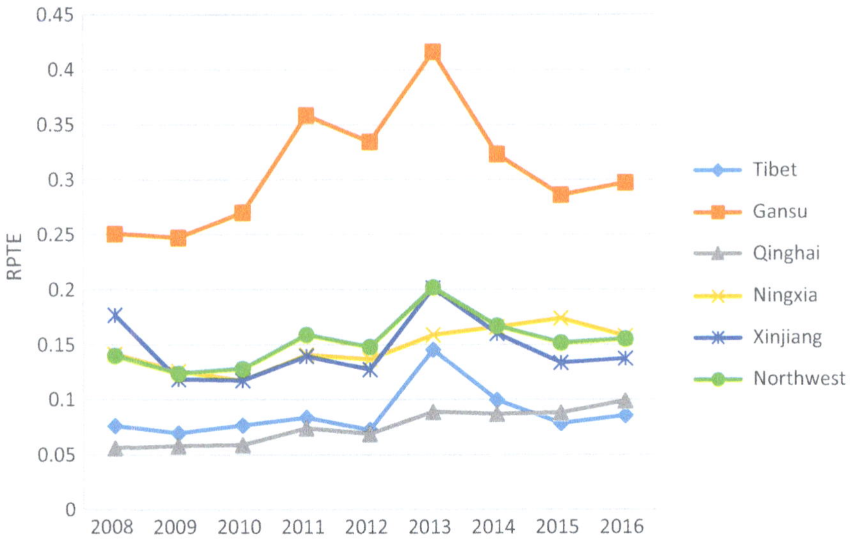

Fig. 6.35 The trend of RPTE in Northwest from 2008 to 2016. *Data source* Edited by the authors

in the Tibet Plateau, and its geographical location and population density determine its low demand for passenger and freight transport. It is worth noting that the RFTE in Qinghai was significantly higher than that in other provinces; the probable cause of that may be the large amount of outward transport of minerals from Qaidam Basin, which promoted the RFTE of Qinghai (Figs. 6.35, 6.36, 6.37 and 6.38).

6.3 Spatial Autocorrelation Analysis

6.3.1 The Global Moran's I Analysis

In this chapter, we also use the Global Moran's *I* to study the spatial autocorrelation of RPTE and RFTE, and to determine whether it is appropriate to use spatial panel regression (Table 6.6).

Using the RPTE and RFTE of 31 provinces between 2008 and 2016, we compute the Global Moran's I value. The results in Tables 6.7 and 6.8 show the values of the Moran's I were greater than zero and at a significance level of 1%, so we can conclude that both RPTE and RFTE exhibited a positive spatial autocorrelation, meaning that the provinces with high RPTE or RFTE were surrounded by provinces with high RPTE or RFTE, and that the provinces with low RPTE or RFTE were surrounded by provinces with low RPTE or RFTE.

Furthermore, the Moran's I index of RPTE ranged from 0.357 to 0.544, while the Moran's I index of RFTE ranged from 0.244 to 0.439. The significant change in the

Fig. 6.36 The mean value of RPTE in Northwest from 2008 to 2016. *Data source* Edited by the authors

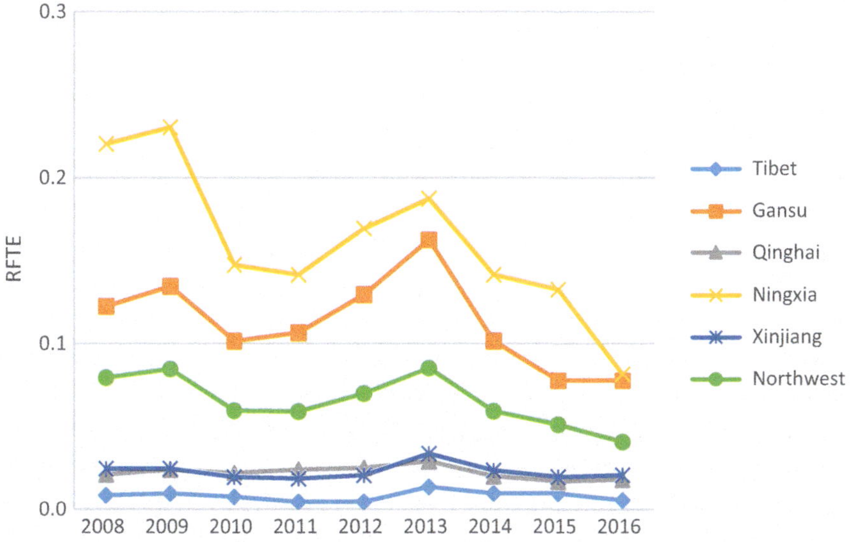

Fig. 6.37 The trend of RFTE in Northwest from 2008 to 2016. *Data source* Edited by the authors

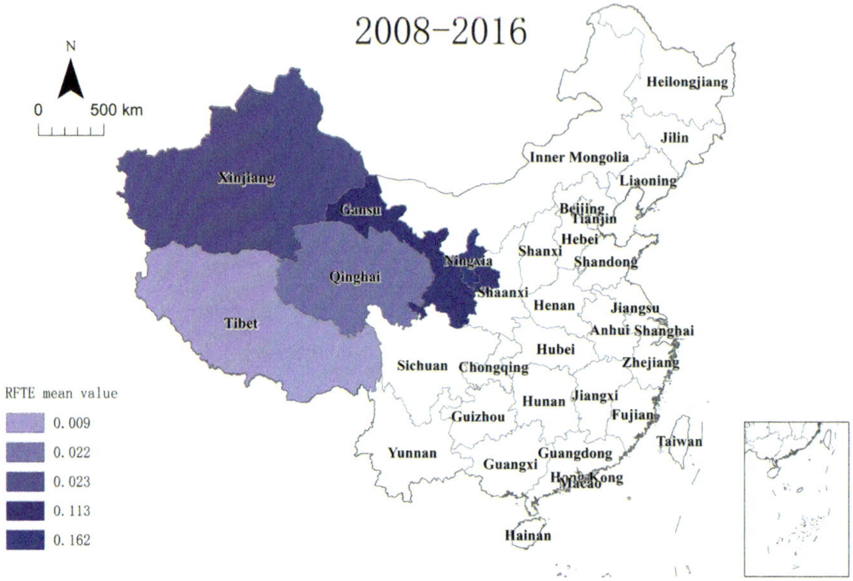

Fig. 6.38 The mean value of RFTE in Northwest from 2008 to 2016. *Data source* Edited by the authors

Table 6.6 The Global Moran's I of RPTEs in China from 2008 to 2016

Year	Moran's I	E(I)	Sd(I)	Z
2008	0.495***	− 0.033	0.115	4.581
2009	0.544***	− 0.033	0.115	4.999
2010	0.440***	− 0.033	0.116	4.097
2011	0.380***	− 0.033	0.117	3.531
2012	0.357***	− 0.033	0.116	3.372
2013	0.480***	− 0.033	0.118	4.335
2014	0.478***	− 0.033	0.118	4.328
2015	0.486***	− 0.033	0.117	4.450
2016	0.509***	− 0.033	0.117	4.652

Note: ***indicates significance at the 1% level
Data source Edited by the authors

Moran's I means that there existed some changes in the global spatial agglomeration impact of RPTE and RFTE.

Table 6.7 The Global Moran's I of RFTEs in China from 2008 to 2016

Year	Moran's I	E(I)	Sd(I)	Z
2008	0.399***	– 0.033	0.101	4.271
2009	0.390***	– 0.033	0.108	3.935
2010	0.332***	– 0.033	0.098	3.730
2011	0.310***	– 0.033	0.094	3.664
2012	0.349***	– 0.033	0.096	3.992
2013	0.439***	– 0.033	0.103	4.562
2014	0.403***	– 0.033	0.098	4.471
2015	0.351***	– 0.033	0.088	4.342
2016	0.244***	– 0.033	0.074	3.765

Note: ***indicates significance at the 1% level
Data source Edited by the authors

Table 6.8 Influencing factors of RPTE

Explanatory variables	Definitions of variables
Economic development level (EDL)	GDP per capita (10^4 RMB)
Railway network density (RND)	Proportion of the length of railways in operation to the regional area (km/km^2)
Urbanization level (UL)	Proportion of urban permanent resident population in the total permanent resident population (%)
Population density (PD)	Ratio of regional permanent resident population to the regional area (person/km^2)
Industrial structure (IS)	Proportion of the tertiary industry in GDP (%)

Note: ***indicates significance at the 1% level
Data source The authors, edited from China Statistical Yearbook (2017)

6.3.2 Local Spatial Autocorrelation Analysis

6.3.2.1 The Local Spatial Autocorrelation Analysis of RPTE

We further use the Moran's I scatterplot and local indicators of spatial association (LISA) to test whether there is spatial agglomeration in local spatial units. Figures 6.39, 6.40, and 6.41 present the Moran's I scatterplot of RPTE in China in 2008, 2012 and 2016, respectively. Most provinces were in the first and third quadrants, and only a few provinces were located in the second and fourth quadrants, indicating that the positive spatial correlations of RPTE above or below the average value were very obvious.

In addition, the Moran scatterplots in 2008, 2012, and 2016 indicate that Jiangsu, Zhejiang, Jiangxi, Anhui, Shandong, Hubei, Hunan and Chongqing always belonged to the HH agglomeration area, meaning that these provinces had high RPTE and

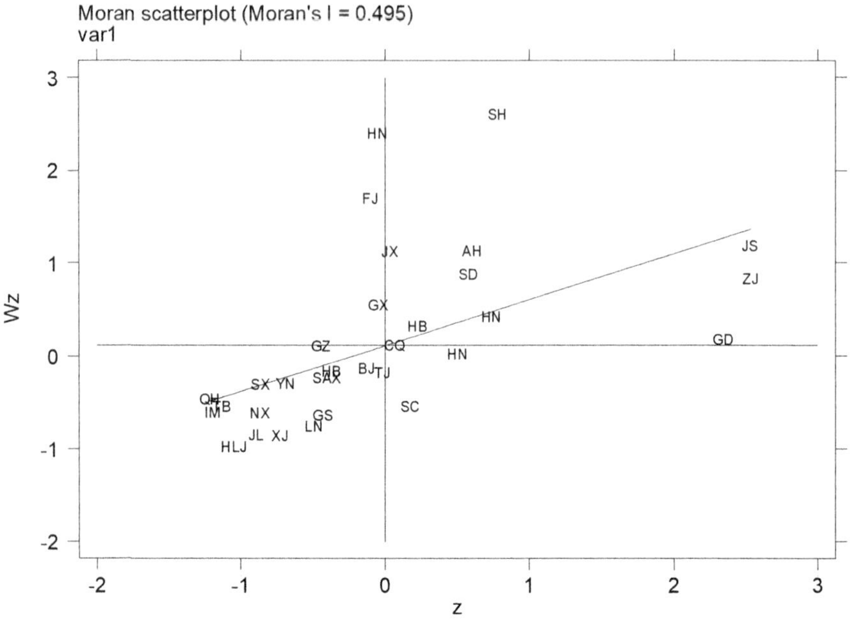

Fig. 6.39 The Moran's I scatterplot of RPTE in 31 provinces in 2008. *Data source* Edited by the authors

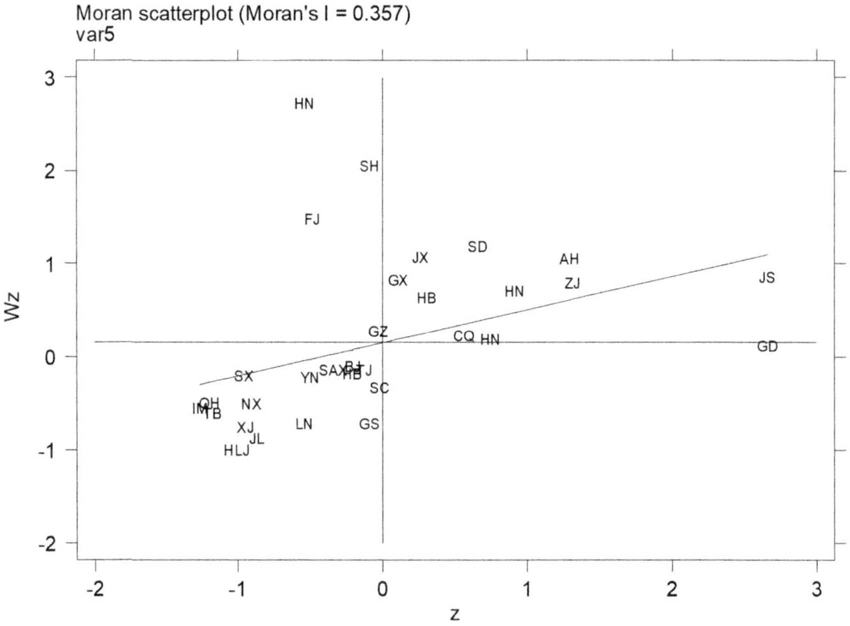

Fig. 6.40 The Moran's I scatterplot of RPTE in 31 provinces in 2012. *Data source* Edited by the authors

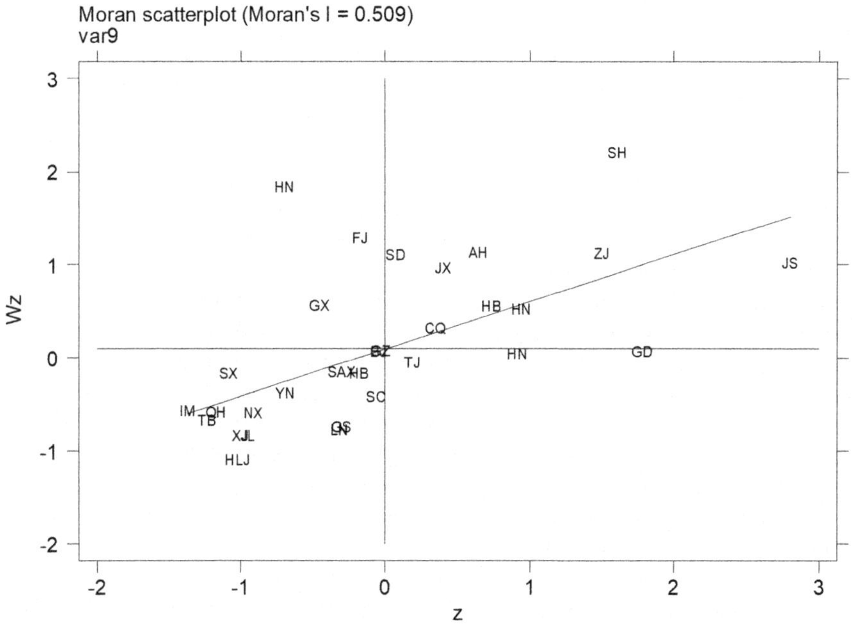

Fig. 6.41 The Moran's I scatterplot of RPTE in 31 provinces in 2016. *Data source* Edited by the authors

were surrounded by provinces with relatively high RPTE; these provinces are mainly located in the Yangtze River Basin of China. In 2008, 2012, and 2016, Beijing, Hebei, Shanxi, Liaoning, Jilin, Heilongjiang, Tianjin, Shaanxi, Yunnan, Ningxia, Qinghai, Gansu, Inner Mongolia, Guizhou and Tibet were always in the LL agglomeration area, meaning that these provinces had low RFTE and were surrounded by provinces with relatively low RFTE; these provinces are mainly located in Northeast, North and Western China.

Figures 6.42, 6.43, and 6.44 depict regions with significant locations color-coded by different types of LISA coefficients of RPTE in China. As can be seen from the figures, the provinces at the significance level of 5% were in the HH-cluster or the LL-cluster.

In 2008, Shanghai, Jiangsu, Zhejiang and Anhui showed the HH-cluster of local spatial autocorrelation of RPTE at the significance level of 5%. In 2012, Shandong and Hunan joined the local HH-cluster area, while Shanghai withdrew from this area. In 2016, the HH-cluster area returned to the situation of 2008, and Shanghai, Jiangsu, Zhejiang and Anhui were in the HH-cluster area.

As for the local LL-cluster area, in 2008 and 2012, Inner Mongolia, Jilin, Heilongjiang and Tibet showed the LL-cluster of local spatial autocorrelation of RPTE. In 2016, Qinghai and Xinjiang joined this area. In sum, the LL-cluster area expanded steadily, while the HH-cluster area changed obviously.

Fig. 6.42 LISA diagram of RPTE in 2008. *Data source* Edited by the authors

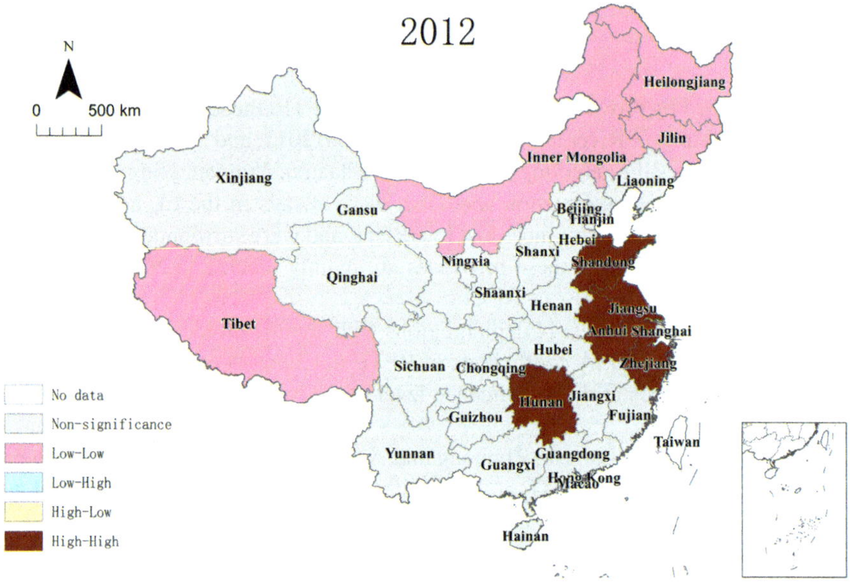

Fig. 6.43 LISA diagram of RPTE in 2012. *Data source* Edited by the authors

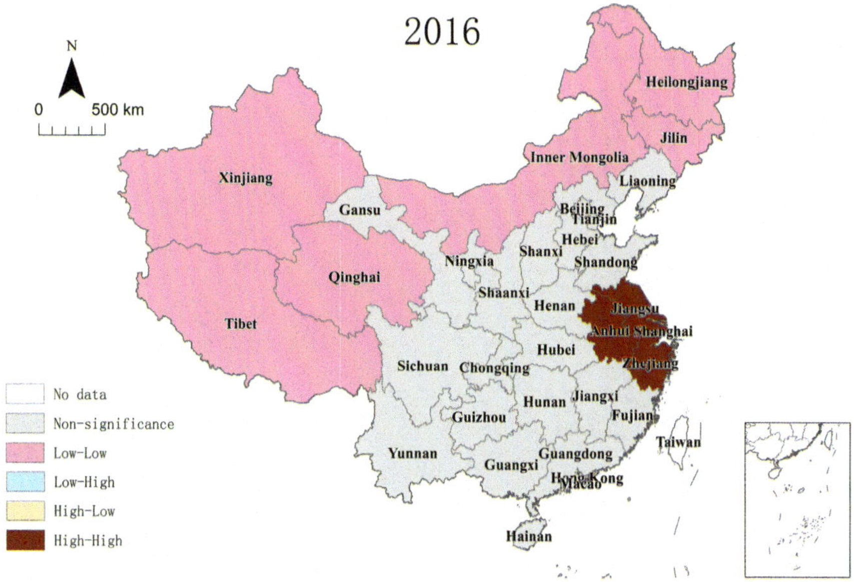

Fig. 6.44 LISA diagram of RPTE in 2016. *Data source* Edited by the authors

6.3.2.2 The Local Spatial Autocorrelation Analysis of RFTE

The local spatial autocorrelation of RFTE will be analyzed here. The following are the Moran's I scatterplots of RFTE. As shown in Figs. 6.45, 6.46 and 6.47, most provinces were in the first and third quadrants (i.e., HH- and LL-cluster areas), indicating that the spatial homogeneity was more significant than the spatial heterogeneity for RFTE. In 2008, 2012 and 2016, there were 22, 21 and 19 provinces located in the first and third quadrants (i.e., HH- and LL-cluster areas), showing a decreasing trend.

Figures 6.48, 6.49, and 6.50 depict regions with significant locations color-coded by different types of LISA coefficients of RFTE in China, and depict the provinces at the significance level of 5%. In 2008, Shanghai, Jiangsu and Shandong showed the HH-cluster of local spatial autocorrelation of RFTE. In 2012 and 2016, Shanghai revealed the HH-cluster of local spatial autocorrelation of RFTE, indicating that Shanghai had a high level of RFTE and was surrounded by provinces with high RFTE.

6.4 Influencing Factors of RPTE and RFTE

The global spatial autocorrelation test shows that there is obvious spatial autocorrelation of RPTE and RFTE in China; therefore, the SDM is also applied to analyze the influencing factors of RPTE and RFTE in China.

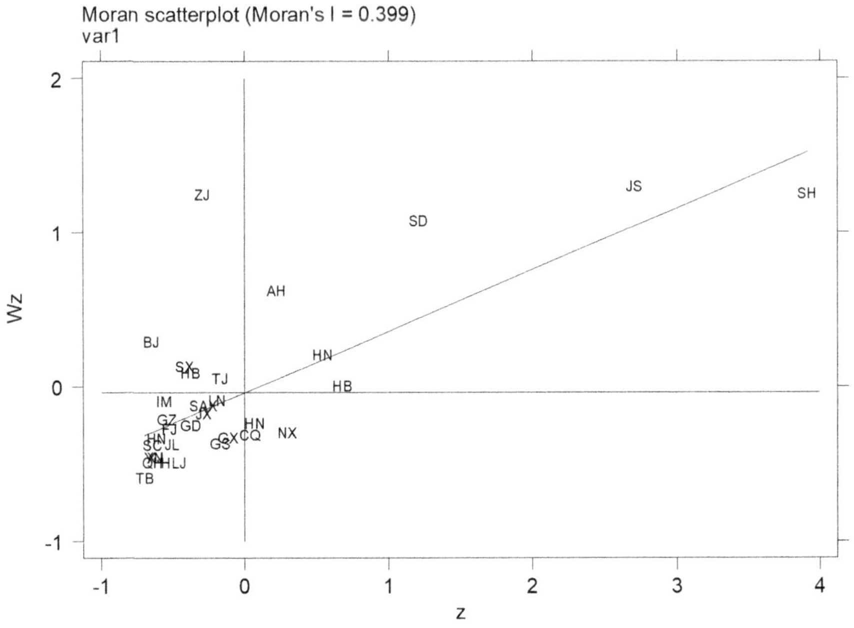

Fig. 6.45 The Moran's I scatterplot of RFTE in 31 provinces in 2008. *Data source* Edited by the authors

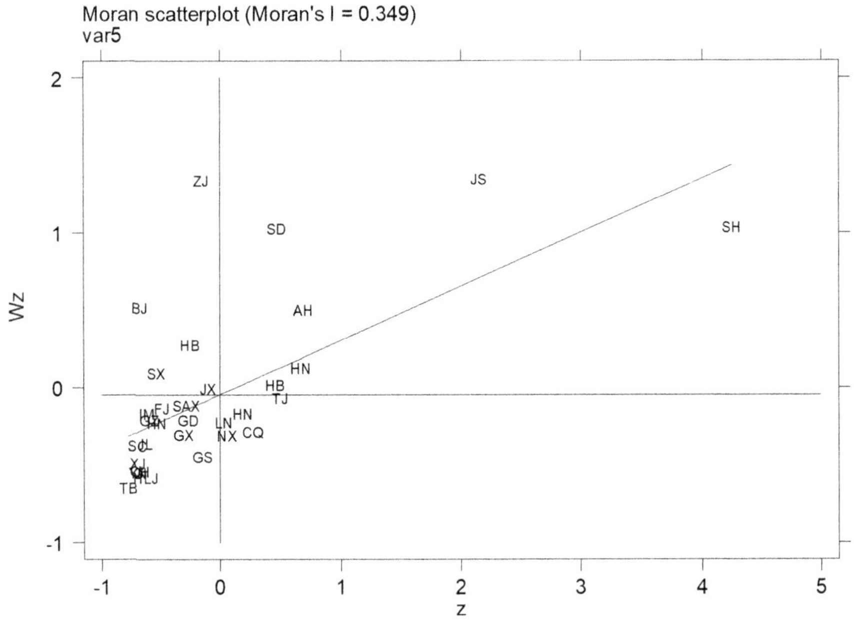

Fig. 6.46 The Moran's I scatterplot of RFTE in 31 provinces in 2012. *Data source* Edited by the authors

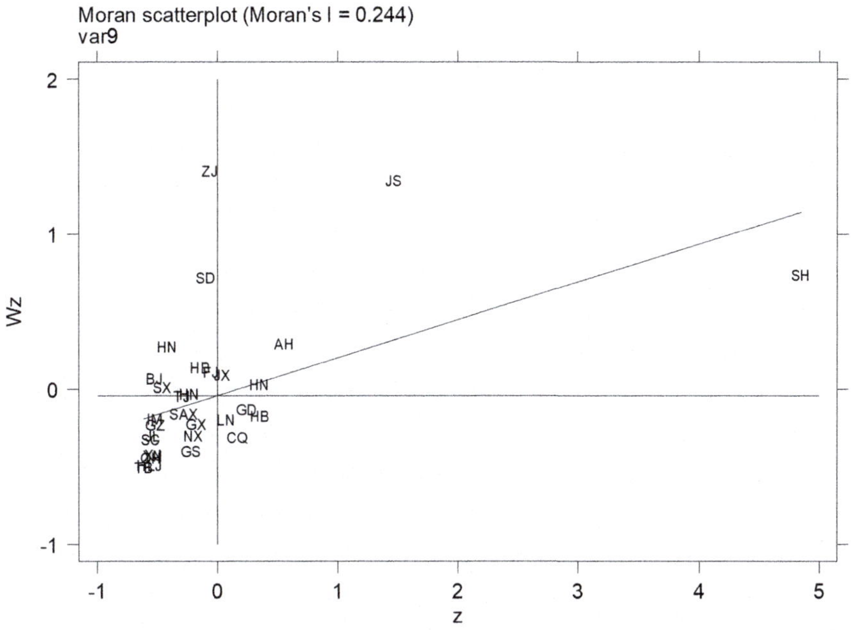

Fig. 6.47 The Moran's I scatterplot of RFTE in 31 provinces in 2016. *Data source* Edited by the authors

Fig. 6.48 LISA diagram of RFTE in 2008. *Data source* Edited by the authors

Fig. 6.49 LISA diagram of RFTE in 2012. *Data source* Edited by the authors

Fig. 6.50 LISA diagram of RFTE in 2016. *Data source* Edited by the authors

6.4.1 Selection of Variables

6.4.1.1 Economic Development Level

The regional economic strength determines whether the region has the ability to improve its transport infrastructure. Provinces with a high level of economic development can more easily attract investment, and are more likely to increase the construction of transport infrastructure and transport equipment manufacturing. Referring to the research results of Alme et al. (2020), Liu et al. (2018), and Yu and Jiang (2019), this study selects the economic development level as an important independent variable for regression analysis on RPTE and RFTE.

6.4.1.2 Railway Network Density

Reasonable railway network density is beneficial to the carrying capacity of railway lines. However, the unreasonable layout or redundant construction of the railway network will inhibit the efficiency of railway transport. Based on the research results of Li (2008) and Ding (2017), this study selects the railway network density as an important regression independent variable affecting RPTE and RFTE, which is equal to the ratio of the length of railways in operation to the land area of each province.

6.4.1.3 Urbanization Level

With the acceleration of urbanization and the continuous economic development in China, residents' need for railway travel and freight transport has shown a growing trend. Referring to the research results of Rong (2001), Ma (2004), and Liu et al. (2018), this study selects the urbanization level as an important regression variable that affects RPTE and RFTE, which is the ratio of urban permanent resident population to the total permanent resident population.

6.4.1.4 Population Density

Population growth will generate new demand for railway passenger and freight transport, and then affect the regional railway transport efficiency. Based on the research results of Wu and Wang (2005), Zhuge et al. (2015), and Liu et al. (2018), this study selects population density as an important regression variable affecting both RPTE and RFTE, which is equal to the ratio of regional permanent resident population to the regional area.

Table 6.9 Influencing factors of RFTE

Explanatory variables	Definitions of variables
Economic development level (EDL)	GDP per capita (10^4 RMB)
Railway network density (RND)	Proportion of the length of railways in operation to the regional area (km/km^2)
Urbanization level (UL)	Proportion of urban permanent resident population in the total permanent resident population (%)
Population density (PD)	Ratio of regional permanent resident population to the regional area (person/km^2)
Industrial structure (IS)	Proportion of the secondary industry in GDP (%)

Data source The authors, edited from China Statistical Yearbook (2017)

6.4.1.5 Industrial Structure

With the improvement of people's living standards, the resident tourism and business travel have been growing in China, resulting in an increasing proportion of the tertiary industry in the social economy. Referring to the research results of Feng (2005) and Guo et al. (2017), this study uses the proportion of the tertiary industry in GDP to measure the impact of industrial structure on RPTE.

The demand for energy and raw materials in industrial production affects the demand for railway freight transport. The technical and economic characteristics of railway transport determine that railway transport is suitable for medium- and long-distance transport of bulk cargoes, such as coal, ore, and building materials, mainly focusing on the secondary industry. Referring to the research results of Geng et al. (2012), and Li et al. (2015a, b), this study uses the proportion of the secondary industry in GDP to measure the impact of industrial structure on RFTE (Tables 6.8 and 6.9).

6.4.2 Regression Analysis of RPTE

6.4.2.1 Construction of the Regression Model

The LM and Wald test results reject the degradation of SDM to a spatial error model or a spatial lag model. The Hausman's test results reject the hypothesis of random effects, so the SDM with fixed effects is selected for regression analysis of RPTE.

With RPTE as the dependent variable and the economic development level, railway network density, urbanization level and population density as the analytic variables, the SDM based on fixed effects is constructed:

$$
\begin{aligned}
RPTEi,t = {} & \rho \sum_{j=1}^{N} Wi,jRPTEi,t + \beta 1DELi,t + \beta 2RNDi,t + \beta 3ULi,t + \beta 4PDi,t \\
& + \beta 5ISi,t + \theta 1 \sum_{j=1}^{N} Wi,jDELi,t + \theta 2 \sum_{j=1}^{N} Wi,jRDi,t + \theta 3 \sum_{j=1}^{N} ULi,t \\
& + \theta 4 \sum_{j=1}^{N} PDi,t + \theta 5 \sum_{j=1}^{N} Wi,jISi,t + \mu + \varepsilon i,t
\end{aligned} \tag{6.1}
$$

The three SDM regression results are shown in Table 6.10. We apply the LR test for selecting the regression model from the three SDMs. The LR test results show that the SDM with spatial and time fixed-effects is more appropriate for regression analysis.

6.4.2.2 The Regression Results

The economic development level and industrial structure have no significant impact. Railway network density has a significant negative impact, while urbanization level and population density have a significant positive impact.

Table 6.10 SDM regression results

	Spatial fixed-effects	Time fixed-effects	Spatial and time fixed-effects
EDL	– 0.0099701	– 0.0545965**	0.733
RND	– 0.0013451***	– 0.0008024***	– 0.0013538***
UL	1.168118***	0.7797608***	1.118614*
PD	0.0011877***	0.0002272 ***	0.0011124***
IS	– 0.2524358**	– 0.4308494***	– 0.0899559
W*EDL	0.060527***	– 0.1258811***	0.0636902*
W*RND	0.0005067*	0.0007803***	0.0003381***
W*UL	– 1.654327***	0.3530895	– 0.6873383
W*PD	– 0.0016791***	0.000628***	– 0.001417***
W*IS	0.2016553	– 0.9667569***	0.5653016*
ρ	0.3476815***	-0.0047762	0.1511282***
Variance sigma2_e	0.0019893***	0.0205164***	0.0018791***
R-squared	0.2919	0.0328	0.1695
Log-likelihood	467.5759	146.2864	479.3409

Note: *** , ** and * indicate significance at the 1%, 5% and 10% levels, respectively
Data source Edited by the authors

Economic Development Level

The regression coefficient of economic development level is not significant In recent years, with the rapid economic development and the increase of personal income, people can afford railway transport; railway passenger transport is not as luxury as air transport, which is no longer easily affected by the macroeconomic environment.

Railway Network Density

Railway network density has a negative effect on RPTE, which indicates that the rapid growth of regional railway network density will inhibit RPTE. With the continuous expansion of railway construction in China, the scale efficiency of railway passenger transport in some provinces of China begins to present a diminishing trend, which indicates that the improvement of RPTE cannot always depend on large-scale investments. Both optimizing infrastructure construction and avoiding the waste of resources are essential. Railway construction should meet the needs of social and economic development for railway transport, and match up with economic development, population growth, land use and other factors. China has a vast territory, and therefore there are great differences of the population size and economic development level among different regions. It is necessary to consider systematically and thoroughly from multiple angles how to optimize the layout of the railway network.

Urbanization Level

Urbanization level shows a significant positive correlation with RPTE as expected. It is worth noting that the regression coefficient of urbanization level is the largest in all the significant variables, which indicates that urbanization level has a greater impact on RPTE. China has entered a critical period of in-depth development of urbanization, and the urbanization level increased from 32.6% in 2000 to 57.35% in 2016. Rapid urbanization has brought heavy passenger traffic. Meanwhile, since the beginning of the twenty-first century, railway transport has always been the most important mode of passenger transport in China among the three major modes of passenger transport (i.e., highway, railway and air transport). The proportion of railway passenger turnover in the total passenger turnover showed an upward trend, rising from 36.7% in 2000 to 40.3% in 2016; therefore, a large amount of passenger traffic brought about by urbanization flowed to railway passenger transport, which significantly improved the utilization efficiency of railway transport infrastructure. In order to give full play to the role of urbanization in improving RPTE, it is necessary to implement differentiated urbanization strategies and railway construction strategies according to the economic development characteristics of different provinces, so as to finally promote the healthy and sustainable development of the whole region.

Population Density

Population density shows a significant positive correlation with RPTE as expected. As mentioned above, railway transport has always been the most important mode of passenger transport in China. A large amount of passenger traffic brought about by population agglomeration flowed to railway transport, which significantly improved PRTE. However, it is worth noting that the regression coefficient of population density is small, which shows that the growth of population density can obviously improve RPTE, but not to a large extent.

Industrial Structure

Industrial structure does not have a significant effect on RPTE, which indicates that the development of the tertiary industry in China at this stage cannot improve RPTE. In order to improve RPTE through the tertiary industry, it is advisable to make full use of the spillover effect of high-speed railways. Since China entered the era of high-speed railways in 2010, the railway passenger transport mode is gradually turning to high-speed railway passenger transport. As a new intercity transport mode, high-speed railways have greatly improved accessibility, accelerated the flow of production factors, contributed to regional economic growth, and greatly impacted industrial structure. The function of a single high-speed railway line is limited. It is crucial to scientifically plan and construct the high-speed railway layout to form a regional high-speed rail network, so that high-speed rail can better play its role in regional economic development. In addition, the overall high-speed rail network planning should be carried out based on the strengthened regional economic ties; meanwhile, relevant supporting facilities of high-speed railways should be improved to better promote the economic development of various regions.

6.4.3 Regression Analysis of RFTE

6.4.3.1 Construction of the Regression Model

The LM test results support the SDM, and the Hausman's test results support fixed effects. With RFTE as the dependent variable and the economic development level, railway network density, urbanization level, population density and industrial structure as the analytic variables, the SDM on fixed effects is constructed:

$$RFTEi,t = \rho \sum_{j=1}^{N} Wi,jRFTEi,t + \beta 1DELi,t + \beta 2RNDi,t + \beta 3ULi,t + \beta 4PDi,t + \beta 5ISi,t + \theta 1 \sum_{j=1}^{N} Wi,jDELi,t + \theta 2 \sum_{j=1}^{N} Wi,jRDi,t + \theta 3 \sum_{j=1}^{N} ULi,t + \theta 4 \sum_{j=1}^{N} PDi,t + \theta 5 \sum_{j=1}^{N} Wi,jISi,t + \mu + \varepsilon i,t \tag{6.2}$$

The LR test is applied for the selection of a fit SDM model. The test results show that the spatial and time fixed-effects model is better than the spatial fixed-effects model and the time fixed-effects model, so the SDM based on spatial and time fixed-effects is selected to analyze the regression results (Table 6.11).

6.4.3.2 The Regression Results

From the results of spatial and time fixed-effects, we can see that the economic development level, railway network density and industrial structure do not have a significant impact on RFTE.

Table 6.11 SDM regression results

	Spatial fixed-effects	Time fixed-effects	Spatial and time fixed-effects
EDL	– 0.0041501	– 0.0195039**	– 0.0032649
RND	– 0.0000706	– 0.0005828***	– 0.0000277
UL	– 0.9247531***	– 0.0721415	– 0.8926591***
PD	– 0.0003514***	0.0003397***	– 0.0003622***
IS	0.0664589	0.0606907	0.0591248
W*EDL	0.0099868	0.038697**	0.0055651
W*RND	– 0.0005679*	0.0002908*	– 0.0004602**
W*UL	0.920415**	– 0.2431737	1.169497*
W*PD	0.0007161**	0.0000769	0.0006094
W*IS	– 0.0319379	– 0.0773225	0.0280523
ρ	0.3160776***	0.1947207**	0.1803202**
Variance sigma2_e	0.0023387***	0.0098176***	0.0022311***
R-squared	0.1191	0.001	0.0633
Log-likelihood	445.7798	247.7671	454.7252

Note: *** , ** and * indicate significance at the 1%, 5% and 10% levels, respectively
Data source Edited by the authors

Urbanization Level and Population Density

The regression coefficients of urbanization level and population density are negative, which is contrary to expectations. For one thing, the regions with a high urbanization level and a high population density in China are mainly located in the economically developed coastal regions. In these regions, the tertiary industry is dominant, and the proportion of the secondary industry in GDP showed a downward trend during the study period, which had a bad impact on RFTE. For another, since the beginning of the new century, the urbanization level in China has risen rapidly. From 2000 to 2016, it rose from 32.6% to 57.3%, and the total population of China also increased from 1.267 billion to 1.383 billion in the same period. The rapid urbanization and population growth have brought huge freight demand. However, the potential of railway freight transport in China has not been fully exploited; the proportion of railway freight turnover in the total freight turnover showed a downward trend, which decreased from 31.1% in 2000 to 12.7% in 2016. Therefore, the huge freight demand brought by urbanization and population growth cannot be fully transformed into the demand for railway freight transport. In order to fully tap the potential of railway transport, it is essential to actively promote the use of railways or waterways for medium- and long-distance freight transport rather than highways. Of course, the role of highways cannot be neglected in the freight transport market; short-distance and scattered freight transport is fit for highway transport.

Economic Development Level and Railway Network Density

The influence of both economic development level and railway network density on RFTE is not significant. Since the beginning of the new century, China's economy has grown rapidly, and more funds have been invested in the construction of railway transport infrastructure, as well as the manufacture and upgrading of transport equipment. However, the long-term large-scale capital investment produces redundant railway construction; there is much unreasonable, repeated railway construction, and the railway network density is too high in some regions (Lu, 2012). The scale effect of railway transport shows a diminishing trend in some provinces.

The threshold for private capital to enter the railway transport sector is too high, which leads to administrative monopoly and market monopoly in railway transport. Therefore, it is of great necessity to make clear the responsibility of property rights, use the modern enterprise system to build Chinese railway enterprises, and lower the entry threshold for private capital in the railway transport sector, to realize the orderly competition between state capital and private capital in the railway transport sector, and increase returns to scale of railway investment. At the same time, according to the development situation and resource endowment of different regions, railway lines should be reasonably arranged to realize the transport network connection between regions. In light of the actual railway transport demand, as well as geographical restrictions, the layout of the railway network should be continuously optimized.

Industrial Structure

Industrial structure has a positive impact on RFTE, but the regression coefficient is not significant, which indicates that the development of the secondary industry does not have a significant impact on RFTE. With the sustained economic development, the freight growth rate may vary at different stages of development. However, the total railway freight volume did not rise but fell from 2013 to 2016. For one thing, China is seeking all-round transformation of the economic growth pattern and building a resource-conserving society, which have affected the transport of bulk commodities. For another, compared with railway transport, highway transport has features of timeliness, efficiency, flexibility, and wide coverage, while railway freight lacks flexibility. To change this situation, it is necessary to continue to deepen railway freight reform, seize the opportunities brought about by industrial upgrading, and integrate industrial upgrading and railway freight transport. The railway sector should also actively respond to the policies of combined rail-road transport, which can reduce the transport cost and make up for the shortcomings of railway freight transport.

6.5 Conclusions

6.5.1 The Overall Spatial Distribution Characteristics of RPTE and RFTE Were Similar

The national RPTE and RFTE were low, and the overall spatial distribution characteristics of RPTE and RFTE were similar: the RPTE and RFTE of the three coastal areas were at a high level, the RPTE and RFTE of Middle reaches of the Yellow River and the Yangtze River, and the Northcast were at an intermediate level, and the RPTE and RFTE of inland areas were at a low level.

6.5.2 The National RPTE First Rose and then Declined, While RFTE Fluctuated in M Shape

The national RPTE fluctuated, with the highest point in 2013. From the 2010s, air transport developed very fast and captured a large share of the passenger transport market, which had a competitive relationship with railway passenger transport. RFTE showed a trend of first rising and then declining, with the highest point in 2013. For one thing, the railway transport sector is impacted by the air transport sector. For another, since 2013, China has increased investment in railway construction to build high-speed railways, which has a bad impact on RPTE and RFTE.

The national RFTE showed an M-shaped trend, which peaked in 2009 and 2013, respectively. The improvement of RFTE from 2011 to 2013 is related to the recovery of the macro-economy of China. In 2012, the Chinese government proposed the establishment of ecological civilization and began to implement strict emission reduction policies (Zhao et al., 2022), resulting in the substantial reduction of the transport volume of many bulk commodities in 2014 and 2015. Through the freight data analysis, it is found that the national railway freight turnover had a negative growth after 2013, which decreased by 13.72% in 2015 compared with the previous year (NBSC, 2021), leading to the gradual decline of RFTE in recent years.

6.5.3 The Impact Mechanism of RPTE and RFTE

We apply the SDM method to analyze the influencing factors of RPTE and RFTE. We found that the economic development level and industrial structure have no significant impact on RPTE. Railway network density has a significant negative impact on RPTE, while urbanization level and population density have a significant positive impact on RPTE. We also found that the economic development level, railway network density and industrial structure do not have a significant impact on RFTE.

References

Alam, K.M., Xuemei, L., Baig, S. et al. (2020) Analysis of Technical, Pure Technical and Scale Efficiencies of Pakistan Railways Using Data Envelopment Analysis and Tobit Regression Model. Netw Spat Econ 20, 989–1014. https://doi.org/10.1007/s11067-020-09510-9

Cao, S. P., He, Y., Zhang, L. F., Chen, Y., Yang, W., Yao, S., & Sun, Q. S. (2021). Spatiotemporal characteristics of drought and its impact on vegetation in the vegetation region of Northwest China. *Ecological Indicators, 133*, 108420. https://doi.org/10.1016/j.ecolind.2021.108420

China Statistical Yearbooks. (2017). China Statistical Publishing House. https://data.cnki.net/yearbook/Single/N2020100004

China Statistical Yearbook. (2009–2017). China Statistical Publishing House. https://data.cnki.net/yearbook/Single/N2020100004

China Transport Statistical Yearbook (2009–2017), China Statistical Publishing House. https://data.cnki.net/yearBook/single?id=N2022010162. Accessed April 24, 2021.

Ding, J. Y. (2017). *Study on the spatial and temporal evolution of regional railway transportation efficiency and its influencing factors in China in the background of 1997–2014 railway acceleration* [Doctoral Dissertation of Anhui Normal University, Wuhu] (In Chinese) https://cdmd.cnki.com.cn/article/cdmd-10370-1017277230.htm

Feng, L. (2005). Analysis of influencing factors of the change of industrial structure on the demand for freight transport in China. *Railway Freight Transport, 2*, 21–23+1 (In Chinese). https://www.cnki.com.cn/Article/CJFDTotal-TDHY200502008.htm

Gao, P. (2021). Estimation of capital stock and capital return rate of China's transportation infrastructure. *Railway Transport and Economy, 3*, 39–41+47 (In Chinese). http://www.cnki.com.cn/Article/CJFDTotal-TDYS200103016.htm

Geng, L. Y., Zhang, T. W., & Zhao, P. (2012). Forecast of railway freight volumes based on LS-SVM with grey correlation analysis. *Journal of the China Railway Society, 34*(03), 1–6 (In Chinese). https://xueshu.baidu.com/usercenter/paper/show?paperid=70adf9a46e86608410265d7e0047d31f&site=xueshu_se

Guo, M. Y., Wang, J., & Shi, Y. T. (2017). Analysis of co-integration relationship between transportation and industrial structure in Shenzhen. *Journal of Shenzhen University (Science and Engineering), 34*(02), 214–220 (In Chinese). https://d.wanfangdata.com.cn/periodical/szdxxb201702014

Li, J. W., & Zhang, G. Q. (2016). Estimation of capital stock and capital return rate of China's transportation infrastructure. *Contemporary Finance & Economics, 6*, 3–14 (In Chinese).https://doi.org/10.13676/j.cnki.cn36-1030/f.2016.06.001

Li, L. B. (2008). Efficiency analysis of the regional railway in China: Based on the application of DEA-Tobit approach. *Soft Science, 4*, 58–63 (In Chinese). http://en.cnki.com.cn/Article_en/CJFDTOTAL-XUXI200804014.htm

Li, R., Dai, M. R., & Li, F. Z. (2015a) Study on selection methods of key influence factors of railway freight transport volume based on grey correlation. *Railway Freight Transport, 33*(11), 11–14+5 (In Chinese). https://xueshu.baidu.com/usercenter/paper/show?paperid=70adf9a46e86608410265d7e0047d31f&site=xueshu_se

Li, Y., He, D., Hu, J., & Cao, J. (2015b) Variability of extreme precipitation over Yunnan province, China 1960–2012. *International Journal of Climatology, 35*, 245–258 (In Chinese). https://doi.org/10.1002/joc.3977

Liu, B. Q., Wei, W., Su, Q., & Zhang, L. L. (2018). Spatio-temporal evolution characteristics and mechanism of railway efficiency in China. *Geographical Research, 38*(5), 26–35 (In Chinese). http://www.dlyj.ac.cn/EN/Y2018/V37/I3/512

Liu, T. Y. (2019). Spatial structure convergence of China's transportation system. *Research in Transportation Economics, 78*. https://doi.org/10.1016/j.retrec.2019.100768

Lu, T. D. (2012). The proposition to avoid the over advance and inappropriate construction of China's transport infrastructures. *Scientia Geographica Sinica, 32*(1), 2–111. http://geoscien.neigae.ac.cn/EN/Y2012/V32/I1/2. Accessed December 7, 2019.

Ma, B. T. (2004). *Urbanization process and inter-city railway passenger transport demand analysis in China* [Doctoral Dissertation of Beijing Jiaotong University, Beijing, 2004] (In Chinese). https://d.wanfangdata.com.cn/thesis/Y586044

Mao, H. Y. (2017). Innovation of mechanism and regional policy for promoting coordinated development of the Beijing-Tianjin-Hebei. *Progress in Geography, 36*(1), 2–14. https://doi.org/10.18306/dlkxjz.2017.01.001

National Bureau of Statistics of China (NBSC). (2021). http://data.stats.gov.cn/easyquery.htm?cn=E0103

Rong, C. H. (2001). From transport development to sustainable transport. *Soft Science, 3*, 4–11 (In Chinese). https://wenku.baidu.com/view/c9d39c0f52ea551810a687ca?fr=xueshu_top

Statistical Yearbook of the Chinese Investment in Fixed Assets (2009–2013, 2015–2017). China Statistical Publishing House. https://data.cnki.net/yearBook/single?id=N2019030174

Wang, J., Yang, F., & Chen, K. (2019). Regional carbon emission evolution mechanism and its prediction approach: A case study of Hebei, China. *Environmental Science and Pollution Research, 26*, 28884–28897. https://doi.org/10.1007/s11356-019-06021-0

Wang, L. H., Zhang, F., Fu, W., Tan, Q., & Chen, J. C. (2021). Analysis of temporal and spatial differences and influencing factors of energy eco-efficiency in energy-rich area of the Yellow River Basin. *Physics and Chemistry of the Earth Parts a/b/c, 121*, 102976. https://doi.org/10.1016/j.pce.2020.102976

Wu, W. Q., & Wang, D. (2005). Researches on coordinated development for all kinds of transport modes. *Railway Economics Research, 2*, 29–33 (In Chinese). https://www.docin.com/p-491341244.html

Xu, X. W., & Xu, Y. Q. (2017). The evaluation of comprehensive competitiveness of coastal ports and spatial evolution of sea port hinterland in China. *Economic Geography, 38*(5), 26–35. https://doi.org/10.18306/dlkxjz.2017.01.001

Yu, L. C., & Jiang, J. Y. (2019). Analysis of the passenger transportation demand-supply condition of Chinese railway and related industrial policies. *Social Sciences in Guangdong, 1*, 26–34+254

(In Chinese). https://wenku.baidu.com/view/5b4115c6a78da0116c175f0e7cd184254b351bdb?fr=xueshu_top

Zeng, L. E., Lu, H. Y., Liu, Y. P., Zhou, Y., & Hu, H. Y. (2019). Analysis of regional differences and influencing factors on China's carbon dioxide emissions efficiency in 2005–2015. *Energie, 12*(16), 3081–4001. https://doi.org/10.3390/en12163081

Zhang, Y., & Chen, P. (2015). A study on the effect of transport improvement to the tourism development in Henan Province. In: M. Li, Q. Zhang, R. Zhang, X. Shi (Eds.), *Proceedings of 2014 1st international conference on industrial economics and industrial security* (Springer, Berlin, Heidelberg). https://doi.org/10.1007/978-3-662-44085-8_68

Zhao, P. J., Zeng, L. E., Lu, H. Y., Zhou, Y., Hu, H. Y., & Wei, X. Y. (2020). Green economic efficiency and its influencing factors in China from 2008 to 2017: Based on the super-SBM model with undesirable outputs and spatial Durbin model. *Science of the Total Environment*, 140026. https://doi.org/10.1016/j.scitotenv.2020.140026

Zhao, P., Zeng, L., Li, P., Lu, H., Hu, H., Li, C., Zheng, M., Li, H., Yu, Z., Yuan, D., Xie, J., Huang, Q., & Qi, Y. (2022). China's transportation sector carbon dioxide emissions efficiency and its influencing factors based on the EBM DEA model with undesirable outputs and Spatial Durbin model. *Energy, 238*, 121934. https://doi.org/10.1016/j.energy.2021.121934

Zhuge, H. Y., Zhou, Y., Wu, W. J., & Liu, B. (2015) Analysis and revelation of market competitiveness of foreign high-speed railways. *Railway Transport and Economy 37*(11), 83–88 (In Chinese). https://doi.org/10.16668/j.cnki.issn.1003-1421.2015.11.17

Chapter 7
Air Transport Efficiency

7.1 Background and Methods

7.1.1 Background

With the development of technology and economy, the elements of social and economic activities flow more frequently around the world. Civil aviation with the characteristics of long-distance and high-speed transport plays an irreplaceable role in all transport networks. China has a vast territory, and the links between various provinces and regions are increasingly close. Air transport has made great contributions to overcoming distance and time constraints. Since the reform and opening-up, the air transport network has gradually improved, and now it has begun to operate at scale (Fig 7.1). However, as China's economy has developed rapidly since the beginning of the twenty-first century, the demand for air transport is increasing day by day (Lu et al., 2019). The Chinese government proposed the basic development principle of "making structural adjustments, enlarging capacity and increasing efficiency" for the air transport network. It intends to develop the air transport network, further adjust and optimize the structure and layout of the civil aviation network, enlarge air transport capacity and improve transport efficiency. Therefore, the study of air transport efficiency has important practical significance.

7.1.2 Methods

In this section, the EBM model is used to measure air transport efficiency (ATE). This study selects 31 provinces in the Chinese mainland as the research area, covering a period from 2008 to 2016. This study selects the capital stock, the number of employees, and the number of flights taking off and landing as the input indicators, and selects passenger throughput and cargo throughput as the output indicators (Table 7.1).

P. Zhao and L. Zeng, *Transport Efficiency and Safety in China*, Population, Regional Development and Transport, https://doi.org/10.1007/978-981-99-1055-7_7

This study estimates the capital stock of the air transport sector based on the perpetual inventory method proposed by Goldsmith. The basic formula of the perpetual inventory method is: $K_{i,t} = I_{i,t} + (1 - \delta_{i,t})K_{i,t-1}$. $K_{i,t}$ represents the capital stock of the air transport sector in t year of i province, $I_{i,t}$ represents the total fixed capital investment in the air transport sector in t year of i province, and δ represents the capital depreciation rate of the air transport sector. Because the total fixed capital investment in the air transport sector in some provinces fluctuated greatly in some years, in order to prevent the occurrence of special deviation values, this study takes the average value of the total fixed asset investment from 2007 to 2009, and then divides the average value by 10% as the capital stock of the air transport sector in the base period (Zhang et al., 2004). The annual capital depreciation rate δ of the air transport sector is 8.76% based on the research results of Li and Zhang (2016). The data are from China Statistical Yearbook (2009–2017), China Fixed Assets Statistical Yearbook (2009–2017), China Transport Statistical Yearbook (2009–2017) and statistical yearbooks of some provinces over the research period.

The data of air transport practitioners are from China Statistical Yearbook (2009–2017).

The data of flights taking off and landing, passengers and cargoes are from the Statistics Bulletin of Civil Airports (2008–2016).

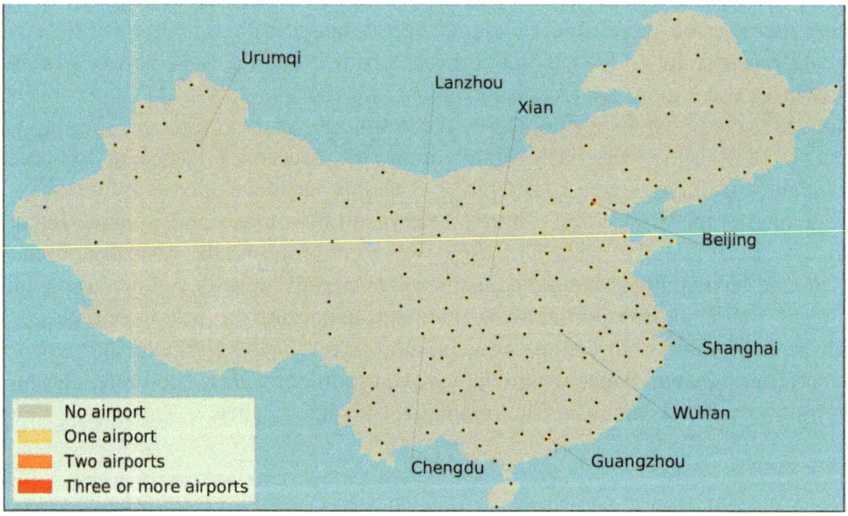

Fig. 7.1 Distribution map of civil airports in China. *Data source* https://doi.org/10.1016/j.jairtraman.2020.102014

Table 7.1 The measurement index system of air transport efficiency

Primary indexes	Secondary indexes
Inputs	The capital stock of the air transport sector (unit: 100 million)
	Total number of employees in the air transport sector (unit: 10,000)
	The number of flights taking off and landing (unit: 10,000)
Outputs	Passenger throughput (unit: 10,000 persons)
	Cargo throughput (unit: 10,000 tons)

Data source: the authors, edited from China Statistical Yearbook (2009–2017), Statistical Yearbook of the Chinese Investment in Fixed Assets (2009–2013, 2015–2017), China Transport Statistical Yearbook (2009–2017), the Statistics Bulletin of Civil Airports (2008–2016), and statistical yearbooks of some provinces over the research period

7.2 Measurement Results

Using maxdea 7.9 ultra software, we calculate the ATEs of 31 provinces in China from 2008 to 2016. The results are listed in Table 7.2 and 7.3. Figure 7.2 is the spatial distribution map of ATE.

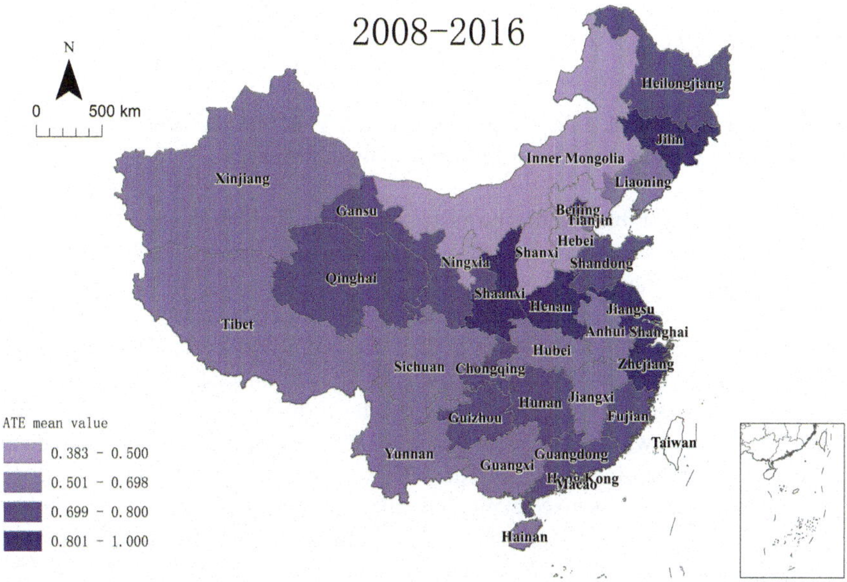

Fig. 7.2 The mean value of ATE intervals in 31 provinces from 2008 to 2016. *Data source* Edited by the authors

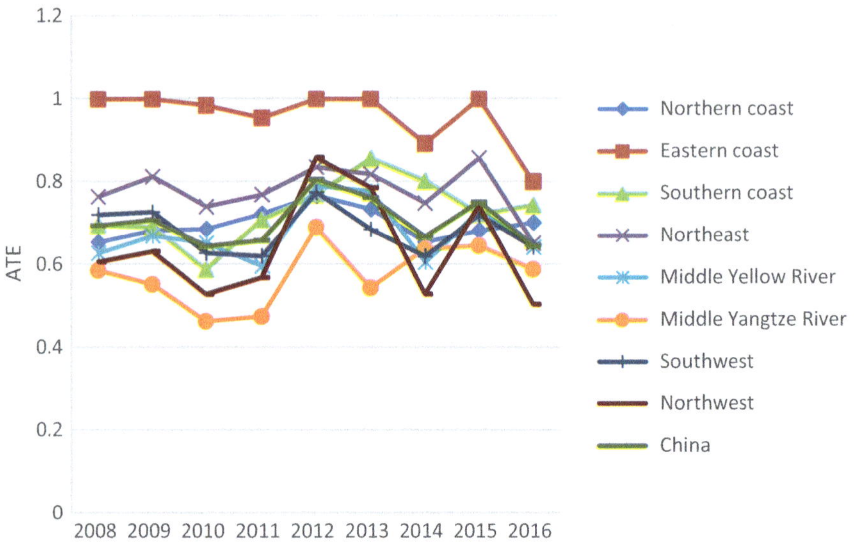

Fig. 7.3 The trend of ATE nationwide and in eight economic zones from 2008 to 2016. *Data source* Edited by the authors

7.2.1 *The Overall Characteristics*

7.2.1.1 The National ATE Fluctuated, Which Was Significantly Affected by the Macroeconomic Environment

It can be seen from Table 7.2 that the mean value of ATE in China's 31 provinces was 0.704, which means that the whole level of ATE needs to increase by 29.6% to reach the production frontier surface. The ATE in 16 provinces was higher than the national level. The national ATE first declined and then rose from 2008 to 2012, with the lowest point in 2010. After 2012, the national ATE showed a significant downward trend. The decline of ATE in 2008, perhaps induced by the financial crisis, and the slow growth rate of passenger and cargo demand in the air transport sector inhibited ATE. After 2010, China was set for economic recovery, the growth rate of air passenger demand significantly increased, and then ATE improved significantly. After 2012, the number of employees, fixed capital investment, number of flights, and passenger volume of the air transport sector all increased rapidly, but the the growth rate of air freight volume increased slowly, which inhibited the overall level of ATE (Fig 7.3).

Table 7.2 ATE of 31 provinces in China from 2008 to 2016

Provinces	2008	2009	2010	2011	2012	2013	2014	2015	2016	Mean
Beijing	1.000	1.000	1.000	1.000	1.000	1.000	1.000	1.000	1.000	1.000
Tianjin	0.654	0.683	0.706	0.765	0.660	0.661	0.616	0.691	0.734	0.685
Hebei	0.212	0.320	0.305	0.408	0.630	0.504	0.338	0.341	0.419	0.386
Shanxi	0.414	0.423	0.385	0.412	0.546	0.521	0.421	0.559	0.505	0.465
Inner Mongolia	0.240	0.256	0.228	0.364	0.615	0.652	0.431	0.702	0.430	0.435
Liaoning	0.785	0.749	0.622	0.639	0.688	0.643	0.498	0.570	0.556	0.639
Jilin	0.839	1.000	1.000	1.000	1.000	1.000	1.000	1.000	0.678	0.946
Heilongjiang	0.671	0.690	0.602	0.670	0.820	0.811	0.750	1.000	0.721	0.748
Shanghai	1.000	1.000	1.000	1.000	1.000	1.000	1.000	1.000	1.000	1.000
Jiangsu	1.000	1.000	1.000	1.000	1.000	1.000	0.806	1.000	0.779	0.954
Zhejiang	1.000	1.000	0.954	0.865	1.000	1.000	0.872	1.000	0.625	0.924
Anhui	0.478	0.485	0.355	0.423	0.577	0.475	0.519	0.582	0.629	0.503
Fujian	0.777	0.772	0.711	0.738	0.835	0.811	0.688	0.779	0.737	0.761
Jiangxi	0.333	0.309	0.294	0.427	0.906	0.489	0.557	0.694	0.651	0.518
Shandong	0.755	0.728	0.735	0.721	0.780	0.773	0.667	0.696	0.650	0.723
Henan	1.000	1.000	1.000	0.611	1.000	0.937	0.572	0.657	0.632	0.823
Hubei	0.541	0.559	0.481	0.474	0.612	0.514	0.479	0.620	0.411	0.521
Hunan	1.000	0.864	0.731	0.583	0.672	0.701	1.000	0.693	0.670	0.768
Guangdong	0.805	0.836	0.737	0.834	0.886	0.761	0.722	0.767	0.826	0.797
Guangxi	0.585	0.614	0.583	0.582	0.769	0.753	0.506	0.599	0.458	0.605
Hainan	0.508	0.465	0.320	0.559	0.593	1.000	1.000	0.622	0.673	0.638
Chongqing	0.666	0.666	0.524	0.682	0.849	0.822	1.000	1.000	0.750	0.773
Sichuan	0.775	0.789	0.566	0.545	0.865	0.584	0.549	0.673	0.666	0.668
Guizhou	1.000	1.000	1.000	0.735	0.747	0.617	0.513	0.647	0.546	0.756
Yunnan	0.581	0.567	0.469	0.560	0.640	0.648	0.534	0.684	0.793	0.608
Tibet	0.440	0.455	0.361	0.554	1.000	1.000	0.656	1.000	0.697	0.685
Shaanxi	0.859	1.000	1.000	1.000	1.000	1.000	1.000	1.000	1.000	0.984
Gansu	0.713	0.808	0.739	0.688	1.000	1.000	0.653	1.000	0.576	0.797
Qinghai	1.000	1.000	0.723	0.647	1.000	0.655	0.435	0.534	0.351	0.705
Ningxia	0.405	0.426	0.402	0.453	0.613	0.578	0.330	0.490	0.401	0.455
Xinjiang	0.479	0.470	0.419	0.502	0.682	0.698	0.572	0.685	0.493	0.556
China	0.694	0.708	0.644	0.659	0.806	0.762	0.667	0.751	0.647	0.704

Data source Edited by the authors

Table 7.3 ATE of eight economic zones from 2008 to 2016

Economic zones	2008	2009	2010	2011	2012	2013	2014	2015	2016	Mean
Northern coast	0.655	0.683	0.687	0.724	0.767	0.734	0.655	0.682	0.701	0.699
Eastern coast	1.000	1.000	0.985	0.955	1.000	1.000	0.893	1.000	0.801	0.959
Southern coast	0.696	0.691	0.589	0.710	0.771	0.857	0.803	0.723	0.745	0.732
Northeast	0.765	0.813	0.741	0.770	0.836	0.818	0.749	0.857	0.652	0.778
Middle Yellow River	0.628	0.670	0.653	0.597	0.790	0.777	0.606	0.730	0.642	0.677
Middle Yangtze River	0.588	0.554	0.465	0.477	0.692	0.545	0.639	0.647	0.590	0.577
Southwest	0.721	0.727	0.628	0.621	0.774	0.685	0.620	0.720	0.642	0.682
Northwest	0.607	0.632	0.529	0.569	0.859	0.786	0.529	0.742	0.504	0.640
China	0.694	0.708	0.644	0.659	0.806	0.762	0.667	0.751	0.647	0.704

Data source Edited by the authors

7.2.1.2 The ATE in Coastal Areas and Northeast China Was High, but It Was Low in the Middle Reaches and Western Areas

Eastern coast had the highest level of ATE. The ATE of Southern coast, Northeast and Middle reaches of the Yangtze River was also high, and the annual average values of ATE in these four regions were all above 0.7. The ATE in Middle reaches of the Yellow River was the lowest.

The air transport sector is a capital and technology intensive industry. Due to the high level of both economic development and technology accumulation in coastal areas, the air transport sector is relatively developed. Meanwhile, coastal areas are densely populated and have high income levels. Therefore, these areas have high demand for air passenger and cargo transport, which has a positive effect on ATE.

The economic development level in Middle reaches of the Yangtze River is lower than that of coastal regions. Middle reaches of the Yangtze River are located in central China. The aviation demand of Middle reaches of the Yangtze River is not as strong as that of the coastal and Northeast regions, and the low output of both air passenger and cargo transport affects the regional ATE. The ATE of Henan and Shaanxi in Middle reaches of the Yellow River was relatively high, but the ATE of Inner Mongolia and Shanxi was relatively low, which inhibited the overall ATE level of Middle reaches of the Yangtze River. Southwest China and Northwest China are the areas with the lowest level economic development in China; the relatively low residential income level affects the ATE of these regions.

7.2.1.3 In Most Regions, the ATEs First Decreased and then Increased in the Early Stage, and Fluctuated in Different Degrees Afterwards

Except for the Northern and Eastern coastal areas, the regional ATEs of the other six regions first fell and then increased in the early research period (2008–2012), and showed different fluctuations afterwards. Under the 4 trillion-yuan fiscal policy, from 2008 to 2010, the aviation capital stock in these regions increased rapidly, while the growth rate of aviation demand has slowly declined, which had bad effects on ATE. After 2010, the economy rebounded and the growth rate of the demand for civil aviation passenger and freight transport picked up, which had a good effect on ATE. However, after 2012, the regional economic development level, air transport investment, civil aviation demand and the impact of high-speed rail in these regions are different, resulting in different trends of ATE. The ATE of Eastern coast was relatively stable, and the regional aviation efficiency was at the production frontier surface in more than half of the study period. The ATE of Northern coast showed significant fluctuations (Fig. 7.4).

7.2.2 *Spatial Variations*

7.2.2.1 Northern Coast

The average value of ATE in Northern coast was 0.699 during the study period, slightly lower than the national average value. The ATE showed an upward trend from 2008 to 2012, and first decreased and then increased after 2012, with the lowest value in 2014.

In terms of provincial differences, the ATE in Beijing was relatively high, and it was at the production frontier surface during the study period. The average annual value of ATE in Hebei was 0.386, one of the lowest in China. During the study period, the change trend of ATE in Tianjin was roughly consistent with that of the region. The ATE of Tianjin showed an upward trend from 2008 to 2011, and first decreased and then increased after 2011, with the lowest value in 2014. The change trend of ATE in Shandong was similar to that of the whole country. The ATE of Shandong first decreased and then increased from 2008 to 2012, but declined again after 2012.

Beijing is the capital of China, and has frequent international exchanges and the largest number of international routes in China. Beijing has close ties with all parts of the world. The air passenger throughput in Beijing is the largest in China, which has a positive effect on ATE. Hebei highly depends on air transport hubs in Beijing and Tianjin, and there is no large regional air transport hub in Hebei, which has a bad effect on the development of ATE (Figs. 7.4 and 7.5).

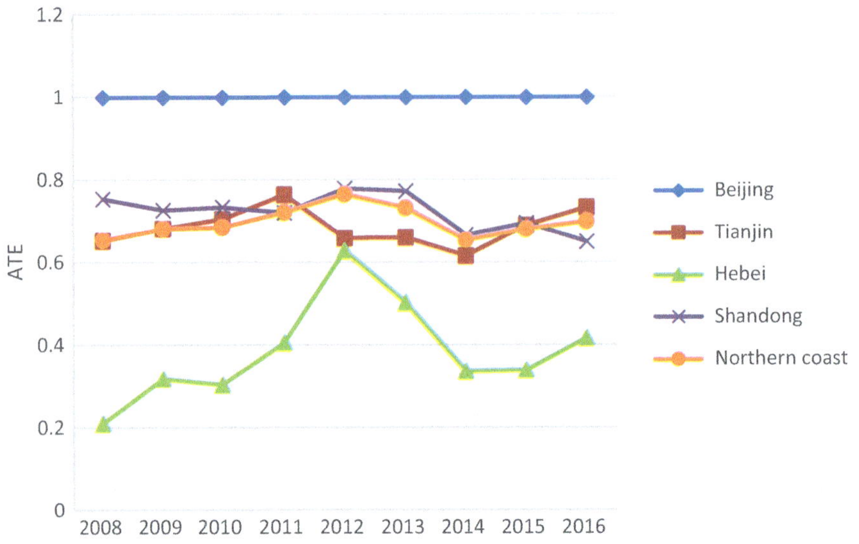

Fig. 7.4 The trend of ATE in Northern coast from 2008 to 2016. *Data source* Edited by the authors

Fig. 7.5 The mean value of ATE in Northern coast from 2008 to 2016. *Data source* Edited by the authors

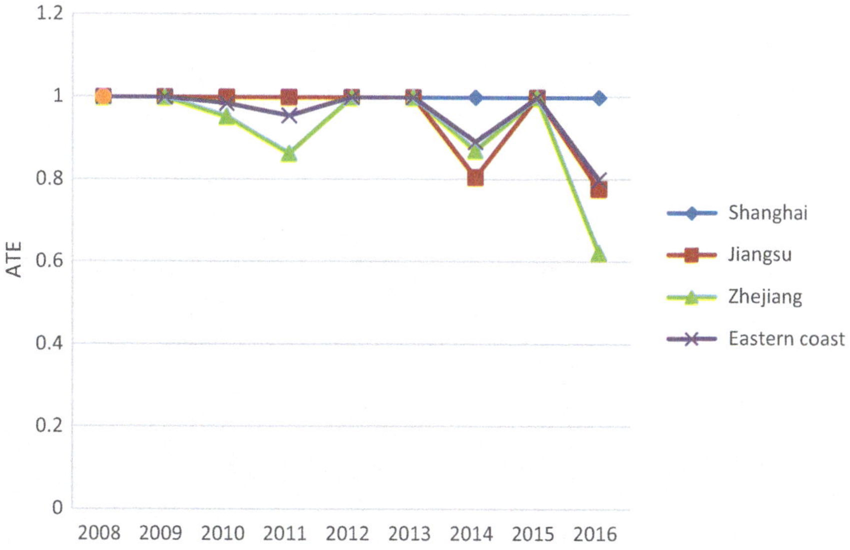

Fig. 7.6 The trend of ATE in Eastern coast from 2008 to 2016. *Data source* Edited by the authors

7.2.2.2 Eastern Coast

During the study period, the mean value of ATE in Eastern coast was 0.959, ranking the first among the eight regions. The regional ATE remained at the forefront in 2008, 2009, 2012, 2013 and 2015, but it was at the lowest level in 2016.

In terms of provincial differences, the ATE of Shanghai always maintained the production frontier surface during the study period. The ATE of Jiangsu was at the production frontier surface during the study period except in 2014 and 2016. The ATE of Zhejiang was at the production frontier surface during the study period except in 2011, 2012, 2014 and 2016.

In the process of economic globalization, the eastern coastal area is one of the most economically developed zones with the fastest economic growth in China. The growth of economy and foreign trade has produced heavy domestic and foreign air transport demand. After entering the era of high-speed rail, many passengers in Eastern coast have chosen high-speed rail over air transport. In 2016, the air passenger throughput of Shanghai and Zhejiang decreased by 33.4% and 57.8% from 2015, respectively, which had a bad effect on ATE, and the ATE in 2016 was at the lowest level during the study period (Figs. 7.6 and 7.7).

7.2.2.3 Southern Coast

During the study period, the mean value of ATE in Southern coast was 0.7, slightly higher than the national average level. The regional ATE first decreased and then

Fig. 7.7 The mean value of ATE in Eastern coast from 2008 to 2016. *Data source* Edited by the authors

increased from 2008 to 2013, with the lowest level in 2010, and it first decreased and then increased from 2013 to 2016, with the lowest level in 2015.

Southern coast is the earliest open area in China, and has frequent domestic and overseas economic exchanges, with a large need for international passenger transport. The airports in Southern coast have a high level of modernization, featuring high efficiency in air cargo transport.

In terms of provincial differences, the ATE of Fujian and Guangdong was higher than the national average level, and the change trends of ATE in these two provinces were roughly consistent with the region. The ATE of Hainan was low. Although Hainan has a high level of air passenger throughput and cargo throughput, the number of aviation employees and capital stock are too high. During the research period, its capital stock ranked fourth in China, while its corresponding air passenger throughput and cargo throughput were ranked below the top 10 in China, which inhibited its ATE (Figs. 7.8 and 7.9).

7.2.2.4 Northeast

During the study period, the average value of ATE in Northeast China was 0.778, which was higher than the national average level. The regional ATE fluctuated between 0.652 and 0.857. Due to the geographical location, air transport has become an important transport mode for trans-regional transport travel in the three northeast

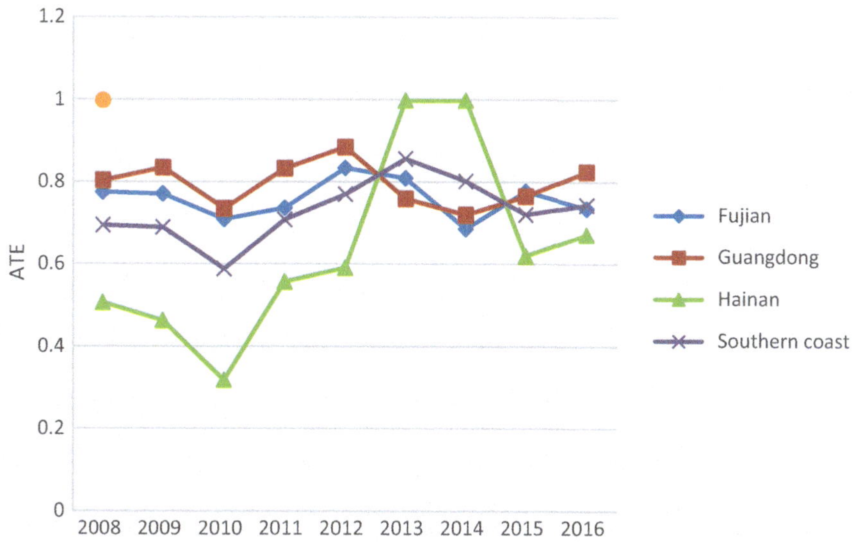

Fig. 7.8 The trend of ATE in Southern coast from 2008 to 2016. *Data source* Edited by the authors

Fig. 7.9 The mean value of ATE in Southern coast from 2008 to 2016. *Data source* Edited by the authors

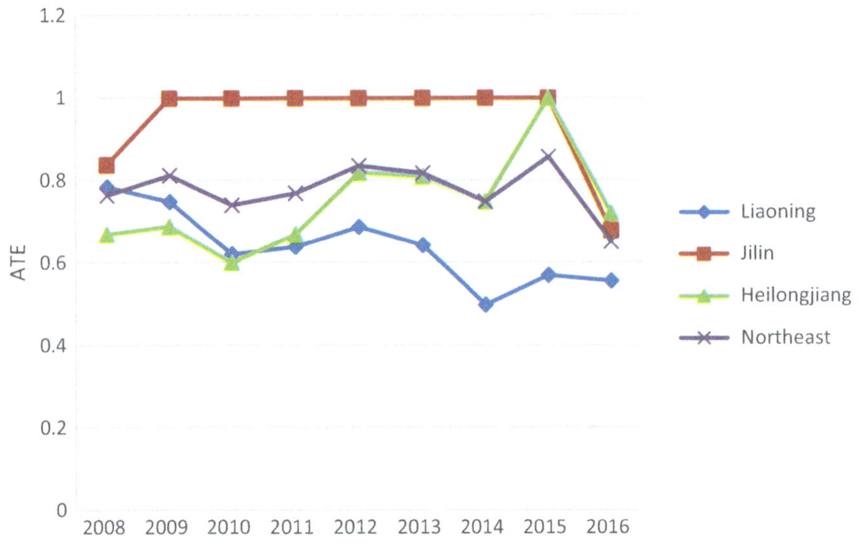

Fig. 7.10 The trend of ATE in Northeast from 2008 to 2016. *Data source* Edited by the authors

provinces, resulting in a large amount of air transport. Jilin had the highest level of ATE in the region; from 2009 to 2015, the ATE in Jilin was at the production frontier surface; but it showed a downward trend from 2015 to 2016. The ATE in Liaoning showed a downward trend during the study period. The ATE in Heilongjiang showed an upward trend from 2008 to 2015 and a downward trend from 2015 to 2016. The three provinces all showed a downward trend from 2015 to 2016, which indicates that high-speed rail have a certain degree of influence to the regional ATE (Figs. 7.10 and 7.11).

7.2.2.5 Middle Reaches of the Yellow River

During the study period, the average annual value of ATE in Middle reaches of the Yellow River was 0.6, which was lower than the national average level. The regional ATE fluctuated between 0.597 and 0.790, with the lowest and highest levels in 2011 and 2013, respectively.

In terms of provincial differences, the ATE of Shaanxi was relatively high, which remained at the production frontier surface during the study period except for in 2008; Shaanxi has been at a higher stage of air transport development, which has numerous airports, and the air passenger and cargo volume of Shannxi is at the forefront of the country. The ATE of Shanxi and Inner Mongolia was low, and the average annual value of ATE in these two provinces was both less than 0.5; Shanxi has the competitive challenges in the air transport industry, and Taiyuan Wusu International Airport is the largest airport in Shanxi, which is competitive with the surrounding airports, such as

Fig. 7.11 The mean value of ATE in Northeast from 2008 to 2016. *Data source* Edited by the authors

the airports of Zhengzhou, Shijiazhuang and Xi'an. The land area of Inner Mongolia is the third largest in China, but its population is only more than 26 million in 2016. Therefore, there are many airports with insufficient passengers, which inhibits the ATE of Inner Mongolia (Figs. 7.12 and 7.13).

7.2.2.6 Middle Reaches of the Yangtze River

The average value of ATE in Middle reaches of the Yangtze River was 0.577, which was the lowest among the eight regions. The regional ATE fluctuated during the study period.

Middle reaches of the Yangtze River are located at the intersection of rail corridors running north–south and those running east–west in China. Compared with air transport, railway transport and highway transport have more advantages, which inhibits ATE.

During the 13th Five-Year Plan period, with the gradual improvement of the high-speed rail network in the central region, the accessibility and scale effect of high-speed rail in the region have further improved, which has attracted many passenger flows from air transport, especially short-distance flights. Therefore, the civil aviation sector should gradually withdraw from the short-distance passenger market and strengthen the development of the long-distance passenger market.

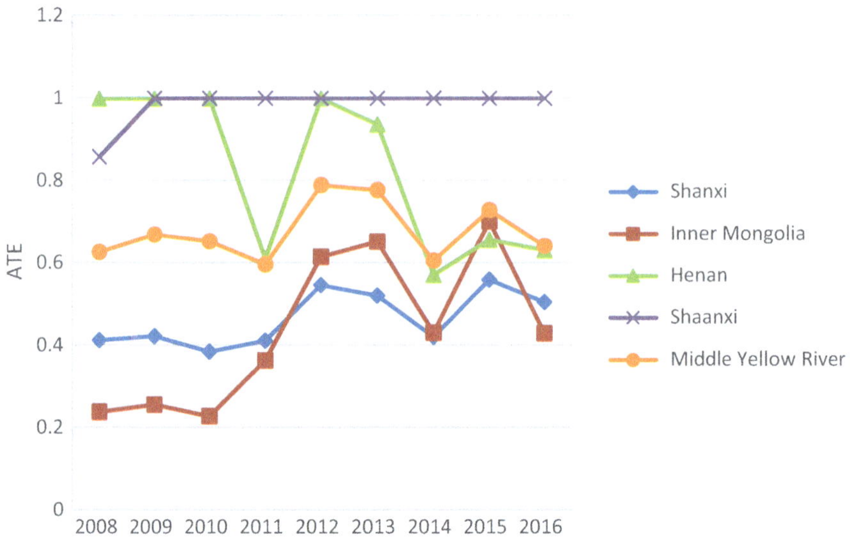

Fig. 7.12 The trend of ATE in Middle reaches of the Yellow River from 2008 to 2016. *Data source* Edited by the authors

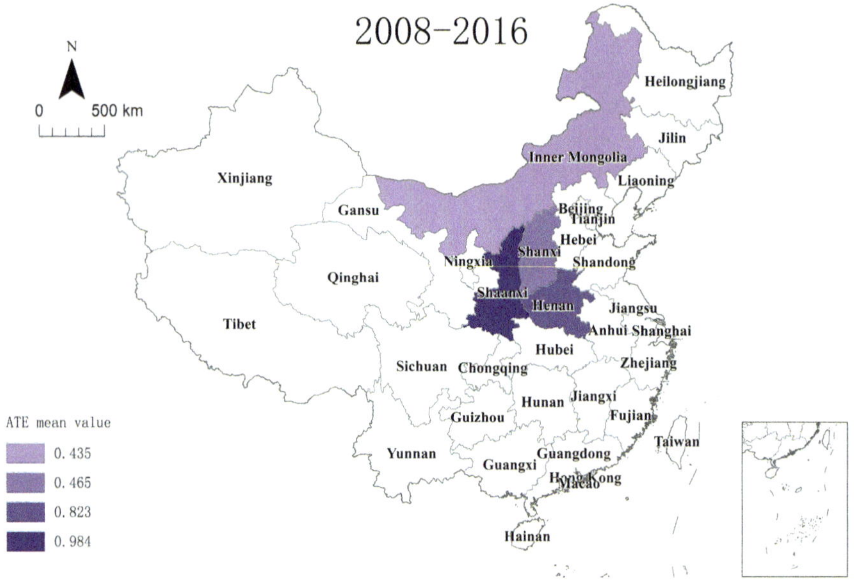

Fig. 7.13 The mean value of ATE in Middle reaches of the Yellow River from 2008 to 2016. *Data source* edited by the authors

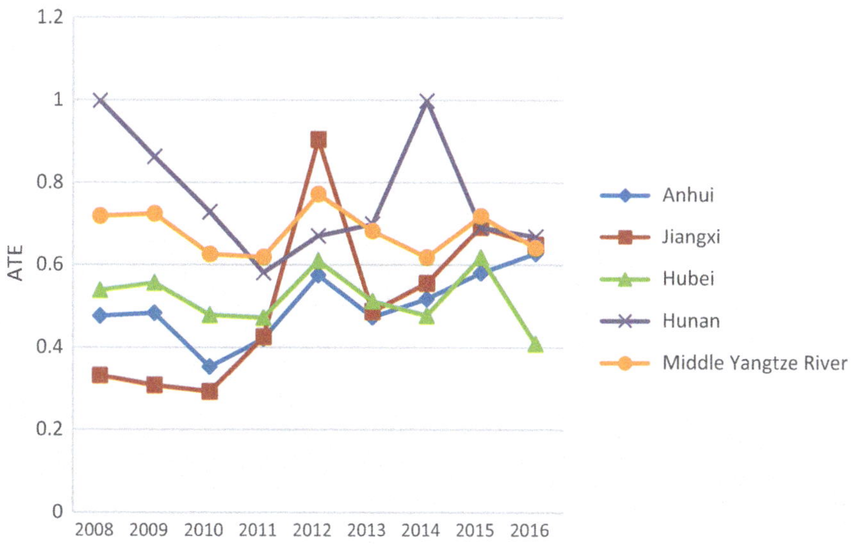

Fig. 7.14 The trend of ATE in Middle reaches of the Yangtze River from 2008 to 2016. *Data source* Edited by the authors

In terms of provincial differences, the ATE of Hunan was the highest, with an average annual value above 0.7, higher than the national average level. The ATE of the other three provinces are close to each other, and the average annual value of ATE in the other three provinces was slightly higher than 0.5 (Figs. 7.14 and 7.15).

7.2.2.7 Southwest

As for Southwest China, the average value of ATE was 0.6, slightly lower than the national average value. The change trend of ATE in Southwest China was similar to that in Middle reaches of the Yangtze River: a fluctuating trend was observed.

In terms of provincial differences, the ATEs of Chongqing and Yunnan were higher than the national average level. The average annual value of ATE in Guangxi was the lowest in the region, which was lower than 0.5. Chongqing and Yunnan are leading tourism provinces, which abound with tourism resources. Yunnan is particularly known for its tourism resources, which attract many tourists at home and abroad and stimulate the improvement of ATE in Yunnan (Figs. 7.16 and 7.17).

7.2.2.8 Northwest

During the study period, the average value of ATE in Northwest China was 0.640, which was one of the lowest in the whole country. The change trend of regional

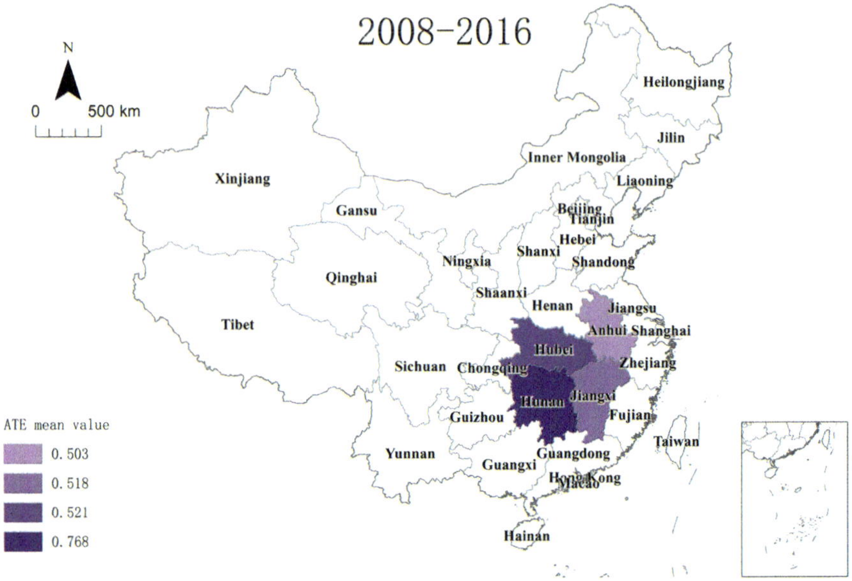

Fig. 7.15 The mean value of ATE in Middle reaches of the Yangtze River from 2008 to 2016. *Data source* Edited by the authors

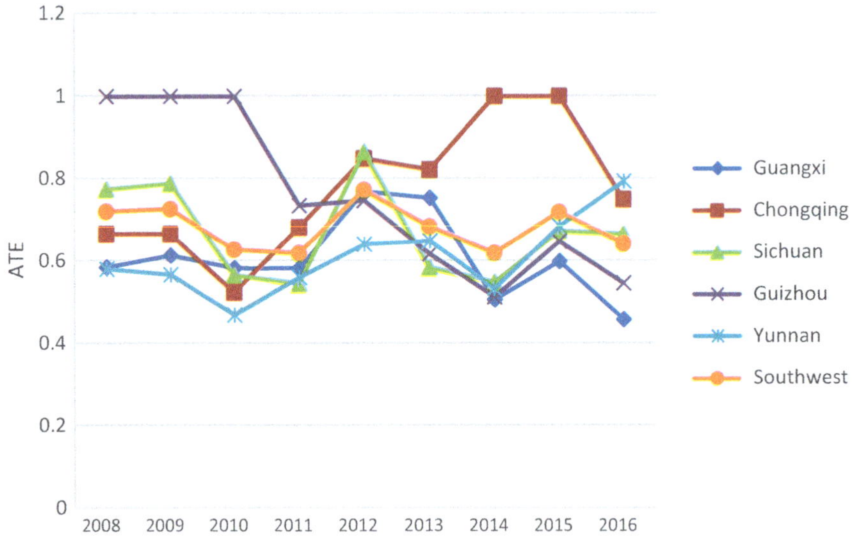

Fig. 7.16 The trend of ATE in Southwest from 2008 to 2016. *Data source* Edited by the authors

Fig. 7.17 The mean value of ATE in Southwest from 2008 to 2016. *Data source* Edited by the authors

ATE was roughly consistent with that of Middle reaches of the Yangtze River and Southwest China.

The implementation of the Western Development Program sped up the development of air transport in Northwest China. However, on the whole, the economic development level of the region is relatively backward, the consumption level of residents is low, and residents lack the ability to afford the air fare of air transport at the present stage, so the development and cultivation of the air transport market still need more time.

In terms of provincial differences, the ATE of Gansu and Qinghai was relatively high comparing with other provinces in Northwest, and the average annual value of ATE in these two provinces was higher than the national average. The ATE of Ningxia was relatively low, and its average annual value was lower than 0.5. However, different from the eastern provinces with higher ATE, the reason for the higher level of ATE in Gansu and Qinghai is not a higher air passenger and cargo volume in these two provinces; although the air passenger and cargo volume in these two provinces is at a low level all the year round, their air transport sector has lower consumption of production factors, resulting in a high level of ATE. However, the air transport volume of Ningxia is relatively low, and Ningxia has a significant number of workforces and flights taking off and landing, having bad effects on ATE (Figs. 7.18 and 7.19).

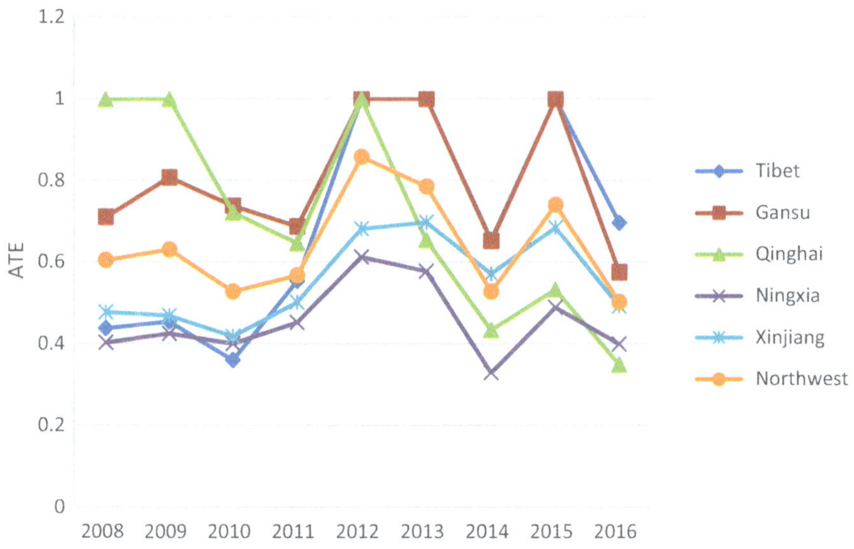

Fig. 7.18 The trend of ATE in Northwest from 2008 to 2016. *Data source* edited by the authors

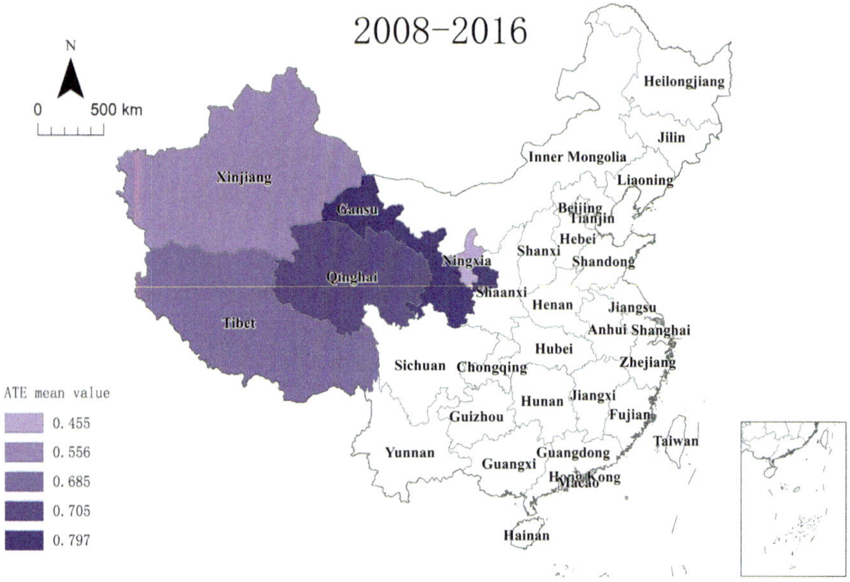

Fig. 7.19 The mean value of ATE in Northwest from 2008 to 2016. *Data source* edited by the authors

7.3 Influencing Factors of ATE

7.3.1 *Selection of Variables*

7.3.1.1 Economic Development Level

Air transport is a bellwether of the economic situation, and the demand for air transport is often positively correlated with the macro economy. When the economic situation is good, the demand for air passenger and cargo transport is larger (Wu & Man, 2018), and the economic crisis often leads to a significant decline in the demand for air transport. Various studies have shown that the correlation between air transport and national economic development level is very high, and that only the increase of resident income can bring about the improvement of air passenger demand. This study selects the economic development level as an important variable to analyze the driving factors of ATE.

7.3.1.2 Population Density

According to the U.S. Census Bureau, the population of the United States increased by 50% to 308 million from 1970 to 2010. In the past 40 years, the population of the United States also has a huge migration change, which leads to the continuous change of aviation demand. Some communities have developed into major business centers and need more large-scale aviation facilities. In other areas, the population is shrinking and the demand for aviation is decreasing. Therefore, the population density can affect the air demand to a certain degree, and then affects ATE. Referring to the research results of Wu and Man (2018), this study selects population density as one of the independent variables.

7.3.1.3 Railway Network Density

Against the backdrop of the China Railway Speed Up Campaign, railway transport has advantages over air transport in price and time when it comes to short- and medium-distance trips. Many passengers choose railway transport instead of air transport, which has a negative impact on airlines. The influence of railways on civil aviation can be called substitution effect. The substitution effect has become more obvious since high-speed rail was put into operation. Based on the research conclusions of Zhang and Zhang (2016), Chen and Jiang (2020), and Liu et al. (2021), this study selects railway network density as an important regression variable.

Table 7.4 Influencing factors of ATE

Explanatory variables	Definitions of variables
Economic development level (EDL)	GDP per capita (10^4 RMB)
Railway network density (RND)	Proportion of the length of railways in operation to the regional area (km/km^2)
Population density (PD)	Ratio of regional permanent resident population to the regional area ($person/km^2$)
Industrial structure (IS)	Proportion of the tertiary industry in GDP (%)

Data source The authors, edited from China Statistical Yearbook (2017)

7.3.1.4 Industrial Structure

Generally speaking, the tertiary industry can bring more aviation demand to the city than the primary and secondary industries, because tourists tend to visit cities with a developed tourism service industry; in other words, cities with a developed tourism service industry will have higher aviation demand. This study selects the industrial structure as an important regression variable, which is equal to the ratio of the tertiary industry to GDP (Table 7.4).

7.3.2 Regression Analysis of ATE

7.3.2.1 Construction of the Regression Model

The ATE value calculated using the EBM model falls between the interval 0 and 1, which is a limited dependent variable. If the Ordinary Least Squares (OLS) model is used to calculate the parameter, the estimated results will be biased but consistent. We utilize the Tobit regression model to analyze the influencing factors of ATE, which can estimate the parameters using maximum likelihood estimation (Table 7.5).

$$\begin{aligned} WTE_{it} &= \beta_0 + \beta_1 EDL_{it} + \beta_2 IL_{it} + \beta_3 PD_{it} \\ &\quad + \beta_4 IS_{it} + \beta_5 UL_{it} + \beta_6 FT_{it} + \beta_7 FI_{it} + u_{it} \end{aligned} \tag{7.1}$$

7.3.2.2 The Regression Results

From the Tobit regression model, we can see that economic development level, population density and industrial structure have a significant positive impact on ATE, while railway network density has a significant negative impact on ATE (Table 7.5).

Table 7.5 Tobit regression results

Variable	Coefficient	Std. Err	Z-statistic	P > |z|
EDL	0.0172605**	0.0079339	1.75	0.040
PD	0.000099***	0.0000237	2.35	0.000
RND	− 0.0002784**	0.0833152	− 2.22	0.014
IS	0.5218317***	0.1699665	2.11	0.002

Note: *** and ** indicate significance at the 1% and 5% levels, respectively
Data source Edited by the authors

Economic Development Level

There is a significant positive correlation between economic development level and ATE, as expected. As China's economy grows, its air transport sector has rapidly developed. Air passenger and cargo throughput has been on the rise overall, which is closely linked to regional socio-economic development (Wu & Man, 2018). At present, the air routes in China are relatively scattered, and local airports tend to be directly connected with top airports of other regions, which is not conducive to the formation of regional hub airports, and harm the maximization of economic benefits for both airlines and passengers, and hinder the formation of a hub-and-spoke network. Under the condition of complete marketization, the aviation network is a self-organizing system; that is to say, the development of the air transport network to a higher level is driven by internal forces. Therefore, with the deregulation of the air transport sector and the continuous growth of air transport demand, the evolution of the aviation network structure is becoming increasingly complex. In the future, the optimization of the air route network will be one of the key issues.

Population Density

The regression coefficient between population density and ATE is significantly positive, which indicates that the higher the population density is, the higher the regional ATE will be. With the rapid development of the economy in China, civil aviation demand is gradually increasing. Due to its large population base and vast land area, China has become one of the fastest growing and largest aviation markets in the world. The spatial distribution of population density in China is quite different, so it is necessary to allocate the aviation network infrastructure reasonably. On a local scale, it is vital to strengthen the complementary functions of airports among the Pearl River Delta, Yangtze River Delta, Beijing-Tianjin-Hebei and other densely populated urban areas by integrating the airport resources of urban agglomerations. On a national scale, efforts are necessary to increase the construction of regional airports in the central and western regions with insufficient airports and high population density, or the regions with low population density but the geographical position leaning. This is very necessary to increase the number of flight destinations, and properly adjust the connecting air routes in these regions, which should be both moderately advanced and within the reach of the region. However, it should be noted that the regression coefficient value of population density is small. The local civil aviation demand depends not only on the local population, but also on the per capita income and economic activities, which are fundamental driving factors.

Railway Network Density

The railway network density has a significant negative impact on ATE at the level of 5%, which shows that the railway factor, mainly the high-speed railway factor, has a restraining effect on ATE. Qian et al. (2016) believed that flights of over 1000 km are less impacted by high-speed rail. It is necessary to analyze the specific impact of high-speed rail on flights in detail. Specifically, some flights with fewer advantages than high-speed rail should be reduced or canceled. Of course, measures can be taken to attract passengers who sway between civil aviation and high-speed rail. In this case, it is not the competition between high-speed rail and civil aviation, but the competition between airlines. If airlines want to attract the passengers, they need to improve their service and reduce travel costs.

Industrial Structure

The regression coefficient of industrial structure is significantly positive, as expected. It is worth noting that the regression coefficient of industrial structure is larger than that of other independent variables, which indicates that the regional industrial structure has a more significant impact on ATE. It is necessary to change the development mode of the air transport sector, coordinate the development of air, railway, and highway transport, and comprehensively enhance the competitiveness of civil aviation. It is better for civil aviation to cooperate with high-speed rail than compete with high-speed rail (Zhang et al., 2019). More specifically, it is advisable to open bus or light rail routes to connect high-speed rail stations with airports, and enable passengers to buy high-speed rail tickets from airlines; and the railway corporation should temporarily increase the high-speed train service frequency when flights are canceled or delayed.

7.4 Conclusions

7.4.1 The National ATE Was High

The mean value of ATE in China's 31 provinces was 0.704 from 2008 to 2016, which means that the whole level of ATE needs to increase by 29.6% to reach the production frontier surface. The ATE in 16 provinces was higher than the national level. The overall level of ATE in China was high.

7.4.2 The National ATE Showed a Fluctuating Trend

The national ATE first declined and then rose from 2008 to 2012, with the lowest point in 2010. After 2012, the national ATE showed a significant downward trend. The decline of ATE in 2008, perhaps induced by the financial crisis, and the slow growth of passenger and cargo demand in the air transport sector inhibited ATE. After 2010, China was set for economic recovery, the air passenger demand significantly increased, and then ATE improved significantly. After 2012, the number of employees, fixed capital investment, number of flights, and passenger volume of the air transport sector all increased rapidly, but the air freight volume increased slowly, which inhibited the overall level of ATE.

7.4.3 The ATE in Coastal Areas and Northeast China Was High, but It Was Low in the Middle Reaches and Western Areas

Eastern coast had the highest level of ATE. The ATE of Southern coast, Northeast and Middle reaches of the Yangtze River was also high, and the annual average values of ATE in these four regions were all above 0.7. The ATE in Middle reaches of the Yellow River was the lowest. The air transport sector is a capital and technology intensive industry. Due to the high level of both economic development and technology accumulation in coastal areas, the air transport sector is relatively developed. Meanwhile, coastal areas are densely populated and have high income levels. Therefore, these areas have high demand for air passenger and cargo transport, which has a positive effect on ATE.

7.4.4 The Impact Mechanism of ATE

We apply the Tobit method to analyze the influencing factors of ATE. We found that economic development level, population density and industrial structure have a significant positive impact on ATE, while railway network density has a significant negative impact on ATE.

References

China Statistical Yearbooks (2017). China Statistical Publishing House. https://data.cnki.net/yearbook/Single/N2020100004

China Statistical Yearbook (2009–2017). China Statistical Publishing House. https://data.cnki.net/yearbook/Single/N2020100004

China Transport Statistical Yearbook (2009–2017). China Statistical Publishing House. https://data.cnki.net/yearBook/single?id=N2022010162. Accessed April 24, 2021.

Chen, Z., & Jiang, H. (2020). Impacts of high-speed rail on domestic air cargo traffic in China. *Transportation Research Part A, 142*, 1–13. https://doi.org/10.1016/j.tra.2020.10.002

Li, J. W., & Zhang, G. Q. (2016). Estimation of capital stock and capital return rate of China's transportation infrastructure. *Contemporary Finance & Economics, 6*, 3–14. (In Chinese). https://doi.org/10.13676/j.cnki.cn36-1030/f.2016.06.001

Liu, S., Wan, Y., & Zhang, A. (2021). Does China's high-speed rail development lead to regional disparities? A network perspective. *Transportation Research Part A: Policy and Practice, 138*, 299–321. https://doi.org/10.1016/j.tranpol.2021.05.026

Lu, W., Park, S. H., Huang, T., & Yeo, G. T. (2019). An analysis for Chinese airport efficiency using weighted variables and adopting CFPR. *The Asian Journal of Shipping and Logistics, 35*, 230–242. https://doi.org/10.1016/j.ajsl.2019.12.010

National Bureau of Statistics of China (NBSC). (2021). http://data.stats.gov.cn/easyquery.htm?cn=E0103. Accessed April 29, 2021.

Qian, Z., Chen, L., Hao, Q J. (2016). High-speed- railway Reshape the Regional Development of Our Country Territory. *Reform of Economic System.198*(03):56-62. https://d.wanfangdata.com.cn/periodical/jjtzgg201603009

Statistical Yearbook of the Chinese Investment in Fixed Assets (2009–2013, 2015–2017). China Statistical Publishing House. https://data.cnki.net/yearBook/single?id=N2019030174

Statistics Bulletin of Civil Airports (2008–2016). Civil Aviation Administration of China. http://www.caac.gov.cn/XXGK/XXGK/index_172.html?fl=11

Zhang, A., Wan, Y., & Yang, H. (2019). Impacts of high-speed rail on airlines, airports and regional economies: A survey of recent research. *Transport Policy, 81*, A1–A19. https://doi.org/10.1016/j.tranpol.2019.06.010

Zhang, J., Wu, G. Y., & Zhang, J. P. (2004). The estimation of China's provincial capital stock: 1952–2000. *Economic Research Journal, 10*, 35–44. http://en.cnki.com.cn/Article_en/CJFDTOTAL-JJYJ200410004.htm

Zhang, Y. F., & Ni, P.F. (2016). Economic growth spillover and spatial optimization of high-speed railway. *China Industrial Economics, 2*, 21–36. http://kns.ccpd.cnki.net/kcms/detail/detail.aspx?filename=GGYY201602003&dbcode=CCPD

Zhang, Y., & Zhang, A. (2016). Determinants of air passenger flows in China and gravity model: Deregulation, LCCs, and high-speed rail. *Journal of Transport Economics and Policy, 50*(3), 287–303. https://doi.org/10.2139/ssrn.2775501

Wu, X., & Man, S. (2018). Air transportation in China: Temporal and spatial evolution and development forecasts. *Journal of Geographical Sciences, 28*, 1485–1499. https://doi.org/10.1007/s11442-018-1557-y

Chapter 8
Water Transport Efficiency

8.1 Background and Methods

8.1.1 Background

Water transport is a relatively low-cost transport mode based on the use of geographic conditions. It has already become an important component of the comprehensive transport system (He et al., 2017). Water transport is an important way of transport for goods, such as coal and coal products, oil, natural gas and natural gas products, metal ore, steel, mineral construction materials, grain, wood, and so on. The freight turnover of water transport in China increased from 502.2 billion tons in 2008 to 8,170.7 billion tons in 2016, accounting for 47% of the total turnover—almost half of the total (Fig. 8.1). It can be seen that water transport has advantages in cargo transport. It is worth studying how to improve the water transport efficiency (WTE) in China.

8.1.2 Methods

In this study, the EBM model is also used for measuring waterway transport efficiency (WTE). Because of the lack of data in some provinces, the research selects 22 provinces in the Chinese mainland as the research area, with a study period from 2008 to 2016. Based on the geographical distribution, 22 provinces are classified into North China waterway transport zone, Central China waterway transport zone, South China waterway transport zone, and Northeast China waterway transport zone, as shown in Table 8.1 and Fig. 8.2.

This study selects the capital stock of the waterway transport sector, the length of waterways, the sum of motor boats and barges, and the number of employees as input indexes, and passenger turnover and cargo turnover as output indexes. The analysis of each index variable is as follows (Table 8.2).

P. Zhao and L. Zeng, *Transport Efficiency and Safety in China*, Population, Regional Development and Transport, https://doi.org/10.1007/978-981-99-1055-7_8

Table 8.1 Four waterway transport zones in China

Regions	Constitution
North China waterway transport zone	Tianjin, Shandong, Henan, Shaanxi
Central China waterway transport zone	Shanghai, Jiangsu, Zhejiang, Anhui, Jiangxi, Hubei, Hunan, Sichuan, Chongqing, Guizhou
South China waterway transport zone	Fujian, Guangdong, Guangxi, Hainan, Yunnan
Northeast China waterway transport zone	Liaoning, Jilin, Heilongjiang

Data source Edited by the authors

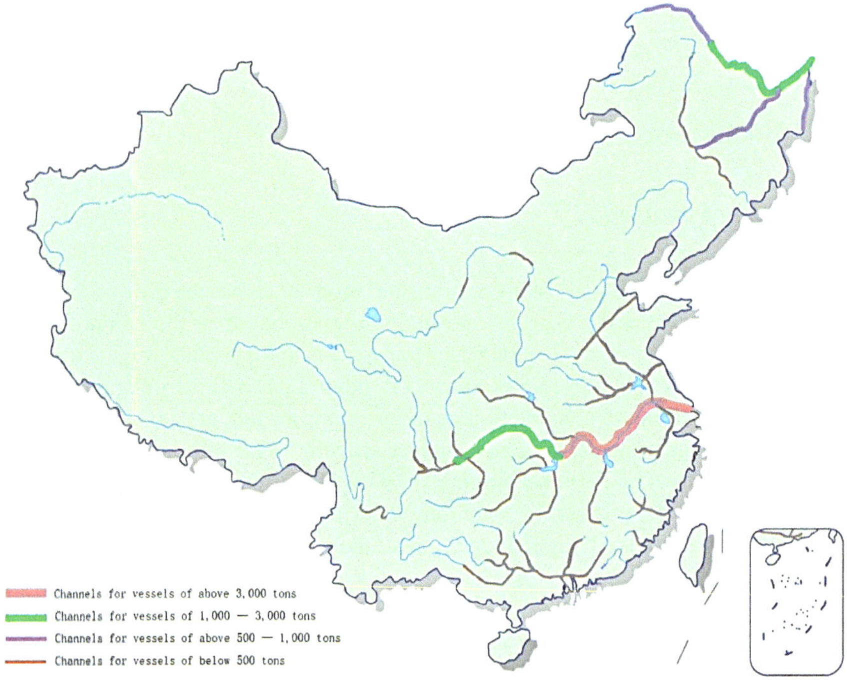

Fig. 8.1 Chart of inland waterways in 2017. *Data source* https://www.51wendang.com/doc/4b0a851fc66beae43d1e2645

This study estimates the capital stock of the water transport sector based on the perpetual inventory method proposed. The basic formula of the perpetual inventory method is: $K_{i,t} = I_{i,t} + (1 - \delta_{i,t})K_{i,t-1}$. $K_{i,\,t}$ represents the capital stock of the waterway transport sector in t year of i province, $I_{i,\,t}$ represents the total fixed capital investment in the waterway transport sector in t year of i province, and δ represents the capital depreciation rate. Based on the research result of Zhang et al. (2004), the capital stock in 2008 is estimated by dividing the total fixed capital investment in 2008 by 10%. The annual depreciation rate δ is 8.76% based on the research results of Li and Zhang (2016). The data are from China Statistical Yearbook (2009–2017),

Fig. 8.2 Four waterway transport zones. *Data source* Edited by the authors

China Fixed Assets Statistical Yearbook (2009–2013, 2014–2017), China Transport Statistical Yearbook (2009–2017) and statistical yearbooks of provinces over the years.

The length of waterways is the material basis for the development of water transport. The relevant data are from China Statistical Yearbook (2009–2017).

The number of commercial civil ships is the sum of both motor boats and barges. The data are from China Statistical Yearbook (2009–2017).

The number of water transport practitioners, passenger turnover, and freight turnover are from China Statistical Yearbook (2009–2017) (Table 8.2).

Table 8.2 The measurement index system of water transport efficiency

Primary indexes	Secondary indexes
Inputs	The capital stock of the waterway transport sector (unit: million RMB)
	Total number of employees in the waterway transport sector (unit: person)
	The length of waterways (unit: kilometer)
	The number of commercial civil ships (unit: ships)
Outputs	Water passenger turnover
	Water freight turnover

Data source China Statistical Yearbook (2009–2017), China Fixed Assets Statistical Yearbook (2009–2013, 2014–2017), China Transport Statistical Yearbook (2009–2017), and statistical yearbooks of provinces over the years

8.2 Measurement Results

According to the EBM model and the variable selection, we calculate the provincial WTEs in China using Maxdea 7.9 ultra software. The results are listed in Tables 8.3 and Table 8.4. Based on the mean values of the provincial WTEs in China from 2008 to 2016. Figure 8.3 was made by ArcGIS 10.0 software. Figure 8.4 shows the trend of WTE nationwide and in four waterway transport zones from 2008 to 2016

Table 8.3 Water transport efficiency of 30 provinces in China from 2008 to 2016

Provinces	2008	2009	2010	2011	2012	2013	2014	2015	2016	Mean
Tianjin	1.000	1.000	1.000	1.000	1.000	1.000	1.000	1.000	0.326	0.925
Liaoning	1.000	1.000	1.000	1.000	1.000	1.000	1.000	1.000	1.000	1.000
Jilin	0.182	0.136	0.110	0.080	0.059	0.067	0.138	0.050	0.046	0.096
Heilongjiang	0.043	0.034	0.035	0.036	0.039	0.056	0.049	0.038	0.033	0.040
Shanghai	1.000	1.000	1.000	1.000	1.000	1.000	1.000	1.000	1.000	1.000
Jiangsu	0.074	0.115	0.111	0.112	0.103	0.274	0.196	0.147	0.132	0.140
Zhejiang	0.530	0.586	0.573	0.531	0.538	0.870	0.549	0.585	0.647	0.601
Anhui	0.157	0.111	0.113	0.105	0.099	1.000	1.000	1.000	1.000	0.510
Fujian	0.445	0.457	0.462	0.483	0.471	0.707	0.570	0.680	0.741	0.557
Jiangxi	0.168	0.218	0.170	0.155	0.147	0.194	0.164	0.135	0.121	0.164
Shandong	0.499	0.664	0.754	0.693	0.544	0.399	0.369	0.376	0.430	0.525
Henan	0.336	0.579	0.583	0.400	0.504	0.474	0.392	0.386	0.432	0.454
Hubei	0.260	0.250	0.241	0.264	0.381	0.417	0.427	0.419	0.421	0.342
Hunan	0.366	0.402	0.456	0.523	0.527	0.656	0.529	0.563	0.570	0.510
Guangdong	0.350	0.310	0.335	0.361	0.421	0.504	0.603	0.624	1.000	0.501
Guangxi	0.170	0.181	0.175	0.185	0.254	0.372	0.317	0.426	0.433	0.279
Hainan	1.000	1.000	1.000	1.000	1.000	1.000	1.000	1.000	1.000	1.000
Chongqing	1.000	1.000	1.000	0.819	0.828	0.863	0.668	0.592	0.587	0.817
Sichuan	0.284	0.218	0.195	0.172	0.137	0.214	0.171	0.140	0.125	0.184
Guizhou	1.000	1.000	1.000	1.000	1.000	1.000	1.000	1.000	1.000	1.000
Yunnan	1.000	0.572	0.728	1.000	0.460	1.000	0.463	1.000	1.000	0.802
Shaanxi	1.000	1.000	1.000	1.000	1.000	1.000	1.000	1.000	1.000	1.000
China	0.539	0.538	0.547	0.542	0.523	0.639	0.573	0.598	0.593	0.566

Data source Edited by the authors

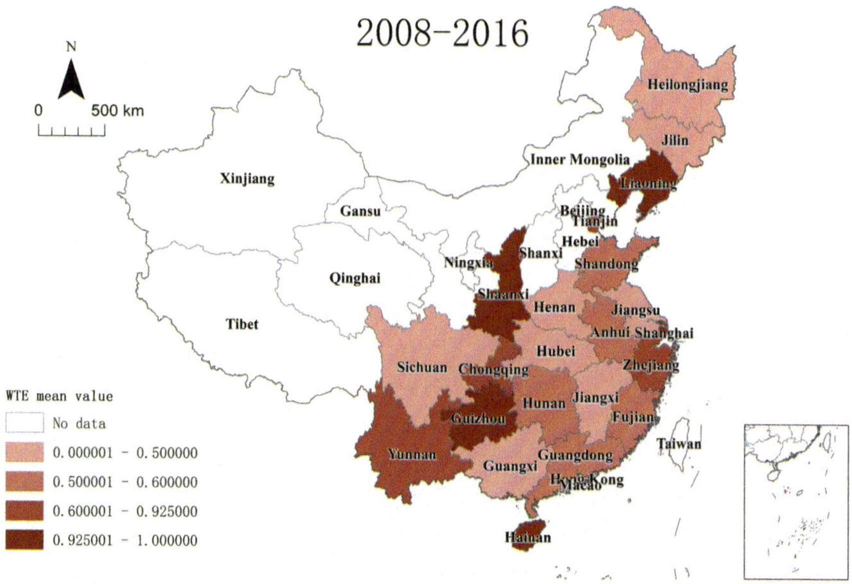

Fig. 8.3 The mean value of WTE intervals in 22 provinces of China from 2008 to 2016. *Data source* Edited by the authors

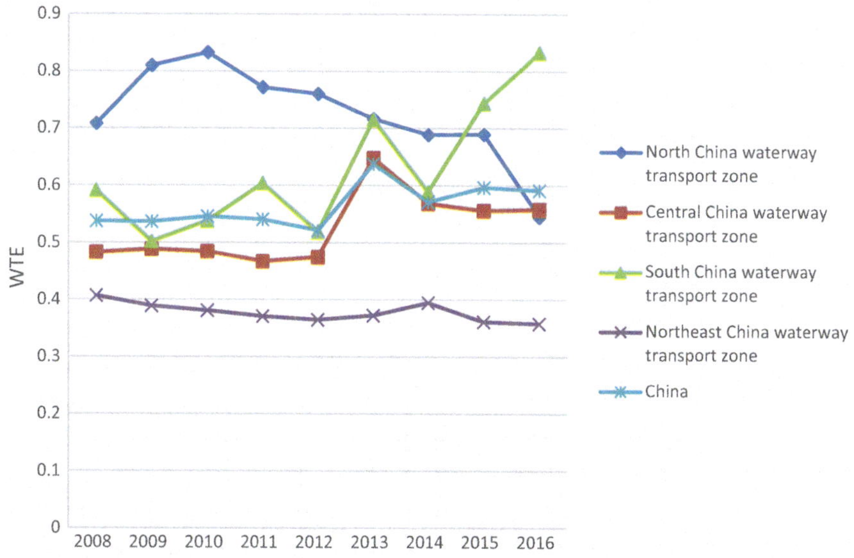

Fig. 8.4 The trend of WTE nationwide and in four waterway transport zones from 2008 to 2016. *Data source* Edited by the authors

Table 8.4 WTE of four waterway transport zones from 2008 to 2016

Waterway transport zones	2008	2009	2010	2011	2012	2013	2014	2015	2016	Mean
North China waterway transport zone	0.709	0.811	0.834	0.773	0.762	0.718	0.69	0.691	0.547	0.726
Central China waterway transport zone	0.484	0.49	0.486	0.468	0.476	0.649	0.57	0.558	0.56	0.527
South China waterway transport zone	0.593	0.504	0.54	0.606	0.521	0.717	0.591	0.746	0.835	0.628
Northeast China waterway transport zone	0.408	0.39	0.382	0.372	0.366	0.374	0.396	0.363	0.36	0.379
China	0.539	0.538	0.547	0.542	0.523	0.639	0.573	0.598	0.593	0.566

Data source Edited by the authors

8.2.1 *The Overall Characteristics*

8.2.1.1 Southeast Coastal Areas and Southwest Provinces Had Higher WTE, but for Different Reasons

The annual average value of WTE in all provinces of China was 0.566, which means that an overall improvement of 43.4% was needed to achieve high WTE simultaneously. Southeast coastal areas and southwest provinces had higher WTE, but for different reasons. The southeast coastal areas and Chongqing have a high level of waterway transport organization and management and a high output of waterway transport, which have a good impact on WTE. Although the output level of water transport is system Yunnan and Guizhou in Southwest China is low, the input level of the water transport system is low, resulting in a relatively high level of water transport efficiency.

From 2008 to 2012, the overall WTE in China was relatively stable, ranging from 0.523 to 0.547. In 2013, it showed a significant improvement, rising to 0.639, which was mainly driven by the significant improvement of WTE in some provinces, such as Zhejiang, Anhui, Fujian and Yunnan. However, in these four provinces, except Anhui, there was a significant decrease in 2014, which made the overall WTE in China drop to 0.573 in 2014. The overall WTE of China rose to 0.593 in 2015, which was rooted in the economic recovery. Specifically speaking, the significant increase in the water passenger volume promoted the improvement of WTE in 2015.

8.2.1.2 The WTE in the South China Waterway Transport Zone Had a Significant Improvement in the End Stage, While that in the North China Waterway Transport Zone Significantly Declined

The WTEs in North and South China waterway transport zones were higher than the national level, while the WTEs in Central and Northeast China waterway transport zones were lower than the national level. The average difference between the WTE of these four areas and the national mean value was 0.16, −0.039, 0.062 and −0.187, respectively. The North China waterway transport zone contains fewer provinces. Boosted by the WTE of Shaanxi and Tianjin, the overall WTE of this region was relatively high. The South China waterway transport zone has a long history of water transport, with a large water transport demand and a high waterway organization level, which are conducive to a high level of WTE. There are many provinces in the Central China waterway transport zone, and the WTE varied from province to province; although there were provinces with high WTE such as Zhejiang and Guizhou, many provinces had low WTE, which reduced the overall regional WTE. Although the WTE of Liaoning in Northeast China was high, Heilongjiang and Jilin have no harbors, restricting the development of waterway transport in the Northeast China waterway transport zone (Table 8.4 and Fig 8.4).

According to the change trend of WTE in four water transport areas, the study period can be divided into two stages: in the first stage from 2008 to 2014, the WTE values of these four regions were in the following order: North China waterway transport zone > South China waterway transport zone > Central China waterway transport zone > Northeast China waterway transport zone; in the second stage from 2015 to 2016, the WTE values of these four regions were in the following order: South China waterway transport zone > North China waterway transport zone > Central China waterway transport zone > Northeast China waterway transport zone.

8.2.2 *Spatial Variations*

8.2.2.1 North China Waterway Transport Zone

The North China waterway transport zone had the highest WTE among the four areas. The average value of WTE in the region showed an inverted U-shaped trend, with the highest value in 2010. The decline was even greater after 2012, which was related to the construction of ecological civilization after 18th CPC National Congress. In order to build ecological civilization, China transformed the economic development pattern, restructure the economy, and the water transport volume of coal and many other bulk goods in North China decreased, which inhibited WTE.

In terms of provincial differences, Shaanxi had the highest WTE not only in the region, but also across China; it maintained the leading position all the year round, followed by Tianjin, Shandong and Henan. Although the waterway passenger and freight turnover in Shaanxi was the low level in China, the input of labor and capital in the water transport industry was relatively low, resulting in the highest WTE level in China. The Port of Tianjin is the largest port in northern China, and its freight turnover is also in the forefront of the country, with higher utilization efficiency of waterway facilities; the WTE of Tianjin was at the production frontier surface from 2008 to 2015, showing a significant decline in 2016; this is because the water freight turnover of Tianjin decreased by 11.5% in 2016, which inhibited the WTE. The trend of WTE in Shandong and Henan was in line with that of the entire region, showing an inverted U-shaped trend, with the highest value in 2010 (Figs. 8.5 and 8.6).

8.2.2.2 Central China Waterway Transport Zone

During the study period, the average value of WTE in the Central China waterway transport zone was 0.5603. The regional WTE was relatively stable from 2008 to 2012, and showed a significant increase in 2013 and a notable decrease afte 2013. This may be due to the reduction of water transport volume of coal and many other bulk goods, which inhibited the WTE.

In terms of provincial differences, Shanghai and Guizhou had the highest WTE, which was in the production frontier surface during the research period. The WTEs

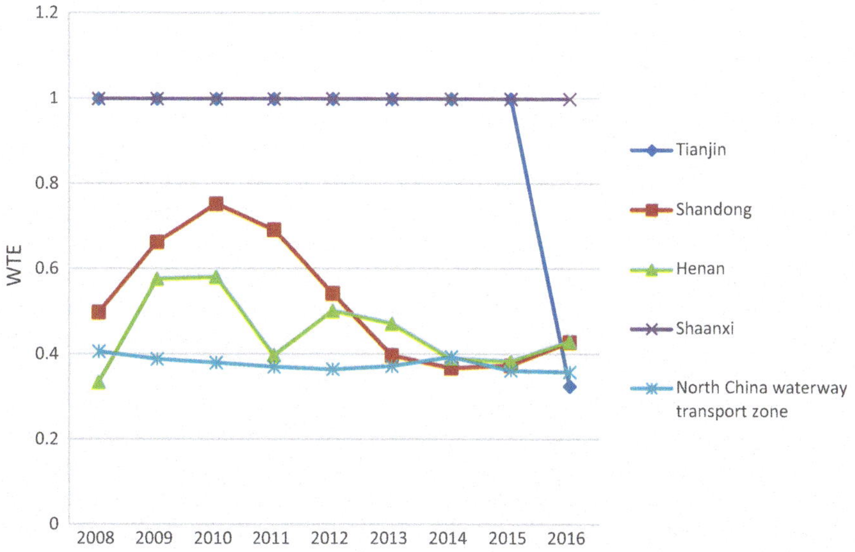

Fig. 8.5 The trend of WTE in the North China waterway transport zone from 2008 to 2016. *Data source* Edited by the authors

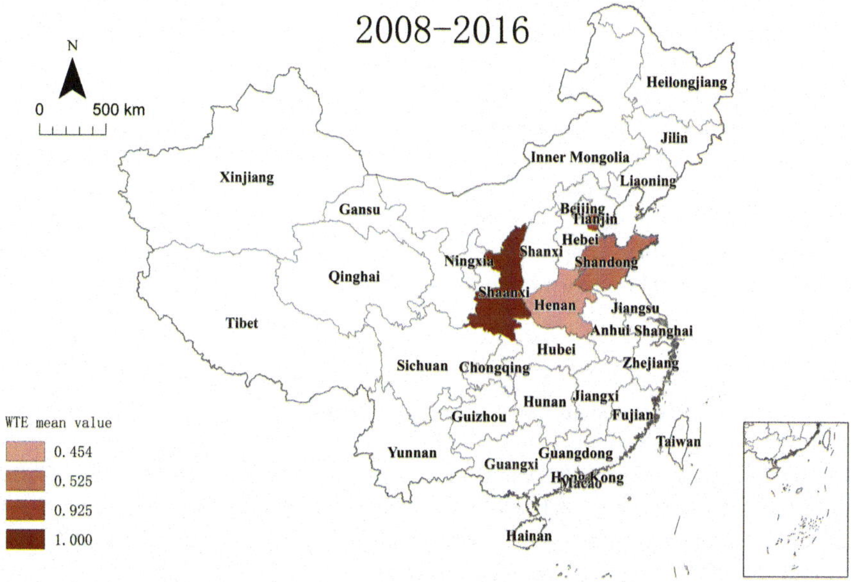

Fig. 8.6 The mean value of WTE in the North China waterway transport zone from 2008 to 2016. *Data source* Edited by the authors

of Zhejiang, Anhui, Hunan and Chongqing were at a high level, with an average annual value above 0.5, while the WTEs of Jiangsu, Jiangxi and Sichuan were at a low level, with an average annual value below 0.2.

Shanghai and Guizhou maintained their leading positions in WTE. The WTE in Guizhou showed a similar pattern to that in Shaanxi; although the waterway passenger and freight turnover was relatively low, the consumption of input factors in Guizhou was relatively low, which had a positive effect on WTE. Shanghai has the largest port in China, with a vast economic hinterland and a huge passenger and freight turnover (Fu and Chen, 2014); it has a well-developed education sector, home to many professional and technical personnel; foreign advanced technology and business philosophies are welcomed here, and it has taken the lead in implementing many electrification transformation projects, such as the logistics park, the Modern Wharf, and the comprehensive transport hub; Since 2010, Shanghai port has become the top container port in the world (Lee & Lam, 2015), and its WTE has been in the production frontier surface for a long time. It is worth noting that although Jiangsu is located in the eastern coast, with a high waterway passenger and freight turnover, its length of waterways, number of employees in the water transport sector and ship investment are too high. For example, the length of waterways and the number of employees in the water transport sector in Jiangsu were the highest in China, which inhibited WTE.

The WTE of Hubei remained stable from 2008 to 2012; after 2013, the WTE in Hubei showed an upward trend, and the significant increase of water freight turnover was an important driving factor of this improvement. The WTE of Anhui showed a slow downward trend from 2008 to 2012, but increased significantly in 2013 and reached the production frontier surface; this is due to the significant increase of waterway freight turnover in Anhui in 2013. In contrast, the WTE of Chongqing was in the production frontier surface from 2008 to 2010, and showed a downward trend after 2010. Especially after 2013, the WTE of Chongqing dropped sharply; the significant decline in the water passenger turnover was an important influencing factor. Other provinces in the region showed a trend of first rising and then declining (Figs. 8.7 and 8.8).

8.2.2.3 South China Waterway Transport Zone

During the study period, the annual average value of WTE in the South China waterway transport zone was 0.628, which was a little higher than that of the North China waterway transport zone. The regional WTE was the lowest in 2009, and showed a fluctuating upward trend after 2009.

In terms of provincial differences, the WTE of Hainan was the highest in the region, which was in the production frontier surface during the research period. The WTE of Yunnan was at a high level, with an average annual value above 0.8, while the WTE of Guangxi was at a low level, with an average annual value below 0.3.

The WTE of Guangdong was the lowest in 2009, and showed a rapid upward trend after 2009; especially in 2016, the WTE of Guangdong was greatly improved

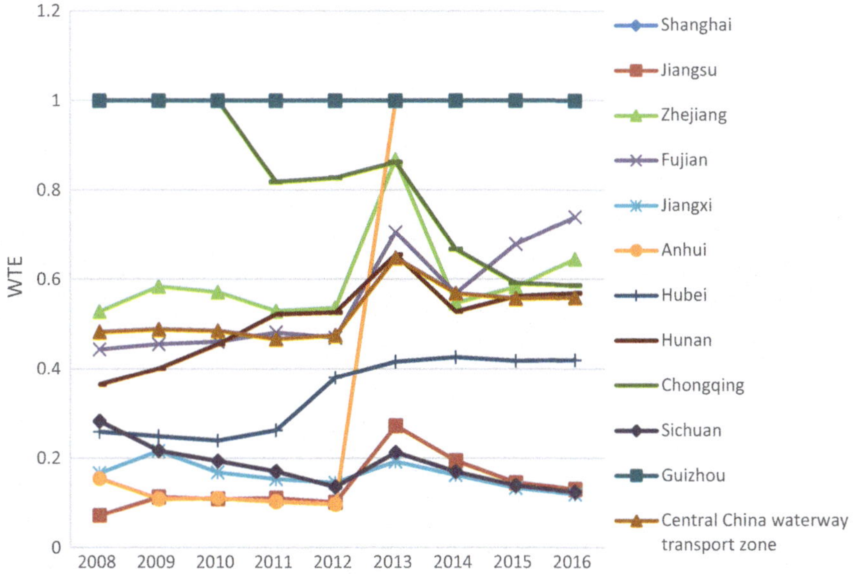

Fig. 8.7 The trend of WTE in the Central China waterway transport zone from 2008 to 2016. *Data source* Edited by the authors

Fig. 8.8 The mean value of WTE in the Central China waterway transport zone from 2008 to 2016. *Data source* Edited by the authors

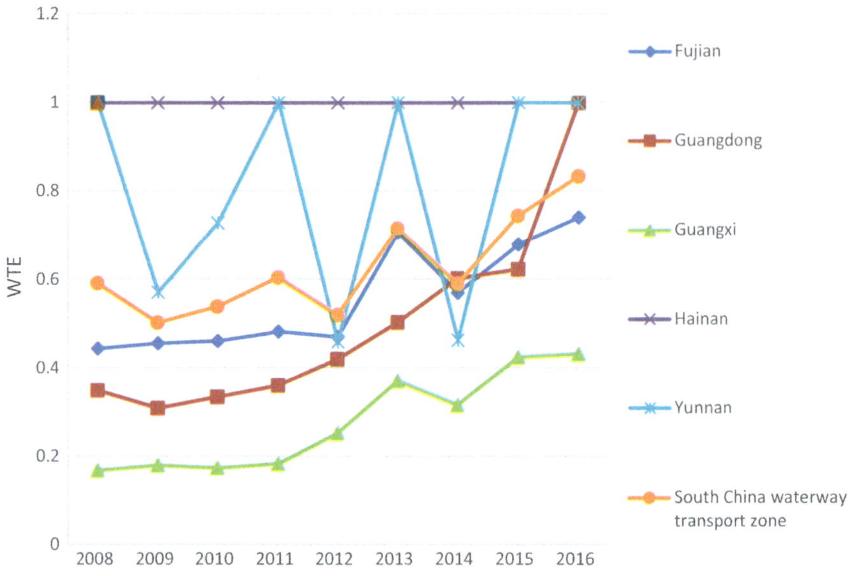

Fig. 8.9 The trend of WTE in the South China waterway transport zone from 2008 to 2016. *Data source* Edited by the authors

and reached the production frontier surface; in 2016, the water freight turnover increased by more than six times as compared to 2008, which is conducive to promoting the WTE of Guangdong. During the study period, the WTEs of Fujian and Guangxi showed a fluctuating upward trend; this is due to the increase in the water passenger and freight turnover. During the study period, the WTE of Yunnan showed a fluctuating trend (Figs. 8.9 and 8.10).

8.2.2.4 Northeast China Waterway Transport Zone

During the study period, the annual average value of WTE in the Northeast China waterway transport zone was 0.379, which was the lowest among the four regions. The regional WTE showed a downward trend from 2008 to 2012, and an inverted U-shaped trend from 2012 to 2016, with the high level in 2014 in the second stage.

During the study period, the WTE of Liaoning was in the production frontier surface, and the water passenger and freight turnover of Liaoning was at a high level in the whole country. The WTEs of Jilin and Heilongjiang were relatively low, and the annual average values of WTE were 0.091 and 0.04, respectively. The WTE in Jilin had a similar change trend to that of the region, showing a downward trend from 2008 to 2012, and inverted U-shaped trend from 2012 to 2016, with the highest value in 2014 in the second stage. The WTE of Heilongjiang fluctuated between 0.03 and 0.04 (Figs. 8.11 and 8.12).

Fig. 8.10 The mean value of WTE in the South China waterway transport zone from 2008 to 2016. *Data source* Edited by the authors

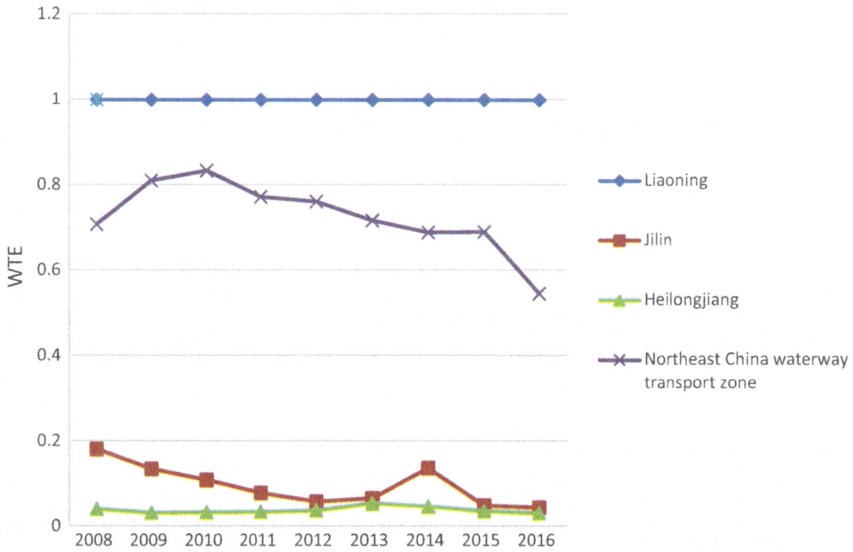

Fig. 8.11 The trend of WTE in the Northeast China waterway transport zone from 2008 to 2016. *Data source* Edited by the authors

Fig. 8.12 The mean value of WTE in the Northeast China waterway transport zone from 2008 to 2016. *Data source* Edited by the authors

8.3 Influencing Factors of WTE

8.3.1 Selection of Variables

8.3.1.1 Economic Development Level

The higher the level of regional economic development, the greater the demand for water passenger and cargo transport and the more the investment in water transport infrastructure construction by local governments. This research selects the regional per capita GDP as one of the important factors (Zhu, 2021).

8.3.1.2 Informatization Level

The development of modern information technology and science and technology has an important impact on container transport. The application of information technology in container ports, such as electronic data interchange (EDI) and Vessel Traffic Services (VTS), simplifies the procedures of water transport and greatly improves the WTE. This research selects the proportion of telecom business in GDP as a measurement index of regional informatization level (Jiang et al., 2017).

8.3.1.3 Population Density

Population density reflects the local transport demand level to a certain extent. The higher the population density is, the higher the transport demand density is. In order to ensure the accuracy of the data, this research selects the regional population density as one of the control variables (Zhu, 2021).

8.3.1.4 Industrial Structure

Regional industrial structure has an impact on water transport. For example, in a tourist city, the tourism industry is relatively developed, and the port will play a more important role as a passenger transport hub. However, in an industrial city with a developed heavy industry, which is dominated by secondary industries, the port has to serve the function of a freight terminal. Based on the studies of Jiang et al., (2017), regional industrial structure can be measured by the proportion of the secondary industry in GDP.

8.3.1.5 Urbanization Level

Focusing on the development of industry, commerce, trade and culture, the city is the engine of economic growth in various regions. The acceleration of urbanization and industrialization will bring steady growth of domestic water transport demand, especially the demand for inland water transport, which will affect WTE. This study selects the regional urbanization level as one of the control variables.

8.3.1.6 Opening-Up Level

Water transport is an important way for developing international economic ties. The development of the water transport sector needs the support from the policy of opening to the outside world. The higher the degree of regional opening-up, the more attention will be paid to the development of water transport (Zhu, 2021). The proportion of regional foreign trade in GDP and the proportion of foreign investment in GDP can be used to measure the regional opening-up level (Zong & Xiao, 2016) (Table 8.5).

8.3.2 Regression Analysis of WTE

The WTE calculated with the EBM model is between 0 and 1, which is the same as ATE. Therefore, the Tobit model is used to analyze the influencing factors of WTE.

Table 8.5 Influencing factors of WTE

Explanatory Variables	Definitions of Variables
Economic development level (EDL)	GDP per capita (10^4 RMB)
Informatization level (IL)	Proportion of telecom business in GDP (%)
Population density (PD)	Ratio of regional permanent resident population to the regional area (person/km^2)
Industrial structure (IS)	Proportion of the secondary industry in GDP (%)
Urbanization level (UL)	Proportion of urban permanent resident population in the total permanent resident population (%)
Foreign trade (FT)	Proportion of regional foreign trade in GDP (%)
Foreign investment (FI)	Proportion of foreign investment in GDP (%)

Data source The authors, edited from China Statistical Yearbook (2017) and China Trade and External Economic Statistical Yearbook (2009–2017)

In this research, the WTE is taken as the explained variable, and the economic development level, industrial structure, informatization level, population density, urbanization level, foreign trade and foreign investment are taken as the influencing factors. The Tobit model is as follows (Table 8.6):

$$
\begin{aligned}
WTE_{it} = {} & \beta_0 + \beta_1 EDL_{it} + \beta_2 IS_{it} + \beta_3 IL_{it} + \beta_4 PD_{it} \\
& + \beta_5 UL_{it} + \beta_6 FT_{it} + \beta_7 FI_{it} + u_{it}
\end{aligned}
\tag{8.1}
$$

From the Tobit regression model, we can see that industrial structure has a significant negative impact on WTE, while informatization level, population density and foreign investment have a significant positive impact on WTE. However, the economic development level, urbanization level and foreign trade have no significant impact on WTE (Table 8.6).

There is a negative correlation between economic development level and WTE, but it is not significant. At present, the development mode of many ports in China is still

Table 8.6 Tobit regression results

Variable	Coefficient	Std. Err	Z-statistic	P > |z|
EDL	−0.0090392	0.0393981	−0.23	0.819
IL	3.768729**	1.619452	2.33	0.021
PD	0.0003918***	0.0001288	3.04	0.003
IS	−1.974922***	0.6079839	−3.25	0.001
UL	−0.4237559	0.8240105	−0.51	0.608
FT	−0.2582474	0.1813147	−1.42	0.156
FI	11.48625***	2.329834	4.93	0.000

Note: ***, **, and * indicate significance at the 1%, 5%, and 10% levels, respectively
Data source Edited by the authors

too extensive. The blind pursuit of economic growth and neglect of the high-quality development of ports seriously affect the sustainable development of ports.

The industrial structure has a significant negative impact on WTE at the level of 1%, which is contrary to expectations. The secondary industry is the main factor affecting freight intensity (freight ton-kilometers per kilometer). The freight intensity of the heavy industry is higher than that of the light industry, and the freight intensity of the light industry is higher than that of the service industry. With the optimization of industrial structure and layout, the proportion of the secondary industry in GDP gradually decreases, and the proportion of water transport in the freight market increases; therefore, industrial structure and WTE show opposite trends. With the optimization and adjustment of the industrial structure, it is necessary to change the development mode of the water transport sector, actively cultivate the shipping market, expand water transport services, vigorously develop the modern shipping industry, and encourage ports and shipping enterprises to extend the industrial chain; the government should gradually raise the market access standards, optimize the market exit mechanism, supervise the market order in accordance with relevant national laws, regulations, rules and policies, and enhance the management of waterway and port facilities.

The regional informatization level has a positive impact on WTE, as expected. Accelerating the informatization of ports and ships will help improve WTE, reduce transport costs, and strengthen the competitiveness of the regional waterway transport sector as a whole. It is advisable to ride the crest of the internet wave to accelerate the informatization process of the waterway transport sector. The Chinese government should increase financial investment to encourage the application of information technology in the water transport sector; informatization cannot do without professionals, so the government needs to increase the investment in higher education and vocational education, and encourage logistics enterprises to strengthen cooperation with universities, scientific research institutions and vocational training institutions, so as to jointly formulate a talent training plan in line with the development needs of the waterway transport sector.

The regression coefficient of population density is significantly positive, which indicates that the higher the population density, the higher the regional WTE. Since the reform and opening-up, some regions with water transport mode have attracted more people to settle permanently; these regions have big demand for water transport. Therefore, it is of great necessity to vigorously promote the adjustment of shipping capacity and structure, facilitate the development of large-scale and specialized ships that match waterway technical standards, and actively develop special cargo ships and container transport.

The regression coefficient of urbanization level is negative, but not significant, which shows that urbanization level has not become a driving factor of WTE. Although urbanization can bring huge demand for water transport, the current extensive high-speed urbanization in China results in unreasonable allocation and low efficiency of production factors in some regions, which restricts the development of many industries—the water transport sector may be one of them. Therefore, it is necessary to improve the construction of various transport and public service facilities in some regions, including those facilities for water transport, and gradually improve the quality of urbanization.

The regression coefficient of foreign trade is negative, which is also not significant. At present, the development of foreign trade has greatly increased the demand for cargo distribution capacity in many ports; however, the coordination of various transport means (such as railways, highways, and waterways) in many ports in China is unreasonable. Therefore, it is necessary to ensure that the transport corridor is unimpeded, strengthen the interconnection of highway, rail and water infrastructure, develop intermodal transport with water transport as the core in the port economic zone, and build a highly efficient water transport corridor.

The regression coefficient of foreign investment is significantly positive. The Chinese government should make effort to formulate and revise the laws and regulations relating to the economic and trade cooperation with foreign capital, to improve the investment mechanism of foreign capital, to attract more foreign investment and to create a more relaxed investment environment.

8.4 Conclusions

8.4.1 Southeast Coastal Areas and Southwest Provinces Had Higher WTE

Southeast coastal areas and southwest provinces had higher WTE, but for different reasons. The southeast coastal areas and Chongqing have a high level of waterway transport organization and management and a high output of waterway transport, which have a good impact on WTE. Although the output level of water transport in Yunnan and Guizhou in Southwest China is low, the input level of the water transport system is low, resulting in a relatively high level of water transport efficiency.

8.4.2 The Overall WTE was Stable First and then Increased

From 2008 to 2012, the overall WTE in China was relatively stable, ranging from 0.523 to 0.547. In 2013, it showed a significant improvement, rising to 0.639, which was mainly driven by the significant improvement of WTE in some provinces, such as Zhejiang, Anhui, Fujian and Yunnan. However, in these four provinces, except Anhui, there was a significant decrease in 2014, which made the overall WTE in China drop to 0.573 in 2014. The overall WTE of China rose to 0.593 in 2015, which was rooted in the economic recovery. Specifically speaking, the significant increase in the water passenger volume promoted the improvement of WTE.

8.4.3 *The Impact Mechanism of WTE*

We apply the Tobit method to analyze the influencing factors of WTE. We found that industrial structure has a significant negative impact on WTE, while informatization level, population density and foreign investment have a significant positive impact on WTE. However, the economic development level, urbanization level and foreign trade have no significant impact on WTE.

References

China Statistical Yearbooks(2017). China Statistical Publishing House, Beijing. https://data.cnki.net/yearbook/Single/N2020100004

China Statistical Yearbook (2009–2017).China Statistical Publishing House, Beijing. https://data.cnki.net/yearbook/Single/N2020100004

China Transport Statistical Yearbook (2009–2017), China Statistical Publishing House, Beijing. https://data.cnki.net/yearBook/single?id=N2022010162 (accessed April 24, 2021)

He, D., Gao, P., Sun, Z., & Lau, Y.-Y. (2017). Measuring Water Transport Efficiency in the Yangtze River Economic Zone. *China. Sustainability., 9*(12), 2278. https://doi.org/10.3390/su9122278

Jiang, Z., Zhu, H., & Cao, Y. (2017). Efficiency pattern and spatial strategy of ports in Yangtze River Delta Region. *Chinese Geographical Science, 27*, 298–310. https://doi.org/10.1007/s11769-017-0864-z

Lee, P. W., Lam, J. (2015). Container Port Competition and Competitiveness Analysis: Asian Major Ports. In: Lee CY., Meng Q. (eds) Handbook of Ocean Container Transport Logistics. International Series in Operations Research & Management Science, vol 220. Springer, Cham. https://doi.org/10.1007/978-3-319-11891-8_4

Li, J.W., Zhang, G.Q. (2016). Estimation of Capital Stock and Capital Return Rate of China's Transportation Infrastructure. Contemporary Finance & Economics (06), 3–14.(In Chinese) DOI:https://doi.org/10.13676/j.cnki.cn36-1030/f.2016.06.001

National Bureau of Statistics of China(NBSC), 2021. http://data.stats.gov.cn/easyquery.htm?cn=E0103 (accessed April 29, 2021).

Statistical Yearbook of the Chinese Investment in Fixed Assets (2009–2013, 2015–2017), China Statistical Publishing House, Beijing. https://data.cnki.net/yearBook/single?id=N2019030174

Zhang, J., Wu, G. Y., Zhang, J. P. (2004).The Estimation of China's Provincial Capital Stock: 1952–2000. Economic Research Journal, 10, 35–44 http://en.cnki.com.cn/Article_en/CJFDTOTAL-JJYJ200410004.htm

Zhu, B. (2021). Analysis of Port Efficiency and Influencing Factors Based on DEA-Tobit. In: Huang, C., Chan, YW., Yen, N. (eds) International Conference on Data Processing Techniques and Applications for Cyber-Physical Systems. Advances in Intelligent Systems and Computing, vol 1379 . Springer, Singapore. https://doi.org/10.1007/978-981-16-1726-3_67

Zong, G., Xiao, X.X., (2016). Research on Interactive Development between Manufacturing and Logistics Industry in Yangtze River Delta. The Theory and Practice of Finance and Economics. 37(03):111–116. (In Chinese) DOI:https://doi.org/10.13676/j.cnki.cn36-1030/f.2016.06.001.

Chapter 9
Urban Transport Efficiency

9.1 Background and Methods

9.1.1 Background

The urban transport sector is an important urban industrial economic sector. An unreasonable and low-efficient transport sector will affect the overall organization and operation of the city. On the contrary, a high-performing urban transport sector will optimize the overall organization and operation of the city. By critically evaluating the efficiency of the urban transport sector, we can find out which cities have redundant input of transport infrastructure resources or insufficient output, so as to improve the utilization rate of urban basic resources and avoid resource waste.

At present, there are some achievements in the research on urban transport efficiency, but most of them are aimed at urban public transport efficiency or highway transport efficiency in city clusters. Therefore, this study enriches the research content of urban transport efficiency and analyzes urban transport efficiency (UTE), so that the study of UTE is more complete and systematic.

9.1.2 Methods

In this section, the EBM model is used to measure UTE. This study selects 274 cities in the Chinese mainland as the research area, with a study period from 2008 to 2014. Labor and the length of highways are selected as input indicators, and the passenger volume and freight volume are selected as output indicators. (Table 9.1).

(1) Labor: the labor force engaged in the urban transport sector (unit: 1000 persons).
(2) Length of highways: it can only be calculated with the actual length of highways having been completed, checked and accepted or put into operation (unit: 10,000 km).

P. Zhao and L. Zeng, *Transport Efficiency and Safety in China*, Population, Regional Development and Transport, https://doi.org/10.1007/978-981-99-1055-7_9

Table 9.1 Descriptive statistics of the data

Variable Obs	Variable Obs	Min	Max	Mean	Std. Dev
Labor	1918	0.4	66.9	2.43	5.3
Length of highways	1918	605	127,392	12,114.3	8905.8
Passenger volume	1918	155	201,722	11,086.6	16,956
Freight volume	1918	552	110,136	12,354.4	11,940.6

Data source The authors, edited from China City Statistical Yearbook (2009–2015)

(3) The transport sector is a service sector, and its most representative output is the passenger volume and freight volume. In this study, the passenger volume (unit: 10,000 persons) and freight volume (unit: 10,000 tons) in the urban transport sector are selected as output indicators.

Using available data, this study selects 274 prefecture-level cities and centrally administered municipalities from 2008 to 2014 as research objects. The data come from China City Statistical Yearbook (2009–2015).

9.2 Analysis of UTE

The data of 274 prefecture-level cities and centrally administered municipalities directly from 2008 to 2014 are put into the EBM model for calculating UTE, and the calculation results are shown in Appendix Table 1 and Fig. 9.1.

9.2.1 The Overall Characteristics

The annual average of UTE from 2008 to 2014 in China was 0.254, which indicates that the national UTE only reached 25.5% of the ideal level. The cities studied can be divided into five categories: (1) high-level cities (with an annual average UTE value between 0.6 and 1): only 17 cities reached this stage, accounting for 6.2% of all cities studied; (2) relatively high-level cities (with an annual average UTE value between 0.35 and 0.6): there were 29 cities at this level, accounting for 10.6% of all cities studied; (3) mid-level cities (with an annual average UTE value between 0.254 and 0.35): 51 cities were at this level, accounting for 18.6% of all cities studied; (4) relatively low-level cities (with an annual average UTE value between 0.15 and 0.254): there were 100 cities at this level, accounting for 36.5% of all cities studied; (5) low-level cities (with an annual average UTE value less than 0.15): 77 cities remained at this stage, which accounted for 28.1% of all cities studied (Fig. 9.1).

To sum up, China's UTE development is still at the primary stage at present, which has plenty of room for growth.

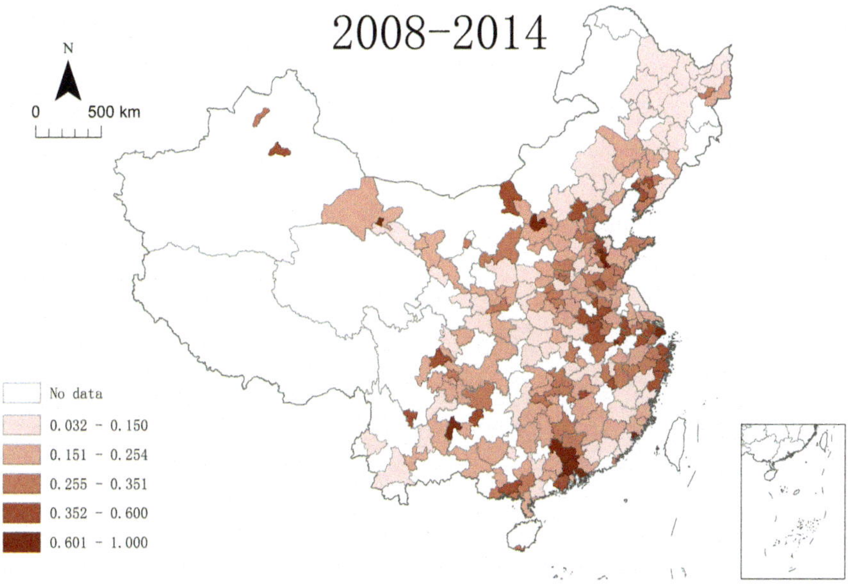

Fig. 9.1 The mean value of UTE intervals in 274 cities of China from 2008 to 2014. *Data source* Edited by the authors

9.2.2 *The Time Difference Analysis of UTSE*

It can be seen that the overall level of UTE fluctuated between 0.214 and 0.278, with the highest value in 2009 and the lowest value in 2013. During the study period, the coefficient of variation of UTE fluctuated between 0.752 and 0.855, with an average value of 0.702. It first declined, then increased, and then fell again. The coefficient of variation was the smallest in 2011, indicating that the national level remained stable in 2011, with not too much change change (Fig. 9. 2).

With the argic 10 software, we draw the UTE in 2008, 2011 and 2014 (Figs. 9.3, 9.4 and 9.5). According to the calculation results of Appendix Table 1. As can be seen from the figure, compared with 2008, the UTE in 152 out of 274 cities improved significantly in 2011, accounting for 55.5% of all cities studied. The UTE in nine cities remained unchanged from 2008, accounting for only 3.3% of all cities studied, while the UTE in 114 cities significantly declined, accounting for 41.6% of all cities studied. Therefore, the overall level of UTE in China improved, which is consistent with the improvement in the average value of UTE from 0.251 in 2008 to 0.262 in 2011.

Compared with 2008, the UTEs in Wuhai, Ordos, Zhoushan, Tai'an, Foshan, Dongguan, Liupanshui, Lanzhou and Jiayuguan remained unchanged in 2011. These cities are in either East or West China. Besides Tai'an (0.144) and Lanzhou (0.148), other cities were kept at the production frontier level.

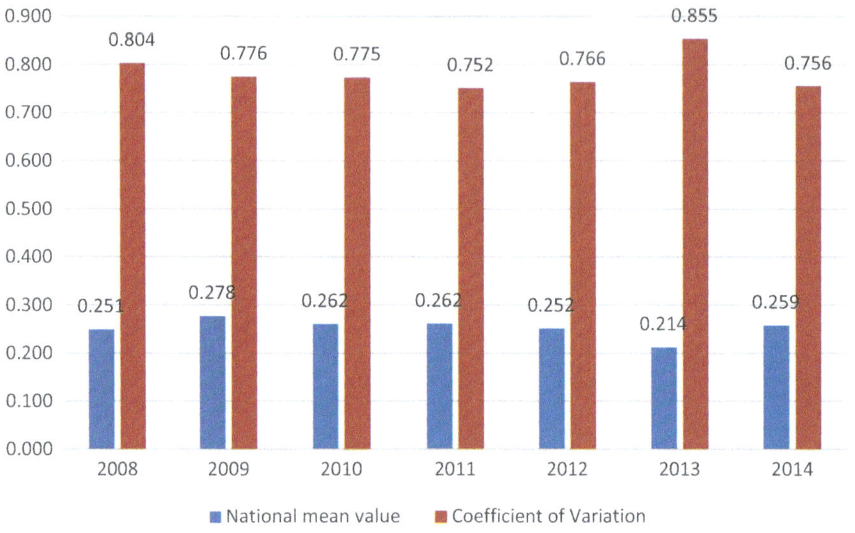

Fig. 9.2 The change trend of the coefficient of variation of UTE in China. *Data source* Edited by the authors

From 2008 to 2011, among the 114 cities with declining UTE, 25 cities (Lincang, Yulin, Laibin, Heihe, Tongliao, Laiwu, Sanmenxia, Qujing, Yangquan, Xiangtan, Jincheng, Taiyuan, Ankang, Zigong, Liaoyuan, Shuozhou, Qinhuangdao, Jinzhong, Chaoyang, Yuxi, Luzhou, Lu'an, Qingdao, Binzhou and Bazhong) saw a drop in UTE of more than 40%. These cities are mainly located in the Midwest (Fig. 9.3).

As we have seen, the UTE in 141 of 274 cities improved significantly in 2014 as compared to 2011, accounting for 51.5% of all cities studied. This proportion is significantly lower than that (55.5%) in 2011 compared with 2008. In 2014, two cities saw no change in UTE, while 131 cities saw a significant decline in UTE as compared with 2011, accounting for 47.8% of all cities studied. This proportion is significantly higher than that (41.6%) in 2011 compared with 2008.

In 2014, the improvement rate of 25 cities was more than 100%, including Yulin, Zunyi, Wuwei, Zhuzhou, Lu'an, Lincang, Xuancheng, Yangjiang, Bayannur, Chuzhou, Lijiang, Baoshan, Zhangye, Jiuquan, Heyuan, Simao, Zhangjiajie, Chongzuo, Guangan, Huangshan, Xiangfan, Yuxi, Wuzhou, Shaoyang and Suihua. These cities are in eastern, central and western regions, but not in the northeast.

From 2011 to 2014, only two cities remained at the production forefront level, which are Zhoushan and Jiayuguan. The number of cities remaining at the production forefront level significantly reduced during this period.

Compared with 2011, 53 cities saw a drop of more than 40% in UTE in 2014, while only 25 cities saw a drop of more than 40% in UTE in 2011 compared with 2008 (Figs. 9.4 and 9.5).

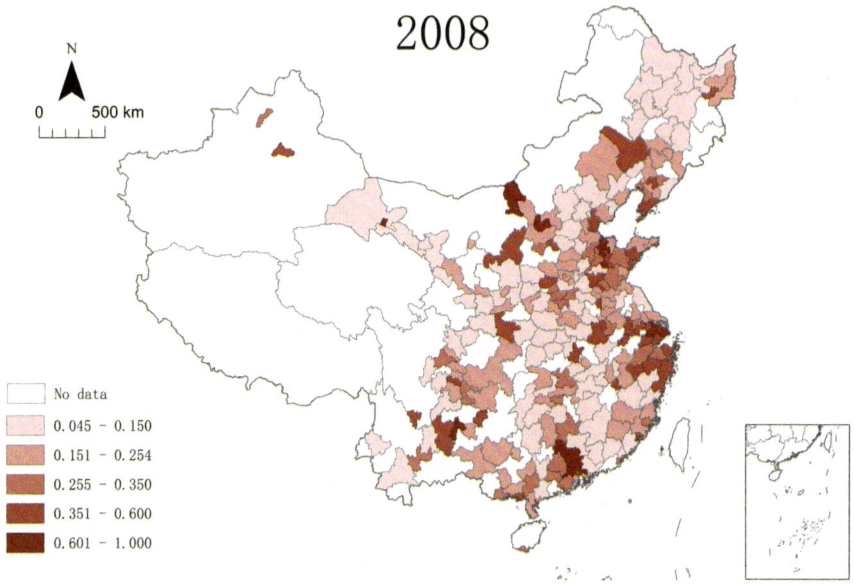

Fig. 9.3 The UTE intervals in 274 cities of China in 2008. *Data source* Edited by the authors

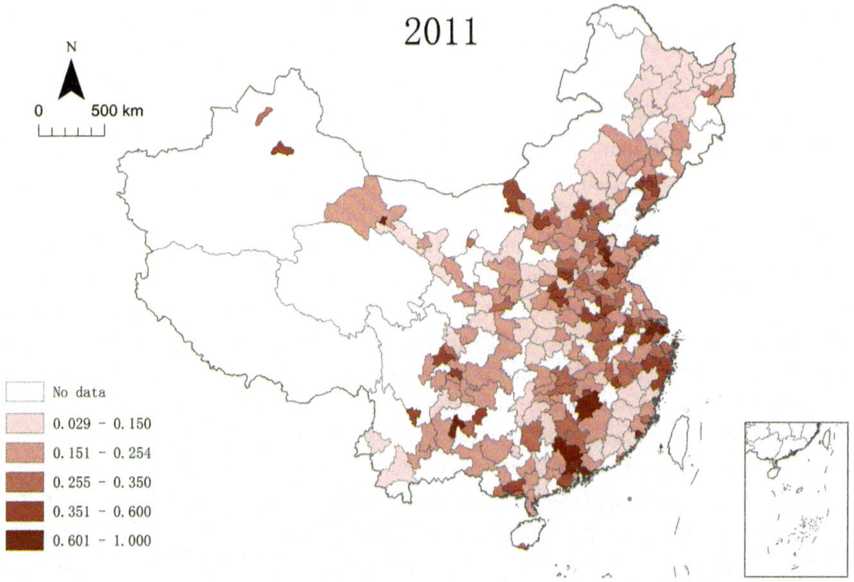

Fig. 9.4 The UTE intervals in 274 cities of China in 2011. *Data source* Edited by the authors

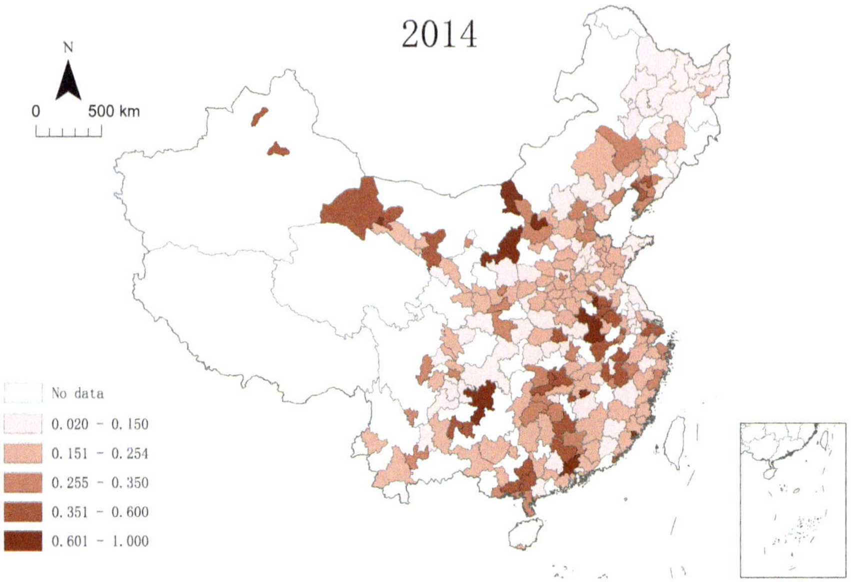

Fig. 9.5 The UTE intervals in 274 cities of China in 2014. *Data source* Edited by the authors

9.2.3 The Spatial Difference Analysis of UTE

During the study period, the UTE of Zhoushan and Jiayuguan was at the production frontier level, while that of Heihe was at the lowest level, with an annual value of only 0.032, indicating that the spatial difference of UTE was large. It can be found that the cities with a higher level of economic development are mainly concentrated in the eastern region, while the cities with a lower level of economic development are mostly distributed in central and western regions. A possible explanation is that the eastern region has some advantages in social economy and transport location, and has absolute advantages in urban infrastructure construction and investment attraction; the industrial structure is also relatively reasonable, the application of multimodal transport in the eastern region is more mature, and the production cost of the transport sector can be greatly reduced through multimodal transport. However, the economic development of central, western and northeast regions is backward, the urban infrastructure construction is insufficient, these three regions have limited ability to attract transport investment, the industrial structure is still uneven, and multimodal transport is not well developed, resulting in the low level of UTE in these regions.

The average annual values of UTE in eastern, central, western and northeast regions were 0.295, 0.246, 0.247 and 0.179, respectively. The value in the eastern region was higher than that in the whole nation, and also that in central, western and northeast regions, which indicates that the input–output conversion ability of the urban transport industry in the eastern region was stronger than that in the other three regions. More specifically, under a certain scale of production factor inputs, the output level in the eastern region was significantly higher than that in the other regions. The value in the western region was slightly higher than that in the central region and slightly lower than the national level. The value in the northeast region was the lowest among the four regions, and its gap with the production frontier surface was the largest.

9.2.4 The Production Frontier Surface Analysis of UTE

If the efficiency value is 1, it means that the DMU is in the production frontier surface. According to Appendix Table 9.6, the number of cities in the production frontier surface is shown in the following Table 9.2.

Table 9.2 The regional characteristics of cities in the production frontier surface

	East China	Central China	West China	Northeast China	The number of cities
2008	Zhoushan, Foshan, Dongguan, Qingyuan	Shuozhou, Chizhou	Wuhai, Ordos, Liupanshui, Jiayuguan	–	10
2009	Dongguan, Zhoushan, Zhuhai, Shenzhen, Qingyuan	Maanshan	Liupanshui, Jiayuguan, Ordos, Wuhai	–	10
2010	Zhoushan, Zhuhai, Dongguan	Tongling	Wuhai, Ordos, Liupanshui, Jiayuguan	–	8
2011	Zhoushan, Zhuhai, Dongguan, Foshan	Tongling	Wuhai, Ordos, Liupanshui, Jiayuguan	–	9
2012	Zhoushan, Zhuhai, Dongguan, Foshan	Tongling	Wuhai, Ordos, Liupanshui, Jiayuguan	–	9
2013	Zhoushan, Zhuhai, Foshan	–	Wuhai, Liupanshui, Jiayuguan	–	6
2014	Zhoushan, Guangzhou	Lu'an	Zunyi, Yulin, Jiayuguan	–	6

Data source Edited by the authors

Zhoushan and Jiayuguan remained in the production frontier surface during the whole research period. In 2008 and 2009, the number of cities in the production frontier surface reached 10, and in the remaining years, the number was less than 10.

From a regional perspective, the cities in the production frontier surface were mainly distributed in eastern and western areas. In 2009, five cities in the eastern area were in the production frontier surface, accounting for 50% of the total number of cities in the production frontier surface. From 2008 to 2012, four cities in the western area were in the production frontier surface. Two cities of the central region were in the production frontier level in 2008, as opposed to only one from 2009 to 2012 and in 2014. During the whole research period, there was no city in Northeast China that remained in the production frontier surface, which indicates that the UTE in Northeast China desperately needed to be improved.

9.3 Urban Transport Accessibility Analysis

In 2016, the urbanization rate in China was 59.6%, and the urban population showed a rapid growth trend. The rapid increase of urban population has brought great pressure to urban transport. At the same time, as traffic congestion and serious traffic pollution are very prominent, it is particularly important to study urban transport accessibility.

The urban transport accessibility index (UTAI) used in this research is defined as: the ratio of free-flow travel time to the total travel time in the urban area. Based on existing data, we select some Chinese cities from 2015 to 2017 as research objects: 45 cities in 2015, 60 cities in 2016, and 100 cities in 2017. The data come from the analysis report on urban transport in major cities of China (2015–2017). The calculation results are shown in Tables 9.3, 9.4 and 9.5. The spatial distribution maps are Figs. 9.6, 9.7, and 9.8.

In 2015, the cities with a UTAI value above 0.7 are Wuxi, Nantong, Jiaxing, Zhongshan, Ningbo, Dongguan, Suzhou, Quanzhou, Weifang, Changzhou, Huizhou and Taizhou; these cities are located in eastern coastal areas. The UTAI in Wuxi was the highest, which reached 0.777, meaning that congested travel time accounted for 22.3% of the total travel time in Wuxi; according to the analysis report on urban transport in major Chinese cities (2015), in Wuxi, there was no traffic jam in the urban area during the whole day; during the rush hour, the average traffic speed was 32.44 km/h in Wuxi, about 46% faster than that in Beijing, the city with the most free-flow travel time in 2015.

In 2015, the UTAI of Hangzhou was 0.582, which is the lowest among the cities studied. In the analysis report, it is mentioned that Hangzhou was in a state of traffic congestion for more than six hours every day in 2015, which means a total of above 1600 h spent in traffic jams all year round.

In 2016, the cities with a UTAI value above 0.7 are Nantong, Wuxi, Zhenjiang, Weifang, Zhongshan, Quanzhou, Jiaxing, Changzhou, Taizhou, Shaoxing, Xuzhou, Datong, Suzhou and Ningbo; except Datong, consistent with the spatial distribution characteristics in 2015, these cities are also located in eastern coastal areas. In 2016,

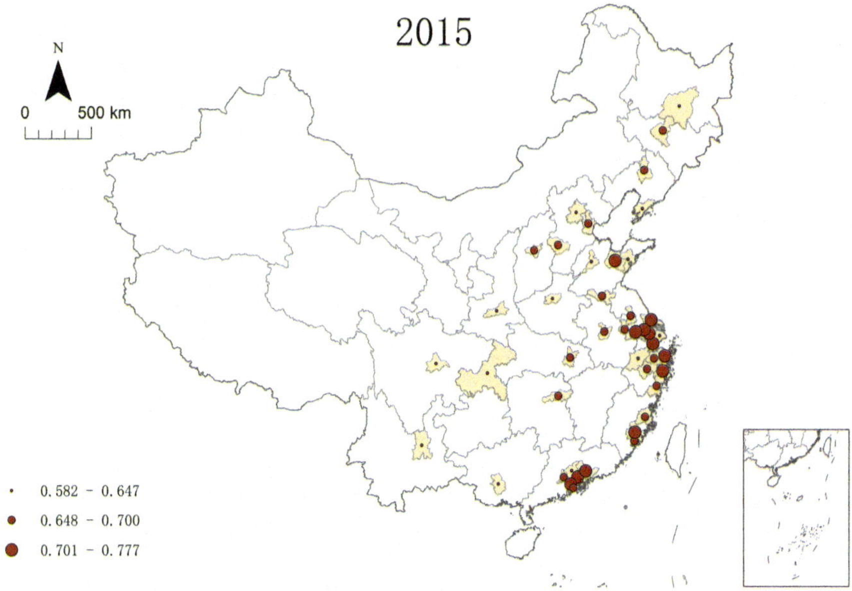

Fig. 9.6 The UTAI intervals in 45 cities of China in 2015. *Data source* Edited by the authors

Table 9.3 The UTAI of Chinese cities (2015)

City	UTAI	City	UTAI	City	UTAI
Wuxi	0.777	Wenzhou	0.693	Xi'an	0.647
Nantong	0.749	Zhuhai	0.691	Chengdu	0.647
Jiaxing	0.745	Yangzhou	0.689	Chongqing	0.638
Zhongshan	0.735	Xiamen	0.687	Shanghai	0.638
Ningbo	0.733	Changchun	0.685	Qingdao	0.636
Dongguan	0.732	Taiyuan	0.684	Shenzhen	0.629
Suzhou	0.719	Xuzhou	0.683	Dalian	0.628
Quanzhou	0.716	Tianjin	0.675	Kunming	0.624
Weifang	0.708	Jinhua	0.673	Nanning	0.619
Changzhou	0.707	Hefei	0.672	Zhengzhou	0.617
Huizhou	0.706	Shenyang	0.671	Beijing	0.596
Taizhou	0.706	Foshan	0.668	Guangzhou	0.596
Nanjing	0.698	Fuzhou	0.664	Jinan	0.592
Shaoxing	0.697	Wuhan	0.662	Harbin	0.585
Shijiazhuang	0.693	Changsha	0.661	Hangzhou	0.582

Data source Edited by the authors

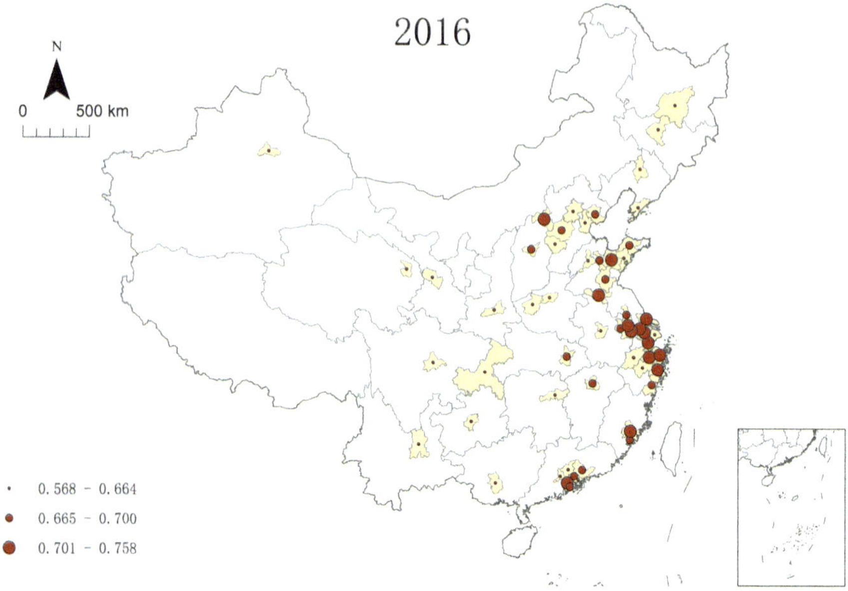

Fig. 9.7 The UTAI intervals in 60 cities of China in 2016. *Data source* Edited by the authors

the UTAI in Nantong was the highest, which reached 0.758, meaning that congested travel time only accounted for 24.2% of the total travel time. The UTAIs of Beijing, Lanzhou, Harbin and Jinan were lower than 0.6, which means that congested travel time accounted for more than 40% of the total travel time in these cities (Fig. 9.8 and Table 9.5). The UTAI of Jinan was 0.568, the lowest in the cities studied (Fig 9.7). In the analysis report on urban transport in major Chinese cities (2016), Jinan was rated the "most congested city" in the Chinese mainland; the traffic congestion in Jinan is mainly related to the many governmental urban road construction projects in Jinan in 2016, which may bring inconvenience to travel in the short term.

In 2017, the cities with a UTAI value above 0.7 are Taizhou, Nantong, Yancheng, Zhenjiang, Wuhu, Changzhou, Wuxi, Huzhou, Suzhou, Langfang, Ningbo, Zhaoqing, Weifang, Jiangmen, Huai'an, Taizhou, Zibo, Shaoxing, Chuzhou, Quanzhou, Linyi and Qinhuangdao; except Huai'an and Chuzhou, these cities are located in eastern coastal areas. Among these cities, Taizhou had the highest level of UTAI, and the value was 0.794, meaning that free-flow travel time accounted for 79.4% of the total travel time. The UTAIs of Guangzhou, Beijing, Harbin and Jinan were lower than 0.6, which means that congested travel time of these four cities accounted for more than 40% of the total travel time.

The UTAI of Jinan was 0.568, the lowest in the cities studied, which decreased by 2.5% compared with 2016. According to the analysis report on urban transport in major Chinese cities (2017), there were 2,078 h of congested travel time in Jinan in 2017, with an average of 5.7 h of congested travel time per day, which significantly

Table 9.4 The UTAI of Chinese cities (2016)

City	UTAI	City	UTAI	City	UTAI
Jinan	0.568	Nanning	0.625	Taiyuan	0.667
Harbin	0.571	Xining	0.621	Shaoxing	0.709
Beijing	0.578	Zhuhai	0.667	Fuzhou	0.685
Lanzhou	0.578	Chengdu	0.633	Ningbo	0.704
Chongqing	0.645	Changchun	0.662	Dongguan	0.694
Guiyang	0.613	Shijiazhuang	0.649	Linyi	0.690
Shenzhen	0.610	Nanjing	0.676	Jiaxing	0.714
Kunming	0.610	Nanchang	0.667	Baoding	0.699
Hangzhou	0.641	Luoyang	0.654	Xuzhou	0.709
Dalian	0.637	Wenzhou	0.671	Changzhou	0.714
Guangzhou	0.606	Tianjin	0.662	Taizhou	0.714
Shanghai	0.629	Yantai	0.690	Ordos	0.680
Hefei	0.654	Hong Kong	0.621	Zhongshan	0.725
Zhengzhou	0.617	Huizhou	0.680	Quanzhou	0.719
Changsha	0.641	Xiamen	0.685	Weifang	0.730
Xi'an	0.617	Tangshan	0.685	Urumqi	0.629
Shenyang	0.637	Zibo	0.699	Datong	0.709
Wuhan	0.667	Suzhou	0.704	Wuxi	0.752
Qingdao	0.649	Jinhua	0.658	Nantong	0.758
Foshan	0.633	Yangzhou	0.699	Zhenjiang	0.752

Data source Edited by the authors

decreased from 6.6 h in 2016. However, compared with other cities, the problem of traffic congestion in Jinan was more serious.

9.4 Influencing Factors of UTE

9.4.1 Reasonable Road Grading System

Referring to international experience, a reasonable road grading system should be a "pyramid" with a small top (expressways) and a big bottom (branch roads), which can ensure the convergence of traffic flows from low-grade roads to high-grade roads, and the diversion of traffic flows from high-grade roads to low-grade roads.

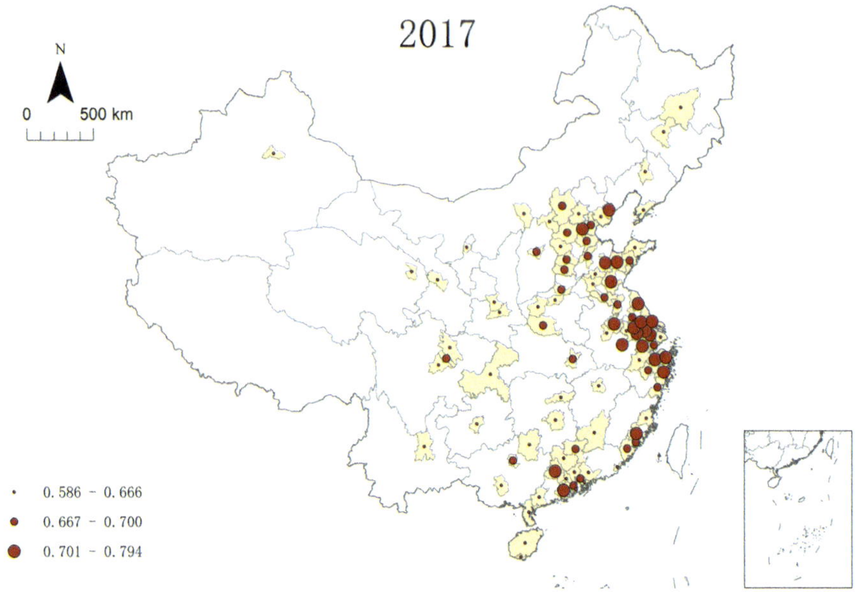

Fig. 9.8 The UTAI intervals in 100 cities of China in 2017. *Data source* Edited by the authors

9.4.2 Reasonable Transport Space Per Capita and Road Network Density

Urban traffic congestion may be caused by the low level of road area per capita. The rapid growth of urban population makes traffic flows increase rapidly, which increases the risk of traffic congestion with unchanged road supply. In addition, reasonable road network density is also an important factor affecting urban traffic congestion. The density of the road network is equal to the total length of roads per unit area (km/km^2). In the urban built-up area, the road network density in Tokyo is 19.04, while that in Beijing is 4.85. The East Third Ring Road and the East Fourth Ring Road in Beijing are very crowded during busy hours.

9.4.3 Control System of Urban Traffic Lights

At present, the urban traffic signal control system in China features a unified traffic light conversion cycle, ignoring the real-time monitoring of road traffic flows. Therefore, optimizing the traffic signal conversion cycle becomes an important research direction to alleviate traffic congestion. The design of an intelligent traffic control system that considers real-time traffic flows is an important measure to reduce traffic congestion.

Table 9.5 The UTAI of Chinese cities (2017)

City	UTAI	City	UTAI	City	UTAI
Taizhou	0.794	Wuhan	0.685	Changchun	0.643
Nantong	0.777	Qingdao	0.682	Qingyuan	0.642
Yancheng	0.742	Xinxiang	0.682	Hangzhou	0.641
Zhenjiang	0.736	Ili	0.681	Foshan	0.640
Wuhu	0.734	Taiyuan	0.679	Shantou	0.640
Changzhou	0.724	Nanyang	0.678	Datong	0.639
Wuxi	0.723	Wenzhou	0.677	Mianyang	0.639
Huzhou	0.719	Jinhua	0.676	Hefei	0.638
Suzhou	0.715	Deyang	0.675	Guiyang	0.637
Langfang	0.714	Baoding	0.674	Hengyang	0.636
Ningbo	0.713	Ordos	0.670	Tai'an	0.634
Zhaoqing	0.713	Zhangjiakou	0.669	Shanghai	0.633
Weifang	0.709	Liuzhou	0.669	Lanzhou	0.631
Jiangmen	0.708	Zhangzhou	0.668	Haikou	0.625
Huai'an	0.708	Cangzhou	0.668	Chongqing	0.625
Taizhou	0.708	Suqian	0.667	Chengdu	0.624
Zibo	0.707	Zhuhai	0.665	Xianyang	0.623
Shaoxing	0.707	Jining	0.665	Urumqi	0.621
Chuzhou	0.706	Tangshan	0.661	Xining	0.620
Quanzhou	0.705	Nanchang	0.660	Yinchuan	0.619
Linyi	0.700	Fuzhou	0.659	Sanya	0.618
Qinhuangdao	0.700	Shijiazhuang	0.658	Luoyang	0.618
Xingtai	0.699	Guilin	0.656	Ganzhou	0.617
Jiaxing	0.698	Nanjing	0.654	Xi'an	0.615
Dezhou	0.697	Yantai	0.654	Nanyun	0.615
Dongguan	0.696	Shenyang	0.652	Kunming	0.611
Zhongshan	0.695	Zhanjiang	0.652	Hohhot	0.610
Xiamen	0.694	Zhengzhou	0.648	Maoming	0.608
Yangzhou	0.694	Hong Kong	0.646	Nanning	0.605
Xuzhou	0.693	Changsha	0.646	Guangzhou	0.597
Tianjin	0.691	Huizhou	0.644	Beijing	0.591
Handan	0.691	Shenzhen	0.644	Harbin	0.586
Lianyungang	0.687	Dalian	0.643	Jinan	0.582
Shaoguan	0.686				

Data source Edited by the authors

9.4.4 *The Non-linear Rate of Urban Roads*

The non-linear rate of urban roads can reflect the "detour degree". Because of their distribution function, policy factors and environmental factors, highways cannot connect the starting and ending points in a straight line, showing a tortuous form between two points. The non-linear rate of urban roads refers to the ratio of the actual traffic distance between the starting and ending points of the road to the spatial linear distance between the two points. In 2017, the average value of the non-linear rate of urban roads of 100 cities in China was 1.59, and that of the top 10 cities with traffic congestion was 1.71, which is slightly higher than the national average level.

9.5 Conclusions

9.5.1 *The Overall UTE is Still at a Low Level and Has Plenty of Room for Growth*

China's UTE development is still at the primary stage at present, which has plenty of room for growth. The overall level of UTE fluctuated between 0.214 and 0.278, with the highest value in 2009 and the lowest value in 2013. During the study period, the coefficient of variation of UTE fluctuated between 0.752 and 0.855, with an average value of 0.702. It first declined, then increased, and then fell again. The coefficient of variation was the smallest in 2011, indicating that the national level remained stable in 2011, with not much change.

9.5.2 *The Spatial Difference of UTE is Large*

It can be found that the cities with a higher level of economic development are mainly concentrated in the eastern region, while the cities with a lower level of economic development are mostly distributed in central and western regions. A possible explanation is that the eastern region has some advantages in social economy and transport location, and has absolute advantages in urban infrastructure construction and investment attraction; the industrial structure is also relatively reasonable, the application of multimodal transport in the eastern region is more mature, and the production cost of the transport sector can be greatly reduced through multimodal transport. However, the economic development of central, western and northeast regions is backward, the urban infrastructure construction is insufficient, these three regions have limited ability to attract transport investment, the industrial structure is still uneven, and multimodal transport is not well developed, resulting in the low level of UTE in these regions.

9.5.3 The Cities with a High Level of UTAI are Mainly Located in Eastern Coastal Areas

We analyze the characteristics of UTAI in the main cities of China in 2015, 2016 and 2017. We have found that the cities with a high level of UTAI are mainly located in eastern coastal areas. As for the economic loss due to traffic congestion, we analyze the data of 2016 and have found that although the overall level of UTAI in central cities was lower than that in eastern cities, the per capita transport spending was higher than that of eastern cities. In 2016, the average economic loss caused by traffic congestion in central cities was RMB 6,866.9, while that in eastern cities and western cities was RMB 6,524.375 and RMB 6,281.1, respectively.

Appendix

See the Table 9.6.

References

China urban statistical yearbook. (2009–2015). China Statistical Publishing House, Beijing. https://data.cnki.net/Yearbook/Single/N2016030128. Accessed April 24, 2020.

The analysis report of ruban transportation in major Cities in China. (2015). https://www.docin.com/p-1944032378.html. Accessed April 24, 2020.

The analysis report of ruban transportation in major Cities in China. (2016). https://www.docin.com/p-2072970162.html. Accessed April 24, 2020.

The analysis report of ruban transportation in major Cities in China. (2017). https://max.book118.com/html/2018/0523/168241832.shtm. Accessed April 24, 2020.

Table 9.6 Measurement results of urban transport efficiency

City	2008	2009	2010	2011	2012	2013	2014	Mean
Beijing	0.146	0.551	0.530	0.452	0.408	0.178	0.318	0.369
Tianjin	0.439	0.406	0.303	0.264	0.262	0.262	0.276	0.316
Shijiazhuang	0.130	0.162	0.169	0.172	0.180	0.186	0.132	0.161
Tangshan	0.227	0.295	0.278	0.287	0.284	0.259	0.204	0.262
Qinghuangdao	0.145	0.101	0.089	0.080	0.077	0.079	0.078	0.093
Handan	0.242	0.258	0.280	0.281	0.275	0.265	0.228	0.261
Xingtai	0.172	0.151	0.145	0.171	0.180	0.127	0.216	0.166
Baoding	0.127	0.162	0.164	0.176	0.175	0.169	0.173	0.164
Zhangjiakou	0.077	0.066	0.080	0.089	0.086	0.070	0.094	0.080
Chengde	0.063	0.064	0.062	0.076	0.081	0.068	0.118	0.076
Cangzhou	0.205	0.278	0.231	0.284	0.286	0.278	0.301	0.266
Langfang	0.186	0.198	0.194	0.169	0.173	0.178	0.222	0.189
Hengshui	0.065	0.068	0.072	0.080	0.078	0.075	0.085	0.075
Taiyuan	0.348	0.279	0.237	0.177	0.169	0.185	0.215	0.230
Datong	0.164	0.314	0.267	0.259	0.170	0.204	0.250	0.233
Yangquan	0.372	0.234	0.207	0.187	0.173	0.169	0.193	0.219
Changzhi	0.234	0.182	0.149	0.142	0.142	0.129	0.178	0.165
Jincheng	0.431	0.232	0.210	0.218	0.198	0.149	0.184	0.232
Shuozhou	1.000	0.701	0.543	0.540	0.575	0.378	0.966	0.672
Jinzhong	0.254	0.160	0.162	0.141	0.146	0.118	0.214	0.171
Yuncheng	0.135	0.142	0.153	0.091	0.093	0.085	0.176	0.125
Xinzhou	0.184	0.163	0.169	0.152	0.154	0.112	0.255	0.170
Linfen	0.115	0.110	0.102	0.094	0.088	0.088	0.098	0.099
Luliang	0.199	0.176	0.164	0.155	0.152	0.079	0.172	0.157
Hohhot	0.220	0.182	0.187	0.163	0.214	0.198	0.260	0.204
Baotou	0.880	0.535	0.538	0.532	0.589	0.473	0.654	0.600
Wuhai	1.000	1.000	1.000	1.000	1.000	1.000	0.764	0.966
Chifeng	0.188	0.138	0.131	0.113	0.136	0.073	0.217	0.142
Tongliao	0.356	0.162	0.232	0.156	0.151	0.093	0.288	0.205
Ordos	1.000	1.000	1.000	1.000	1.000	0.499	0.578	0.868
Hulunbuir	0.120	0.106	0.099	0.108	0.110	0.095	0.144	0.112
Bayannur	0.112	0.068	0.073	0.089	0.056	0.028	0.273	0.100
Ulanqab	0.159	0.092	0.093	0.099	0.100	0.079	0.142	0.109
Shenyang	0.216	0.271	0.243	0.229	0.231	0.192	0.216	0.228
Dalian	0.374	0.390	0.321	0.274	0.286	0.278	0.320	0.321
Anshan	0.233	0.345	0.363	0.366	0.372	0.241	0.278	0.314

(continued)

Table 9.6 (continued)

City	2008	2009	2010	2011	2012	2013	2014	Mean
Fushun	0.200	0.192	0.194	0.227	0.217	0.168	0.202	0.200
Benxi	0.314	0.349	0.304	0.289	0.292	0.255	0.256	0.294
Dandong	0.127	0.190	0.154	0.144	0.139	0.121	0.132	0.144
Yingkou	0.328	0.456	0.438	0.343	0.336	0.346	0.406	0.379
Fuxin	0.123	0.174	0.151	0.164	0.171	0.128	0.245	0.165
Liaoyang	0.442	0.522	0.459	0.511	0.551	0.464	0.598	0.507
Panjin	0.255	0.345	0.338	0.372	0.432	0.410	0.403	0.365
Tieling	0.218	0.213	0.202	0.210	0.184	0.119	0.228	0.196
Chaoyang	0.148	0.098	0.088	0.083	0.092	0.062	0.137	0.101
Huludao	0.169	0.276	0.231	0.189	0.238	0.199	0.314	0.231
Changchun	0.107	0.120	0.110	0.105	0.098	0.069	0.061	0.096
Jilin	0.135	0.154	0.147	0.164	0.162	0.080	0.156	0.143
Siping	0.153	0.134	0.131	0.133	0.149	0.070	0.194	0.138
Liaoyuan	0.164	0.101	0.102	0.087	0.080	0.063	0.117	0.102
Tonghua	0.183	0.207	0.196	0.214	0.183	0.222	0.093	0.185
Changbai Mountains	0.165	0.159	0.147	0.174	0.173	0.175	0.110	0.157
Songyuan	0.137	0.170	0.197	0.207	0.185	0.094	0.214	0.172
Baicheng	0.051	0.054	0.055	0.064	0.055	0.040	0.092	0.059
Harbin	0.094	0.092	0.080	0.079	0.074	0.059	0.062	0.077
Qiqihar	0.071	0.085	0.080	0.075	0.116	0.058	0.052	0.077
Jixi	0.201	0.204	0.185	0.163	0.137	0.102	0.099	0.155
Hegang	0.120	0.200	0.158	0.133	0.112	0.053	0.090	0.124
Shuangyashan	0.193	0.236	0.116	0.122	0.121	0.065	0.053	0.130
Daqing	0.066	0.349	0.091	0.079	0.071	0.099	0.069	0.118
Yichun	0.068	0.141	0.127	0.114	0.101	0.043	0.057	0.093
Jiamusi	0.079	0.081	0.072	0.062	0.075	0.057	0.110	0.077
Qitaihe	0.448	0.481	0.298	0.278	0.283	0.135	0.153	0.297
Heihe	0.067	0.032	0.037	0.029	0.023	0.016	0.020	0.032
Suihua	0.045	0.041	0.042	0.049	0.050	0.041	0.098	0.052
Shanghai	0.685	0.783	0.693	0.632	0.592	0.482	0.589	0.637
Nanjing	0.566	0.429	0.382	0.388	0.394	0.202	0.242	0.372
Wuxi	0.431	0.356	0.352	0.375	0.349	0.185	0.184	0.319
Xuzhou	0.205	0.197	0.228	0.216	0.178	0.194	0.129	0.192
Changzhou	0.449	0.318	0.492	0.284	0.270	0.153	0.134	0.300
Suzhou(in jiangsu)	0.605	0.481	0.510	0.691	0.614	0.354	0.385	0.520
Nantong	0.187	0.232	0.219	0.219	0.207	0.118	0.122	0.186

(continued)

Table 9.6 (continued)

City	2008	2009	2010	2011	2012	2013	2014	Mean
Lianyungang	0.247	0.230	0.216	0.239	0.217	0.140	0.117	0.201
Huai'an	0.108	0.160	0.171	0.178	0.164	0.109	0.109	0.143
Yancheng	0.145	0.160	0.154	0.161	0.160	0.095	0.108	0.141
Yangzhou	0.226	0.180	0.172	0.183	0.170	0.104	0.102	0.163
Zhenjiang	0.267	0.284	0.262	0.261	0.255	0.133	0.130	0.227
Taizhou(in Jiangsu)	0.231	0.259	0.253	0.250	0.232	0.249	0.169	0.235
Suqian	0.128	0.282	0.285	0.258	0.256	0.119	0.175	0.215
Hangzhou	0.284	0.286	0.259	0.246	0.231	0.224	0.169	0.243
Ningbo	0.503	0.547	0.473	0.343	0.332	0.329	0.333	0.409
Wenzhou	0.539	0.555	0.514	0.499	0.390	0.406	0.297	0.457
Jiaxing	0.434	0.340	0.360	0.312	0.265	0.250	0.209	0.310
Huzhou	0.506	0.489	0.414	0.394	0.316	0.209	0.163	0.356
Shaoxing	0.349	0.321	0.295	0.277	0.240	0.283	0.162	0.275
Jinhua	0.357	0.370	0.379	0.378	0.320	0.337	0.124	0.323
Quzhou	0.395	0.307	0.302	0.308	0.275	0.155	0.275	0.288
Zhoushan	1.000	1.000	1.000	1.000	1.000	1.000	1.000	1.000
Taizhou(in Zhejing)	0.378	0.405	0.448	0.423	0.350	0.384	0.220	0.372
Lishui	0.131	0.165	0.138	0.125	0.113	0.086	0.185	0.135
Hefei	0.255	0.325	0.303	0.271	0.256	0.278	0.228	0.274
Wuhu	0.412	0.462	0.447	0.300	0.286	0.288	0.304	0.357
Bengbu	0.178	0.393	0.417	0.417	0.434	0.417	0.528	0.398
Huainan	0.211	0.379	0.388	0.360	0.366	0.360	0.457	0.360
Ma'anshan	0.413	1.000	0.846	0.517	0.531	0.362	0.452	0.589
Huaibei	0.412	0.453	0.453	0.459	0.424	0.444	0.691	0.477
Tongling	0.374	0.989	1.000	1.000	1.000	0.973	0.352	0.813
Anqing	0.129	0.347	0.302	0.320	0.384	0.253	0.417	0.307
Huangshan	0.215	0.207	0.183	0.190	0.223	0.186	0.396	0.229
Chuzhou	0.168	0.147	0.136	0.156	0.217	0.244	0.476	0.221
Fuyang	0.138	0.427	0.393	0.394	0.398	0.425	0.698	0.410
Suzhou(In Anhui)	0.189	0.315	0.292	0.314	0.365	0.263	0.516	0.322
Lu'an	0.483	0.491	0.446	0.284	0.208	0.269	1.000	0.454
Bozhou	0.167	0.427	0.369	0.396	0.432	0.212	0.389	0.342
Chizhou	1.000	0.632	0.658	0.628	0.477	0.186	0.332	0.559
Xuancheng	0.239	0.380	0.396	0.224	0.220	0.203	0.725	0.341
Fuzhou	0.232	0.312	0.255	0.234	0.208	0.207	0.186	0.233
Xiamen	0.481	0.741	0.665	0.663	0.683	0.630	0.980	0.692

(continued)

Table 9.6 (continued)

City	2008	2009	2010	2011	2012	2013	2014	Mean
Putian	0.266	0.289	0.256	0.335	0.245	0.186	0.197	0.254
Sanming	0.162	0.141	0.131	0.124	0.126	0.094	0.204	0.140
Quanzhou	0.159	0.195	0.183	0.169	0.159	0.150	0.171	0.169
Zhangzhou	0.141	0.125	0.112	0.109	0.107	0.081	0.186	0.123
Nanping	0.076	0.072	0.068	0.065	0.068	0.034	0.059	0.063
Longyan	0.236	0.183	0.171	0.142	0.139	0.071	0.144	0.155
Ningde	0.123	0.119	0.113	0.144	0.120	0.119	0.170	0.130
Nanchang	0.092	0.128	0.114	0.109	0.097	0.113	0.105	0.108
Jingdezhen	0.055	0.115	0.108	0.083	0.093	0.078	0.059	0.084
Pingxiang	0.182	0.275	0.360	0.289	0.353	0.213	0.366	0.291
Jiujiang	0.085	0.154	0.159	0.156	0.147	0.122	0.161	0.140
Xinyu	0.097	0.587	0.519	0.576	0.672	0.379	0.811	0.520
Yingtan	0.371	0.456	0.523	0.390	0.460	0.217	0.309	0.390
Ganzhou	0.108	0.203	0.189	0.178	0.180	0.107	0.224	0.170
Ji'an	0.072	0.105	0.098	0.684	0.091	0.063	0.120	0.176
Yichun	0.103	0.202	0.139	0.155	0.175	0.130	0.199	0.158
Fuzhou	0.086	0.188	0.191	0.205	0.203	0.109	0.206	0.170
Shangrao	0.167	0.268	0.312	0.289	0.313	0.182	0.405	0.276
Jinan	0.271	0.262	0.226	0.199	0.190	0.148	0.140	0.205
Qingdao	0.375	0.271	0.240	0.221	0.208	0.202	0.146	0.238
Zibo	0.589	0.762	0.709	0.694	0.581	0.630	0.240	0.601
Zaozhuang	0.313	0.767	0.601	0.565	0.449	0.358	0.129	0.455
Dongying	0.157	0.155	0.136	0.123	0.109	0.124	0.158	0.137
Yantai	0.254	0.360	0.350	0.314	0.259	0.254	0.105	0.271
Weifang	0.336	0.346	0.247	0.209	0.227	0.157	0.173	0.242
Jining	0.370	0.363	0.285	0.252	0.209	0.182	0.240	0.272
Tai'an	0.144	0.196	0.144	0.144	0.143	0.108	0.064	0.135
Weihai	0.249	0.318	0.327	0.349	0.325	0.325	0.104	0.285
Rizhao	0.221	0.363	0.342	0.305	0.266	0.220	0.157	0.268
Laiwu	0.603	0.500	0.494	0.267	0.202	0.186	0.172	0.346
Linyi	0.321	0.439	0.395	0.347	0.346	0.194	0.191	0.319
Dezhou	0.275	0.285	0.267	0.283	0.269	0.126	0.121	0.232
Liaocheng	0.086	0.201	0.175	0.163	0.139	0.118	0.135	0.145
Binzhou	0.631	0.671	0.376	0.373	0.272	0.132	0.203	0.380
Heze	0.109	0.393	0.311	0.285	0.263	0.087	0.157	0.229
Zhengzhou	0.307	0.354	0.370	0.387	0.295	0.304	0.165	0.312

(continued)

Table 9.6 (continued)

City	2008	2009	2010	2011	2012	2013	2014	Mean
Kaifeng	0.269	0.156	0.163	0.182	0.148	0.156	0.186	0.180
Luoyang	0.121	0.159	0.164	0.173	0.162	0.168	0.153	0.157
Pingdingshan	0.176	0.238	0.221	0.246	0.202	0.211	0.223	0.217
Anyang	0.173	0.302	0.282	0.372	0.318	0.258	0.123	0.261
Hebi	0.175	0.414	0.387	0.391	0.391	0.343	0.331	0.347
Xinxiang	0.108	0.129	0.124	0.149	0.138	0.129	0.191	0.138
Jiaozuo	0.308	0.376	0.376	0.429	0.407	0.246	0.202	0.335
Puyang	0.118	0.121	0.126	0.150	0.165	0.150	0.167	0.142
Xuchang	0.310	0.375	0.331	0.393	0.442	0.346	0.153	0.336
Luohe	0.128	0.163	0.157	0.166	0.212	0.169	0.117	0.159
Sanmenxia	0.238	0.109	0.101	0.108	0.128	0.109	0.124	0.131
Nanyang	0.094	0.098	0.106	0.130	0.132	0.120	0.107	0.112
Shangqiu	0.088	0.194	0.199	0.230	0.228	0.170	0.152	0.180
Xinyang	0.068	0.086	0.085	0.116	0.097	0.110	0.111	0.096
Zhoukou	0.127	0.169	0.162	0.185	0.212	0.128	0.174	0.165
Zhumadian	0.191	0.201	0.185	0.216	0.222	0.190	0.202	0.201
Wuhan	0.366	0.401	0.349	0.250	0.284	0.284	0.302	0.320
Huangshi	0.241	0.251	0.187	0.192	0.225	0.194	0.135	0.204
Shiyan	0.050	0.063	0.060	0.069	0.073	0.071	0.079	0.067
Yichang	0.117	0.090	0.078	0.080	0.086	0.088	0.054	0.085
Xiangyang	0.117	0.101	0.108	0.105	0.116	0.116	0.217	0.126
Ezhou	0.134	0.171	0.121	0.121	0.141	0.131	0.129	0.135
Jingmen	0.150	0.178	0.137	0.126	0.136	0.141	0.099	0.138
Xiaogan	0.149	0.165	0.108	0.116	0.099	0.115	0.056	0.115
Jingzhou	0.059	0.092	0.091	0.110	0.096	0.096	0.170	0.102
Huanggang	0.120	0.103	0.120	0.174	0.147	0.113	0.207	0.140
Xianning	0.077	0.080	0.095	0.125	0.111	0.101	0.185	0.111
Suizhou	0.124	0.178	0.169	0.208	0.201	0.156	0.392	0.204
Changsha	0.242	0.434	0.355	0.300	0.244	0.270	0.175	0.289
Zhuzhou	0.265	0.263	0.227	0.248	0.253	0.249	0.950	0.351
Xiangtan	0.332	0.186	0.184	0.167	0.185	0.183	0.079	0.188
Hengyang	0.138	0.215	0.186	0.219	0.246	0.249	0.347	0.229
Shaoyang	0.114	0.144	0.130	0.134	0.149	0.143	0.270	0.155
Yueyang	0.339	0.372	0.249	0.262	0.187	0.162	0.416	0.284
Changde	0.225	0.193	0.213	0.177	0.170	0.153	0.273	0.201
Zhangjiajie	0.171	0.140	0.143	0.196	0.192	0.164	0.480	0.213

(continued)

Table 9.6 (continued)

City	2008	2009	2010	2011	2012	2013	2014	Mean
Yiyang	0.135	0.357	0.268	0.213	0.189	0.154	0.357	0.239
Chenzhou	0.260	0.314	0.263	0.289	0.300	0.244	0.480	0.307
Yongzhou	0.235	0.120	0.120	0.142	0.119	0.357	0.225	0.188
Huaihua	0.099	0.078	0.074	0.111	0.083	0.112	0.190	0.107
Loudi	0.230	0.367	0.397	0.216	0.255	0.180	0.287	0.276
Guangzhou	0.850	0.866	0.724	0.693	0.785	0.805	1.000	0.818
Shenzhen	0.177	1.000	0.938	0.850	0.764	0.745	0.162	0.662
Zhuhai	0.375	1.000	1.000	1.000	1.000	1.000	0.596	0.853
Shantou	0.267	0.310	0.265	0.274	0.300	0.292	0.370	0.297
Foshan	1.000	0.926	0.868	1.000	1.000	1.000	0.676	0.924
Shaoguan	0.145	0.310	0.311	0.320	0.351	0.459	0.285	0.311
Heyuan	0.061	0.070	0.063	0.087	0.088	0.101	0.220	0.099
Meizhou	0.137	0.115	0.095	0.104	0.104	0.063	0.133	0.108
Huizhou	0.120	0.169	0.171	0.175	0.183	0.182	0.195	0.171
Shanwei	0.178	0.107	0.164	0.274	0.263	0.206	0.096	0.184
Dongguan	1.000	1.000	1.000	1.000	1.000	0.821	0.113	0.848
Zhongshan	0.208	0.195	0.171	0.253	0.275	0.360	0.207	0.238
Jiangmen	0.289	0.528	0.525	0.522	0.483	0.410	0.235	0.427
Yangjiang	0.079	0.069	0.066	0.074	0.072	0.083	0.228	0.096
Zhanjiang	0.207	0.266	0.251	0.245	0.231	0.239	0.305	0.249
Maoming	0.121	0.094	0.090	0.102	0.087	0.081	0.132	0.101
Zhaoqing	0.334	0.201	0.198	0.215	0.181	0.189	0.115	0.205
Qingyuan	1.000	1.000	0.753	0.733	0.573	0.626	0.503	0.741
Chaozhou	0.083	0.123	0.114	0.111	0.127	0.128	0.189	0.125
Jieyang	0.153	0.120	0.123	0.624	0.149	0.158	0.115	0.206
Yunfu	0.276	0.157	0.184	0.224	0.218	0.206	0.182	0.207
Nanning	0.176	0.212	0.244	0.197	0.223	0.227	0.248	0.218
Liuzhou	0.158	0.147	0.139	0.137	0.135	0.152	0.168	0.148
Guilin	0.129	0.199	0.207	0.265	0.226	0.140	0.184	0.193
Wuzhou	0.090	0.116	0.165	0.122	0.120	0.135	0.249	0.142
Beihai	0.380	0.300	0.273	0.271	0.280	0.276	0.285	0.295
Fangchenggang	0.286	0.430	0.419	0.342	0.439	0.230	0.278	0.346
Qinzhou	0.191	0.482	0.523	0.529	0.544	0.520	0.563	0.479
Guigang	0.169	0.223	0.224	0.224	0.226	0.257	0.372	0.242
Yulin	0.311	0.251	0.226	0.233	0.214	0.228	0.387	0.264
Baise	0.190	0.185	0.187	0.209	0.234	0.090	0.191	0.184

(continued)

Table 9.6 (continued)

City	2008	2009	2010	2011	2012	2013	2014	Mean
Hechi	0.224	0.262	0.214	0.162	0.182	0.075	0.168	0.184
Laibin	0.732	0.245	0.226	0.242	0.214	0.064	0.177	0.271
Chongzuo	0.082	0.115	0.097	0.105	0.131	0.089	0.240	0.123
Sanya	0.310	0.349	0.381	0.306	0.312	0.413	0.176	0.321
Chongqing	0.184	0.192	0.186	0.194	0.138	0.159	0.071	0.160
Chengdu	0.343	0.622	0.652	0.572	0.495	0.342	0.113	0.448
Zigong	0.429	0.220	0.233	0.227	0.308	0.225	0.105	0.250
Panzhihua	0.443	0.518	0.487	0.507	0.635	0.387	0.286	0.466
Luzhou	0.288	0.265	0.140	0.167	0.152	0.183	0.136	0.190
Deyang	0.176	0.235	0.260	0.274	0.410	0.247	0.178	0.254
Mianyang	0.114	0.108	0.117	0.122	0.106	0.107	0.162	0.120
Guangyuan	0.133	0.138	0.156	0.217	0.233	0.172	0.147	0.171
Suining	0.180	0.194	0.238	0.249	0.206	0.139	0.329	0.219
Neijiang	0.250	0.289	0.331	0.444	0.301	0.264	0.268	0.307
Leshan	0.122	0.189	0.190	0.196	0.182	0.159	0.206	0.178
Nanchong	0.233	0.139	0.176	0.155	0.143	0.125	0.177	0.164
Meishan	0.304	0.240	0.163	0.210	0.192	0.186	0.372	0.238
Yibin	0.157	0.152	0.176	0.200	0.205	0.164	0.146	0.171
Guang'an	0.189	0.286	0.179	0.163	0.272	0.246	0.342	0.240
Dazhou	0.176	0.156	0.178	0.183	0.199	0.138	0.209	0.177
Ya'an	0.131	0.156	0.126	0.161	0.201	0.129	0.282	0.169
Bazhong	0.217	0.136	0.125	0.129	0.114	0.100	0.211	0.148
Ziyang	0.159	0.174	0.168	0.185	0.168	0.106	0.193	0.164
Guiyang	0.423	0.407	0.429	0.516	0.544	0.659	0.803	0.540
Liupanshui	1.000	1.000	1.000	1.000	1.000	1.000	0.506	0.929
Zunyi	0.152	0.142	0.127	0.206	0.236	0.204	1.000	0.295
Anshun	0.176	0.143	0.162	0.263	0.154	0.142	0.446	0.212
Kunming	0.102	0.129	0.114	0.154	0.148	0.147	0.143	0.134
Qujing	0.389	0.174	0.184	0.182	0.194	0.091	0.195	0.202
Yuxi	0.273	0.146	0.135	0.154	0.137	0.076	0.317	0.177
Baoshan	0.082	0.062	0.064	0.064	0.060	0.037	0.181	0.078
Zhaotong	0.094	0.063	0.063	0.076	0.080	0.043	0.150	0.081
Lijiang	0.056	0.069	0.068	0.095	0.086	0.083	0.284	0.106
Simao	0.090	0.069	0.067	0.075	0.076	0.045	0.185	0.087
Lincang	0.376	0.034	0.041	0.038	0.039	0.024	0.125	0.097
Xi'an	0.163	0.369	0.316	0.297	0.316	0.327	0.285	0.296

(continued)

Table 9.6 (continued)

City	2008	2009	2010	2011	2012	2013	2014	Mean
Tongchuan	0.199	0.187	0.177	0.187	0.150	0.147	0.288	0.191
Baoji	0.106	0.106	0.102	0.109	0.099	0.109	0.209	0.120
Xianyang	0.094	0.145	0.141	0.193	0.178	0.181	0.211	0.163
Weinan	0.123	0.159	0.159	0.178	0.233	0.143	0.204	0.171
Yan'an	0.062	0.132	0.119	0.179	0.178	0.100	0.123	0.128
Hanzhong	0.070	0.096	0.102	0.132	0.120	0.108	0.125	0.108
Yulin	0.427	0.080	0.075	0.068	0.080	0.071	1.000	0.257
Ankang	0.380	0.198	0.175	0.201	0.163	0.109	0.283	0.216
Shangluo	0.056	0.027	0.251	0.067	0.053	0.061	0.121	0.091
Lanzhou	0.148	0.175	0.159	0.148	0.139	0.132	0.134	0.148
Jiayuguan	1.000	1.000	1.000	1.000	1.000	1.000	1.000	1.000
Jinchang	0.119	0.154	0.197	0.213	0.195	0.128	0.201	0.172
Baiyin	0.165	0.134	0.124	0.269	0.247	0.194	0.478	0.230
Tianshui	0.075	0.069	0.072	0.168	0.103	0.125	0.183	0.114
Wuwei	0.148	0.064	0.068	0.133	0.113	0.089	0.510	0.161
Zhangye	0.077	0.078	0.072	0.087	0.085	0.082	0.241	0.103
Pingliang	0.242	0.159	0.150	0.153	0.168	0.082	0.168	0.160
Jiuquan	0.082	0.075	0.133	0.159	0.142	0.109	0.410	0.159
Yinchuan	0.203	0.405	0.377	0.316	0.303	0.325	0.328	0.322
Zhongwei	0.181	0.152	0.144	0.130	0.118	0.092	0.254	0.153
Urumqi	0.449	0.523	0.483	0.404	0.419	0.384	0.416	0.440
Karamay	0.320	0.404	0.305	0.265	0.275	0.259	0.440	0.324
Mean	0.251	0.278	0.262	0.262	0.252	0.214	0.259	0.254

Chapter 10
Transport Energy and Climate Change

10.1 Global Transport Energy Consumption and Climate Change

10.1.1 Energy Consumption and Climate Change

10.1.1.1 Hazards of Climate Warming

Global warming is an irrefutable fact. The Fifth Assessment Report of the Intergovernmental Panel on Climate Change (IPCC) pointed out that the global average surface temperature increased by 0.85 °C from 1880 to 2012. According to the report on the Global Climate 2015–2019 released by the World Meteorological Organization (WMO) (2020), the global average temperature from 2015 to 2019 increased by 0.2 °C compared with that from 2011 to 2015, which is 1.1 °C higher than that before industrialization. This period is the warmest times in human history (Fig. 10.1).

The global climate is getting warmer with temperature increasing and ice melting, which will increase the possibility and intensity of extreme weather events and lead to huge negative impacts on natural ecosystems and society. The research of IPCC (2007) shows that climate change will bring great harm. First of all, global warming has caused glacier melt, sea-level rising, and the possibility of extreme weather events. Since 1978, Arctic sea ice has shrunk at a rate of 2.7% every 10 years. From 1961 to 2003, the global average sea level rose at an average annual rate of 1.8 mm. The biological system has also undergone great changes. The occurrence time of spring-specific phenomena has been advanced. The geographical distribution of animal and plant species is moving towards the poles and high-altitude areas, and some species are facing the threat of extinction. Secondly, climate warming has a great impact on the human living environment. For example, the planting of crops is advanced, and abnormal climate conditions greatly increase the uncertainty of agricultural production and destroy the living environment. The report also believes that if the temperature rises by more than 2.5 °C, all regions of the world may

P. Zhao and L. Zeng, *Transport Efficiency and Safety in China*, Population, Regional Development and Transport, https://doi.org/10.1007/978-981-99-1055-7_10

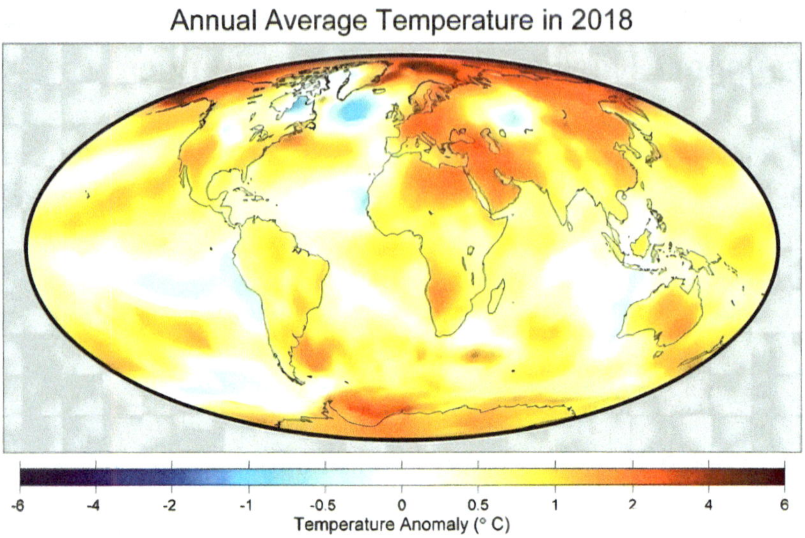

Fig. 10.1 The global average temperature in 2018 compared with the mean value of that from 1951 to 1980. *Data source* https://www.lieqinews.com/a/190208011223659.html

suffer irreversible destruction to the ecological environment, especially developing countries; if the temperature rises by 4 °C, it may cause irreversible damage to the global ecosystem and great losses to the global economy.

10.1.1.2 Impacts of GHGs on Climate Change

Long-lived GHGs caused by human activities are the main causes of global warming. GHGs covered by UNFCCC mainly include carbon dioxide (CO_2), methane (CH_4), nitrous oxide (N_2O), hydrofluorocarbons (HFCs), perfluorocarbons (PFCs) and sulfur hexafluoride (SF6). According to IPCC (2007), since the middle of the twentieth century, anthropogenic GHG emissions have more than a 90% chance of causing global warming.

Due to different radiative properties, various GHGs in the atmosphere have different life cycles. Therefore, they have different impacts on climate warming. Since CO_2 is the most important GHG, CO_2 equivalent is regarded as a standard unit to measure GHG emissions. At present, the global warming potential (GWP) is usually used to measure the impact of different GHGs on climate change. The GWP value of CO_2 is defined as 1, and the ratio of other GHGs to CO_2 is used as the GWP value of the GHG. GWP values of different GHGs are shown in Table 10.1.

The increase of the global average temperature is directly proportional to the increase of GHG emissions (WMO, 2020). In 2019, the content of GHGs in the global atmosphere reached a record high, in which the average concentration of CO_2

Table 10.1 The value of lifetime and GWP of different GHGs

	Lifetime (Year)	GWP		
		20 years	100 years	500 years
CO_2	200–450	1	1	1
CH4	12	62	23	7
N2O	114	275	296	156
HFCs	2600	9400	12,000	10,000
PFCs	50,000	3900	5700	8900
SF6	3200	15,100	22,200	32,400

Data source IPCC (2007)

reached 407.8 ppm (parts per million), which is 0.5% higher than that in 2017 and 147% in 1750. From 2017 to 2018, the growth rate of global CO_2 emissions exceeded the average level of the past 10 years.

Since 1990, the total warming effect of long-lived GHGs has increased by 43%, of which 82% is caused by CO_2. CH_4 is the second longest lived greenhouse gas. Its emissions also hit a record high in the atmosphere in 2018, reaching 1869 ppb (parts per billion), which is more than 259% of the pre-industrial level. The contribution of CH_4 to the warming effect is about 17%, and the growth rate of CH_4 emissions in 2018 is the highest in the past 10 years. In addition, the content of N_2O in the atmosphere reached 331.1 ppb in 2017, which is 123% of the pre-industrial level. Its contribution to the warming effect is about 6%, and it will also destroy the stratosphere and the ozone layer in the atmosphere.

10.1.1.3 Major Types of GHGs

CO_2 is the largest contributor to GHG emissions, followed by CH_4, N_2O, HFCs, PFCs and SF6. IPCC research shows that fossil energy consumption accounted for 74% of anthropogenic CO_2 emissions in 2014. More than 90% of GHG emissions are caused by human activities, and the main source is fossil energy consumption. In 2005, GHG emissions from fossil energy consumption reached 29.4 billion tons of CO_2 equivalent, accounting for 66.5% of anthropogenic GHG emissions. Among them, transport emissions were 6.3 billion tons of CO_2 equivalent (14.3%), electricity and heat emissions were 11 billion tons of CO_2 equivalent (24.9%), industrial emissions were 6.5 billion tons of CO_2 equivalent (14.7%), and the rest of fossil energy consumption emissions were 5.6 billion tons of CO_2 equivalent (12.6%). In 2005, 14.8 billion tons of CO_2 equivalent were emitted from non-fossil energy consumption, accounting for 33.5% of anthropogenic GHG emissions in that year. Among them, 1.9 billion tons of CO_2 equivalent (4.3%) were emitted from industrial processes, 5.4 billion tons (12.2%) from land use change and forestry, 6.1 billion tons (13.8%) from agriculture, and 1.4 billion tons (3.2%) from garbage, sewage and waste.

Since the Industrial Revolution, the rapid growth of fossil energy consumption has led to the rapid increase of CO_2 emissions. The fastest growth has been recorded since 1945. According to the IEA report, in 2017, CO_2 emissions from fossil energy consumption hit a record high. They increased by 235.4% compared with 1971 and reached 32.839 billion tons, with an average annual growth rate of 1.88% from 1917 to 2017. 2010 was the year with the fastest growth of CO_2 emissions from 1971 to 2017, with an increase of 5.98% compared with 2009 (Fig. 10.2). Since 2010, although the CO_2 emissions of global fossil energy consumption have increased year by year, the growth rate has shown a downward trend. In 2015 and 2016, the growth rate was negative, but it rebounded strongly in 2017 (IEA, 2020c). BP believes that the rebound of global carbon emissions in 2017 is mainly due to the increase of carbon emissions in countries such as China, India, Turkey and Iran, which offset the emission reduction of the United States, Ukraine, Mexico, the United Kingdom, and South Africa (BP, 2018).

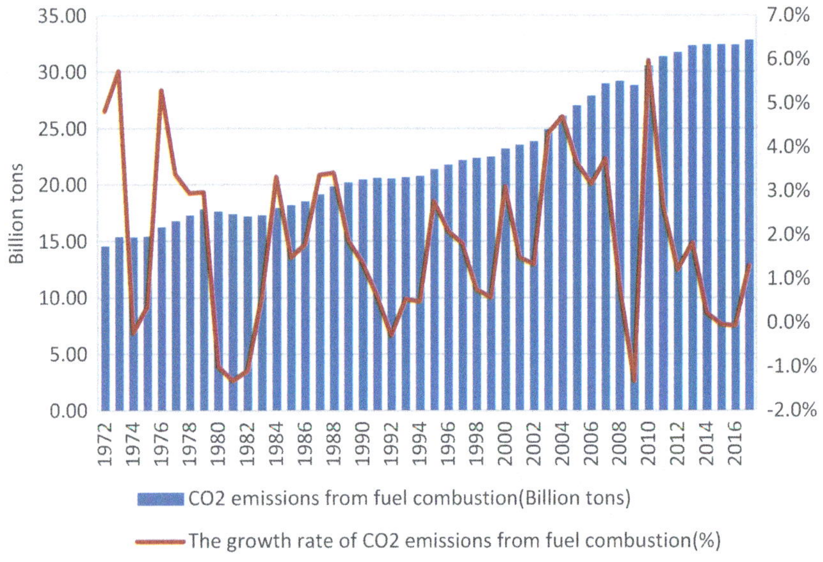

Fig. 10.2 The trend of global CO_2 emissions from fossil energy. *Data source* The authors, edited from Greenhouse Gas Emissions from Energy Highlights 2020edn (IEA, 2020c)

10.1.2 *Transport Energy Consumption*

10.1.2.1 The Transport Sector Has Become the Largest Energy Consuming Sector in the World

Worldwide, with the sustained and rapid development of the transport sector, the end-use energy consumption of the transport sector increased from 964 Mtoe in 1971 to 2809 Mtoe in 2018, with an average annual growth rate of 2.4% (Fig. 10.3). During the same period, as the growth rate of transport energy consumption was higher than that of total energy consumption (1.59% per year), the proportion of transport energy consumption in end-use energy consumption continued to increase. In the two periods 1996–2004 and 2017–2018, the transport sector exceeded the industry sector and became the largest energy consumer. In 2018, transport energy consumption accounted for 29.09% of the total end-use energy consumption (IEA, 2020b). According to the World Energy Balances Highlights 2020edn (IEA, 2020b), the global transport energy demand is expected to grow by more than 25% from 2017 to 2040. With the improvement of residential purchasing power, personal car ownership will continue to grow. Economic development and the increase of personal purchasing power will drive the growth of global trade in goods and services, which will in turn drive the energy demand of commercial transport (heavy-load, air, maritime and railway transport). In addition, the rapid growth of various economic activities and the size of the middle class, especially the middle class in emerging economies, will lead to the highest annual growth rate of aviation demand, which will reach 2.2% from 2017 to 2040 (Exxon Mobil Corporation, 2019).

The total energy consumption of the industrial sector was obviously restrained in the early 1980s and the whole 1990s. The energy demand of the industrial sector rebounded strongly in the 2000s, which was restrained again after 2010. The 2008 global financial crisis may have had an inhibitory effect on that.

Residential energy consumption has slowed down significantly since 1996. Since the 1970s, except for the early 1980s, the demand for transport energy has continued to rise. During the global economic recession in the early 1990s, the Southeast Asian financial crisis in 1997 or the global financial crisis in 2008, the transport energy demand was not suppressed. This shows that the demand for transport energy is relatively sustained and not easily affected by economic recessions. The trend of that is difficult to reverse even through technical progress and policy factors (Figs. 10.3 and 10.4).

Transport energy consumption is closely related to the level of economic development. The higher the per capita GDP, the higher the proportion of transport in energy consumption. When the per capita GDP reaches a high level, the amount of transport energy consumption tends to be stable. As Fig. 10.4 shows, globally, the proportion of transport in end-use energy consumption increased from 27.9% in 2000 to 29.09% in 2018. The main reason is that with the sustained economic growth of China, India and other developing countries, the per capita GDP, transport demand and the number of private cars in these main developing countries increased

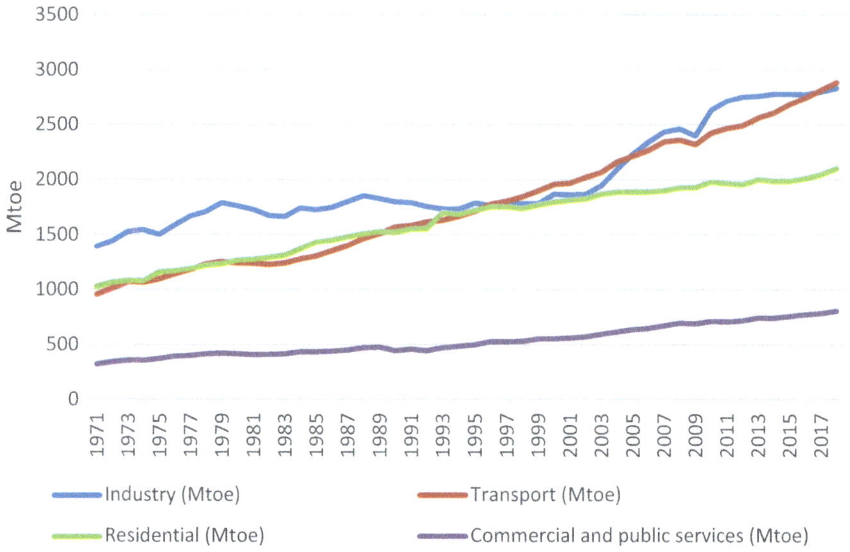

Fig. 10.3 The trend of global fossil energy consumption in major sectors from 1971 to 2018. *Data source* The authors, edited from World Energy Balances Highlights 2020edn (IEA, 2020b)

rapidly, resulting in the rapid expansion of transport energy consumption. During the same period, the average annual growth rate of world transport energy consumption was 2.17%. OECD countries had an average annual growth rate of 0.58%, and the proportion of transport in end-use energy consumption increased slightly from 31.7% to 33.7%; the average annual growth rate of transport energy consumption in non-OECD countries was 4.5%, and the proportion of transport in end-use energy consumption also increased from 17.3% to 20.8%.

10.1.2.2 Transport is the Main Driving Force of the Increase in Oil Consumption

According to the report from IEA (2020b), oil is the main transport energy. In 2018, the global transport energy consumption was about 2.792 billion Mtoe, of which oil accounted for 94.6%. Figure 10.6 shows that transport is the most important factor driving the growth of oil consumption. From 1973 to 2018, the average annual growth rate of global transport oil consumption was 2.24%, far higher than the average annual growth rate of total oil consumption of 1.31% in the same period. During the same period, transport oil consumption increased from 1.02 billion Mtoe to 2.641 billion Mtoe, and its proportion in the total oil consumption also increased from 44.7% to 65.2% (IEA, 2020b). BP also believes that oil is still the primary energy for transport. However, with the increasing use of alternative energy, especially natural gas and electricity, the proportion of oil in transport energy will show a downward trend. In

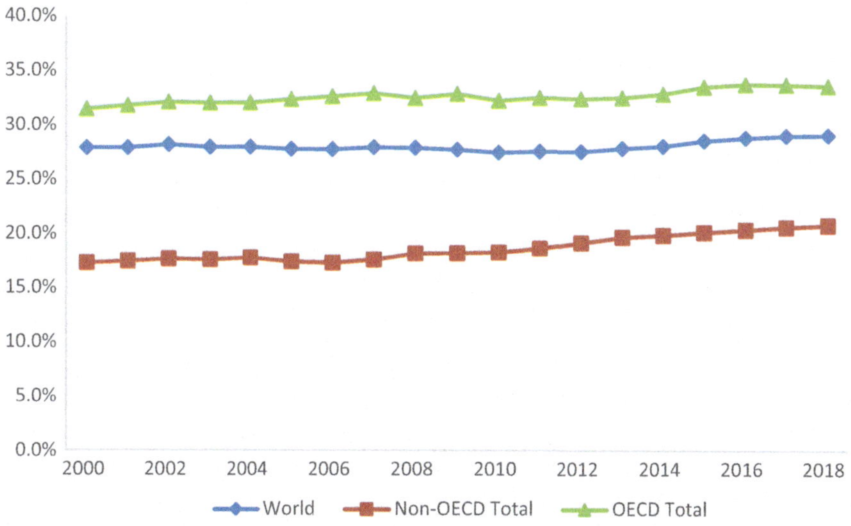

Fig. 10.4 The global share of fossil energy consumption of the transport sector in OECD and non-OECD countries (IEA, 2020b). *Data source* The authors, edited from World Energy Balances (IEA, 2020b)

2018, oil accounted for 94% of transport energy demand. Under the gradual transformation scenario, oil is expected to account for about 85% of that in 2040. According to the prediction of BP (2019), natural gas, electricity and biomass will contribute more than half of the total increment of transport energy demand, accounting for about 15% of the demand of the transport sector by 2040.

Transport oil consumption is still the main source of world oil demand growth. According to the report "Energy outlook for the world and China in 2050 (2019)" issued by the CNPC Economics & Technology Research Institute, the optimization

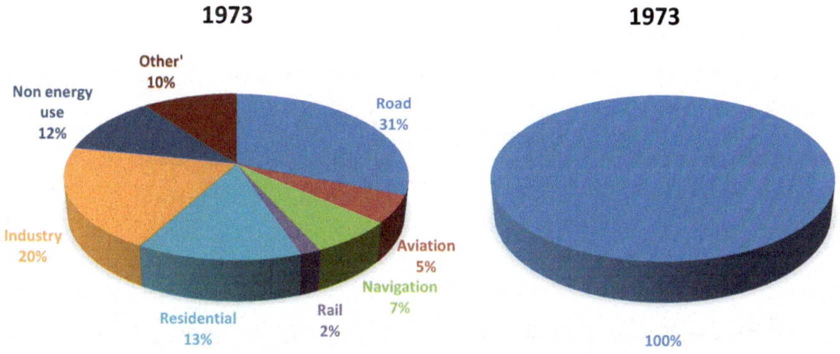

Fig. 10.5 Proportion of major sectors in end-use oil consumption in 1973 and 2018. *Data source* The authors, edited from World Energy Balances Highlights 2020edn (IEA, 2020a)

of the transport network and the growth of demand for transport activities are the main reasons for the rapid growth of transport energy consumption before 2030. The report predicts that oil demand in China will continue to grow due to the increase in transport oil and chemical raw materials before 2030, and it will reach a peak of about 700 million tons in 2030.

In recent years, the increase of transport oil consumption is mainly in non-OECD countries. From 2000 to 2018, the global transport oil consumption increased by 1.94% annually. During these years, the transport oil consumption in non-OECD countries increased by 2.17% annually, while that in OECD countries only increased from 1,115.16 million Mtoe to 117,574 million Mtoe, with an annual average growth rate of only 0.29% (Fig. 10.6). The rapid economic development and growth of vehicle ownership in non-OECD countries stimulated the demand for transport oil to a certain extent. Meanwhile, the new transport energy policies and the technical progress of transport energy utilization in non-OECD countries lagged far behind those of OECD countries. As a result, the proportion of transport oil consumption in non-OECD countries in the global transport oil consumption increased from 29.5% in 2000 to 39.7% in 2018. However, due to the great historical gap between OECD and non-OECD countries, the transport oil consumption in OECD countries was still higher than that in non-OECD countries, and the per capita transport oil consumption in OECD countries was several times higher than that in non-OECD countries. In 2017, the per capita transport oil consumption in OECD countries was 993 kg, 10 times of that in non-OECD countries (99.3 kg) (Fig. 10.7).

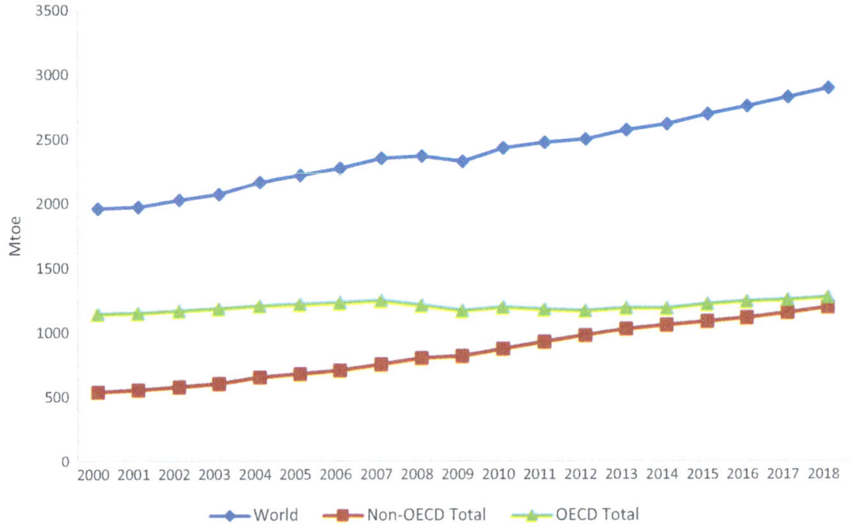

Fig. 10.6 The change trend of transport oil consumption in OECD and non-OECD countries from 2000 to 2018. *Data source* The authors, edited World Energy Balances Highlights 2020edn (IEA, 2020b)

10.1.2.3 Fossil Energy Consumption in the Transport Sector Varies Greatly by Country

The top five countries in transport energy consumption in 2018 are the United States, China, India, Brazil and Japan. The total fossil energy consumption in these countries reached 1.22 billion tons, accounting for 42.8% of the world. The consumption of transport energy in China reached 325 million tons of standard oil, accounting for 11.2% of the world, and the consumption of that in the United States reached 638 million tons, accounting for 22.1% of the world. In 2018, the transport fossil energy consumption of OECD countries was about 1.274 billion tons, accounting for 44.7% of the world.

It can be seen from Fig. 10.8 that the growth rate of transport fossil energy consumption in China, India, Japan and Brazil was basically the same in the 1990s. In the twenty-first century, with the rapid development of Chinese economy, the transport fossil energy consumption in China increased rapidly. In 2002, China overtook Japan as the world's second largest transport fossil energy consumer.

The transport fossil energy consumption in Japan peaked in 2020, which reached 88,303 ktoe, and then showed a slow downward trend. For achieving its commitments made under the Paris Agreement, Japan implemented a series of emission reduction measures in the field of transport, such as improving automobile performance in energy saving and emission reduction, addressing traffic jams, and providing convenient public transport services. Although the number of vehicles is growing, it has successfully restrained transport energy consumption.

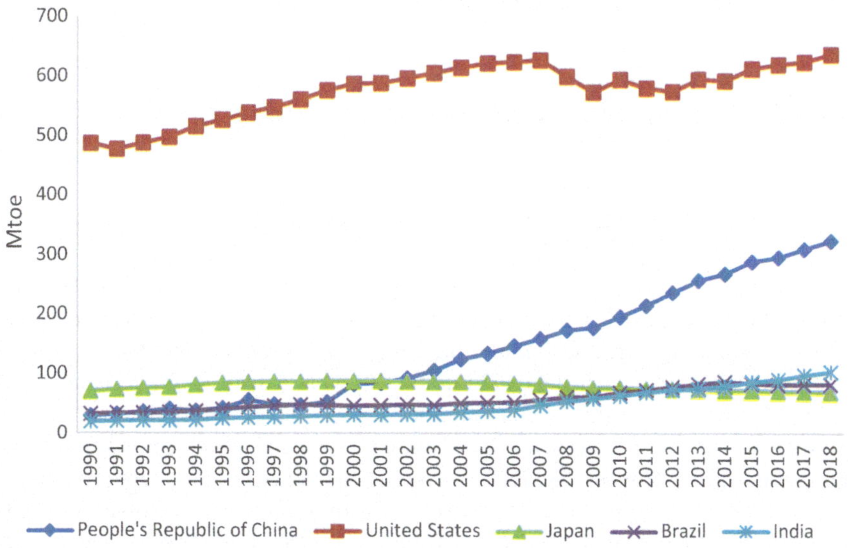

Fig. 10.7 Fossil energy consumption in major countries (IEA, 2020b). *Data source* The authors, edited from World Energy Balances Highlights 2020edn (IEA, 2020b)

From 1991 to 2007, transport energy consumption in the United States continued to grow rapidly. However, affected by the 2008 financial crisis, transport energy consumption declined in the United States from 2008 to 2009. As the economy recovered, transport energy consumption in the United States presented a fluctuating ascending trend after 2010.

Over the years, the world's per capita energy consumption of the transport sector has basically shown a downward trend, from 3512 kgoe in 1971 to 379 kgoe in 2018.

In recent years, although the per capita transport fossil energy consumption in developed countries has declined, it is still far higher than that in developing countries. In 2018, the per capita transport energy consumption in OECD countries was 966 kgoe, five times that of non-OECD countries (189 kgoe).

The gap among the top 10 countries in terms of per capita transport energy consumption is also very large. In 2018, the per capita transport energy consumption in China was 233 kgoe, slightly higher than the world average, while the per capita transport energy consumption in the United States was 1950 kgoe, eight times more than that of China. The United States has a high level of per capita transport energy consumption; the reasons are chiefly as follows: the United States has the most developed highway and civil aviation network in the world, and heavily depends on road transport; civil aviation is the main transport means to undertake long-distance passenger transport in the United States, while the proportion of railway passenger transport in total passenger transport is very low; the energy consumption per unit of freight and passenger volume in highway and air transport is much higher than that in other modes of transport. In addition, the United States has a vast territory and low gasoline prices. All these factors have caused the high level of per capita transport energy consumption in the United States.

10.1.3 Transport CO_2 Emissions and Climate Change

10.1.3.1 The Transport Sector is an Important Source of CO_2 Emissions from Fossil Energy Consumption

Transport CO_2 emissions mainly come from fossil energy consumption in the transport sector. From a global perspective, transport, as a key industry of energy consumption, is naturally a large CO_2 emitter. In 2017, CO_2 emissions from transport fossil energy consumption reached 8.039 billion tons, an increase of 33.2% compared with 6.6325 billion tons in 2007, accounting for 24.5% of the global fossil energy carbon emissions (Fig. 10.9). Transport oil consumption is the main source of transport CO_2 emissions. While global CO_2 emissions are in urgent need of reduction, CO_2 emissions of the transport sector are increasing year by year, which is a bigger cause of concern. The improvement of energy efficiency in transport cannot offset the increase of CO_2 emissions due to the increasing number of transport activities. For example, in the United States, after reaching the peak in 2005, transport-related CO_2 emissions first dropped, and then increased year by year after 2012. In 2016, the transport sector

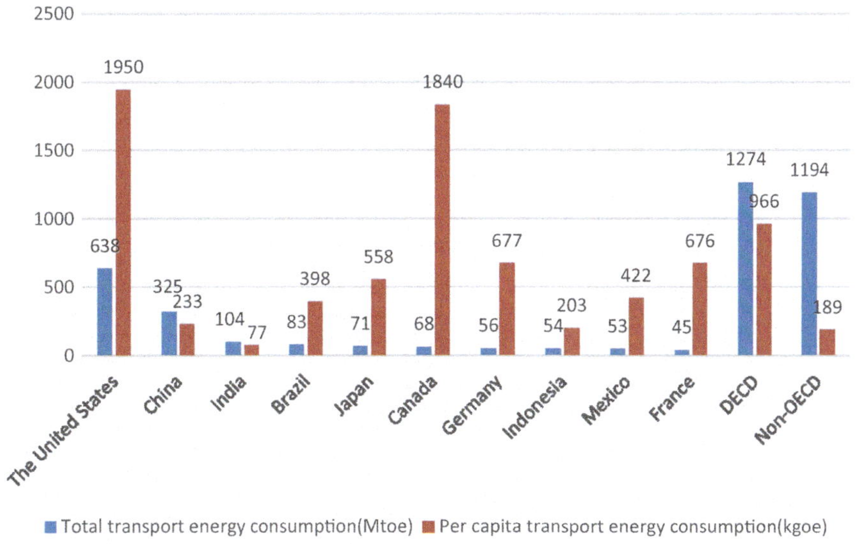

Fig. 10.8 Fossil energy consumption in the transport sector in main countries. *Data source* The authors, edited from World Energy Balances (IEA, 2020b) and World Development Indicators (WB, 2020)

surpassed the power industry for the first time and became the largest source of CO_2 emissions in the United States (Fig. 10.11).

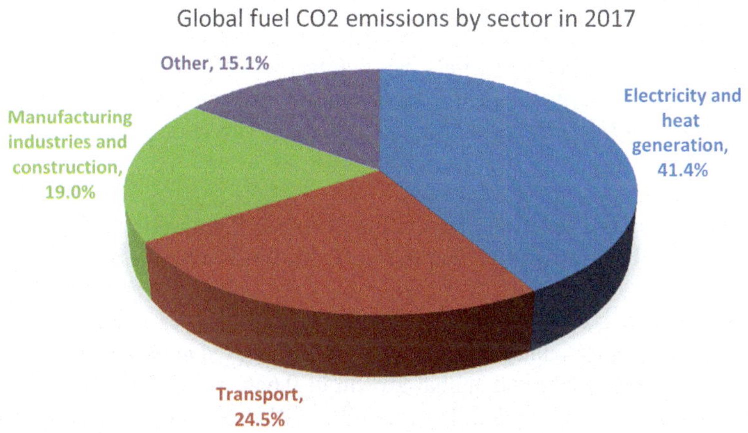

Fig. 10.9 Global CO_2 emissions by sector in 2017. *Data source* CO_2 emissions from fuel combustion 2019 edition (IEA, 2019)

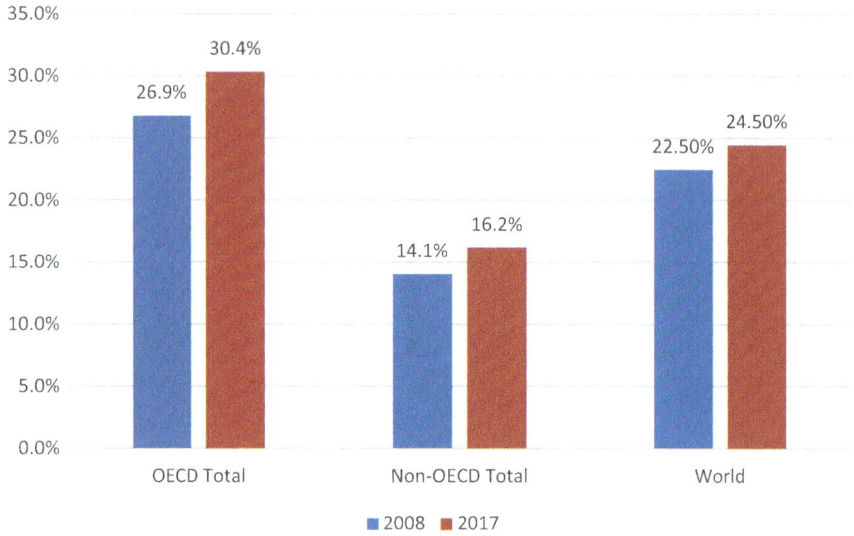

Fig. 10.10 The change trend of global CO_2 emissions by sector from 2002 to 2017. *Data source* CO_2 emissions from fuel combustion 2019 edition (IEA, 2019)

In the case of no significant changes in energy consumption structure, the proportion of transport CO_2 emissions in fossil energy CO_2 emissions is positively correlated with the level of economic development. The proportion of transport CO_2 emissions in OECD countries is much higher than that in non-OECD countries. From the perspective of change trend, transport CO_2 emissions accounted for 30.4% of all fossil energy CO_2 emissions in OECD countries in 2017, compared with 27.3% in 2007, an increase of 3.1 percentage points; while transport CO_2 emissions accounted for 16.2% of all fossil energy CO_2 emissions in non-OECD countries in 2017, compared with 15.4% in 2007, an increase of 0.8 percentage points (Fig. 10.10).

10.1.3.2 There is a Positive Correlation Between Per Capita GDP and Per Capita CO_2 Emissions from the Transport Sector

There is a positive correlation between per capita GDP and per capita transport CO_2 emissions. An average annual increase of 1.5% is observed in global per capita GDP (2007–2017), compared to an annual growth of 0.63% in per capita transport CO_2 emissions over the same period. From 2007 to 2017, the per capita transport CO_2 emissions grew from 994 kg/person to 1,069 kg/person, an increase of 7.6%.

From the perspective of countries, due to the unbalanced development of the world economy, the per capita transport CO_2 emissions vary greatly from country to country. This is mainly reflected in two aspects (Fig. 10.11): the first is the absolute difference of the per capita transport CO_2 emissions. In 2007, the per capita transport CO_2 emissions in OECD countries were 2716 kg, compared to only 521 kg in non-OECD

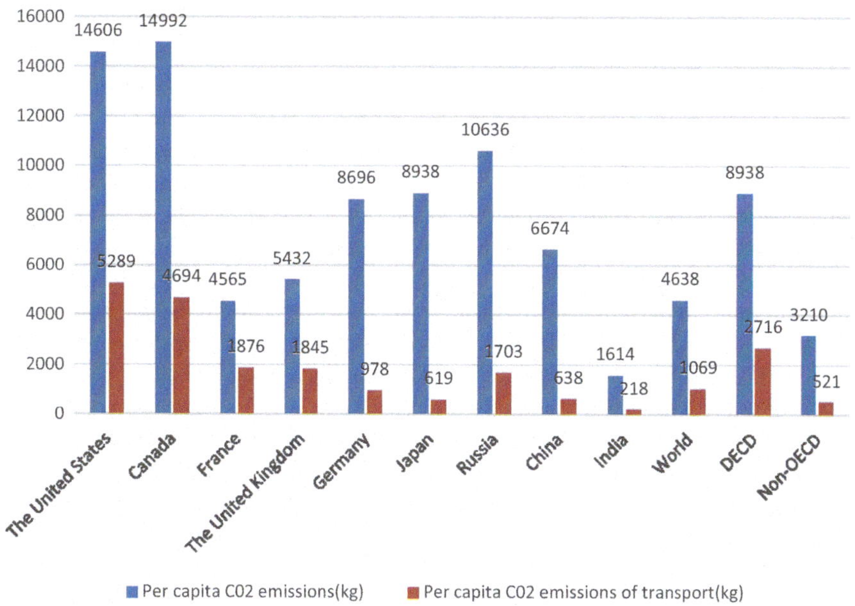

Fig. 10.11 Per capita transport CO_2 emissions in main countries in 2017. *Data source* World Energy Balances (IEA, 2020b)

countries, with a difference of nearly 5.2 times. The United States remained the largest contributor to the total transport CO_2 emissions. The per capita transport CO_2 emissions in the United States were 5289 kg, which is 8.3 times that of China (638 kg) and 24.2 times that of India (218 kg). Secondly, the proportion of transport CO_2 emissions in total CO_2 emissions in OECD countries was 30.4% in 2017, compared to only 16.2% in non-OECD countries. Among OECD countries, the proportion of transport CO_2 emissions in total CO_2 emissions in the United States, Canada, France, the United Kingdom and Germany was 36.2%, 31.3%, 41.9%, 34% and 22.7% respectively, while that in China and India was 9.6% and 14.4% respectively.

10.1.3.3 The Relationship Among Transport Energy Consumption, Transport CO_2 Emissions, and Climate Change

There is a long-standing debate about whether the increase of CO_2 emissions is the cause of global warming. In this section, the relationship between the two is discussed from the perspective of practical statistics.

Comparing the change curve of the global average temperature from 1960 to 2014 with the total CO_2 emissions of the power and heat industry, the transport sector, and the manufacturing and construction industry during the same period, we can see that the change trend of the global average temperature was consistent with that of the power and heat industry and the transport sector. There is a clear correlation between carbon emissions in the two major sectors and global warming. After 1989, the transport sector surpassed the manufacturing and construction industry and became the second largest carbon emission source sector after the power and heat industry (Fig. 10.12).

By calculating the correlation coefficient between the global average temperature and CO_2 emissions from 1960 to 2014, we have found a significant positive correlation between them. The correlation coefficient was 0.8847. The result of linear regression analysis was significant with a probability of 0.000 (Table 10.2).

The global average temperature is regarded as the dependent variable and global CO_2 emissions as the independent variable when we perform simple linear regression analysis. The regression coefficient of the model was 0.0313, with a significance level of 0.01, which shows that CO_2 emissions have a significant impact on the global average temperature, and that the global average temperature may increase by 0.0313 °C when global CO_2 emissions increase by 1 billion tons. In 1960, the global CO_2 emissions from transport were 1.8 billion tons, which rose to 7.39 billion tons in 2014. In the same period, the global average temperature increased 0.72 °C, and CO_2 emissions from transport increased by 5.59 billion tons, indicating that the transport sector contributed 0.175 °C to the global temperature rise.

BP (2019) predicts that the global CO_2 emissions of the transport sector will reach 9.4 billion tons in 2040 in the evolving transition scenario, an increase of 1.1 billion tons compared with 2017, and that the global temperature will rise by 0.034 °C, which will lead to unimaginable consequences (Table 10.2).

Most results have shown that GHGs such as CO_2 were the major cause of the imbalance of the global climate mechanism, while energy consumption was the

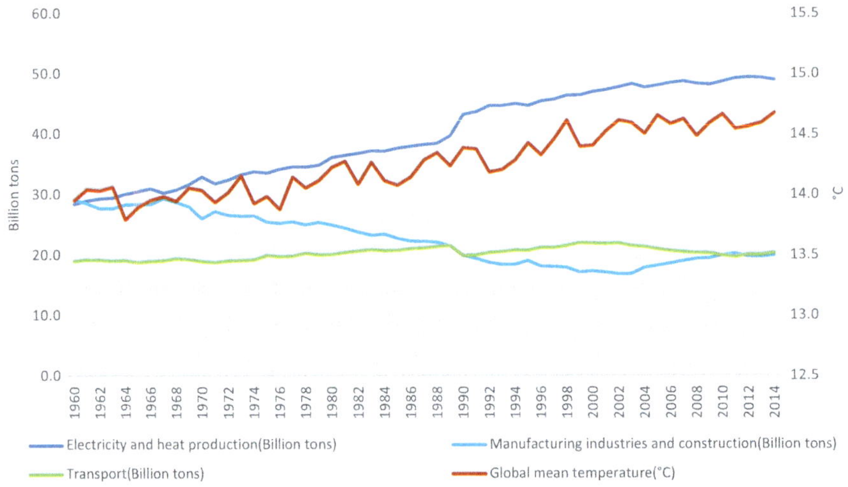

Fig. 10.12 The change trends of global average temperature and CO_2 emissions of different sectors from 1960 to 2014. *Data source* World Energy Balances (IEA, 2020n) and The World Development Indicators (The World Band, 2021)

Table 10.2 Results of Pearson's correlation coefficient

	Industry	Transport	Residential life	Commerce and public services	Other final consumption
Pearson Coef	0.8086 (0.0000)	0.9238 (0.0000)	0.9194 (0.0000)	0.9030 (0.0000)	0.9023 (0.0000)

primary factor resulting in the increase of global CO_2 emissions. The correlation coefficients between global average temperatures and global energy consumption (unit: million tons of oil equivalent) of industry, transport, residential life, commerce and public services, and other industries from 1917 to 2018 are 0.8086 ($P = 0.0000$), 0.9238 ($P = 0.0000$), 0.9194 ($P = 0.0000$), 0.9030 ($P = 0.0000$) and 0.9023 ($P = 0.0000$), respectively. It can be seen that the increase of global energy consumption in industry, transport, residential life, commerce and public services, and other industries has a very significant impact on the global average temperature rise, and the fitting degree of the model is almost 100%. At the same time, it can be seen that the correlation coefficients of the transport sector is the highest. It means that energy consumption of the transport sector has the greatest impact on CO_2 emissions, which confirms the excessive growth of transport energy consumption is directly related to global warming in recent years.

10.1.4 New Trends

10.1.4.1 Prediction of Global Transport Energy

IEA: Global Disposable Energy Consumption of the Transport Sector in 2040 Can Be Reduced by 8.7% Compared with that in 2018 in the Sustainable Development Scenario

The latest IEA outlook presents three scenarios: the current policy scenario, the stated policies scenario and the sustainable development scenario. In the current scenario, governments continue with their current energy policies and do not make a big adjustment to the policies. Of course, IEA believes that energy policy cannot remain unchanged for a long time. Therefore, in 2010, IEA put forward the concept of the new policies scenario that the current energy policies would be adjusted moderately. In the World Energy Outlook 2019 (IEA, 2019b), the new scenario is renamed the Stated Policies scenario. In contrast, the stated policies scenario contains the new policy choice and the object.

According to the IEA World Energy Outlook 2019, in 2030, transport energy consumption will reach 3327 Mtoe in the stated policies scenario and 2956 Mtoe in the sustainable development scenario, which are 16.2% and 3.2% higher than that in 2018 (2863 Mtoe). In 2040, transport energy consumption will be 2606 Mtoe in

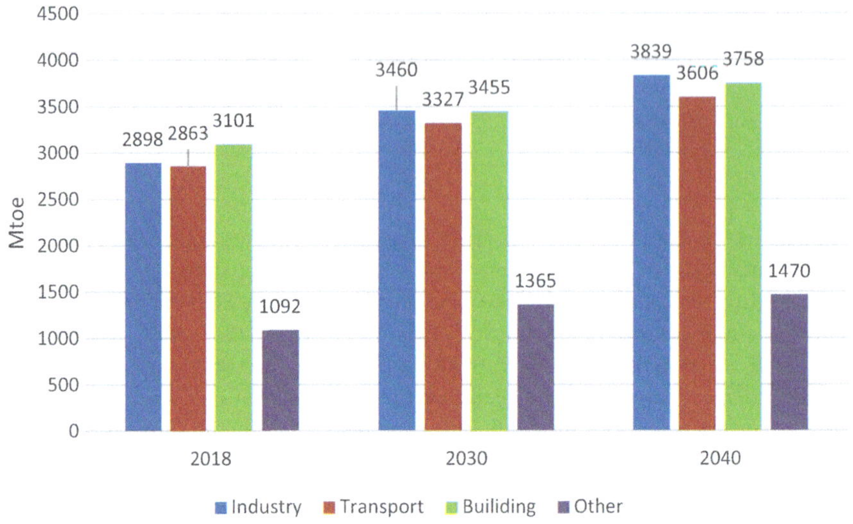

Fig. 10.13 Prediction of global fossil energy consumption of the transport sector in the stated policies scenario. *Data source* The authors, edited from World Energy Outlook 2019 (IEA, 2019b)

the stated policies scenario and 2615 Mtoe in the sustainable development scenario, which are 9% and 8.7% lower than that in 2018 (2863 Mtoe).

IEA (2019b) believes that the reason for the decrease of 8.7% in the sustainable development scenario compared with 2018 lies in higher transport energy efficiency and the reliance of half of the world's cars on electricity (Figs. 10.13 and 10.14).

BP: Compared with the Evolving Transition Scenario, Transport Energy Consumption Will Be Reduced by 227 Mtoe from 2017 to 2040 in the Lower-Carbon Scenario

The BP World Energy Outlook 2019 considers two scenarios: the evolving transition scenario and the lower-carbon scenario. The former assumes that the policy, technology and social preferences of energy utilization will evolve in the same way and at the same pace as in the recent past. The latter integrates all the analyses in the report, and combines the policy measures of the low carbon scenario in industry, construction, transport and power sectors.

According to the BP World Energy Outlook 2019, in 2040, transport energy consumption will be 3521 Mtoe in the evolving transition scenario and 3294 Mtoe in the lower-carbon scenario, increasing by 25% and 16.9%, respectively, compared with 2017 (2817 Mtoe), accounting for 20% of the energy consumption of the whole industry. From 2017 to 2040, the annual growth rate of transport energy consumption will be 1% in the evolving transition scenario and 0.7% in the lower-carbon scenario.

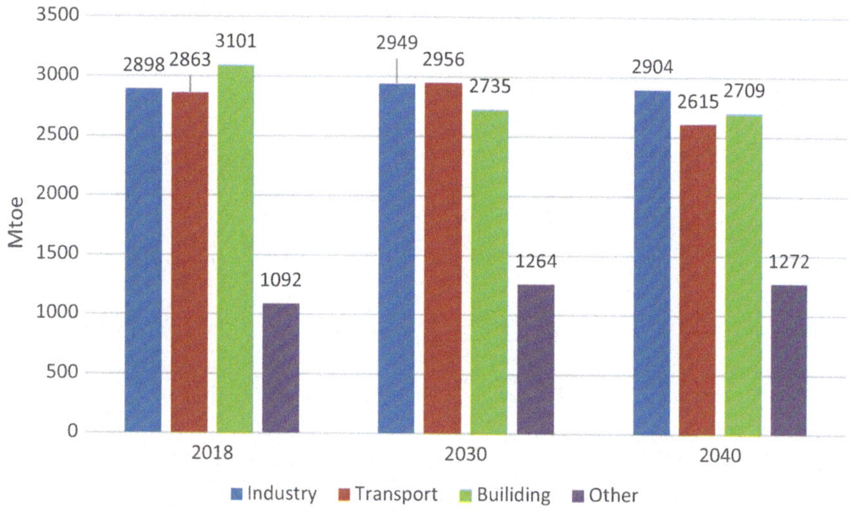

Fig. 10.14 Prediction of global fossil energy consumption of the transport sector in the sustainable development scenario. *Data source* The authors, edited from IEA World Energy Outlook 2019 (IEA, 2019b)

BP (2019) believes that the demand for transport services will almost double from 2017 to 2040 in the evolving transition, but the increase rate of transport energy consumption will only be 25% from 2017 to 2040 due to the rapid improvement of engine efficiency.

The energy consumption growth rate of different transport modes is also affected by differences in energy efficiency improvement. In major global auto markets, the efficiency of diesel locomotives will increase by nearly 50% in 2040 compared with 2017; the energy efficiency of trucks will also increase significantly; therefore, there will be a sharp slowdown in the growth of the demand for energy consumption of highway transport. On the contrary, the energy efficiency of both aviation and shipping is relatively slow to improve. The two transport modes will contribute 50% of energy demand growth of the transport sector from 2031 to 2040, although they currently account for only 20% of the total energy consumption of the transport sector.

BP (2019) believes that the transport energy efficiency of internal combustion engine vehicles, trucks, shipping and aviation can be increased by 54%, 37%, 44% and 32% in 2040 compared with 2017 in the lower-carbon scenario, which is 228 Mtoe less than transport energy consumption (3521 Mtoe) in the evolving transition scenario (Fig. 10.15).

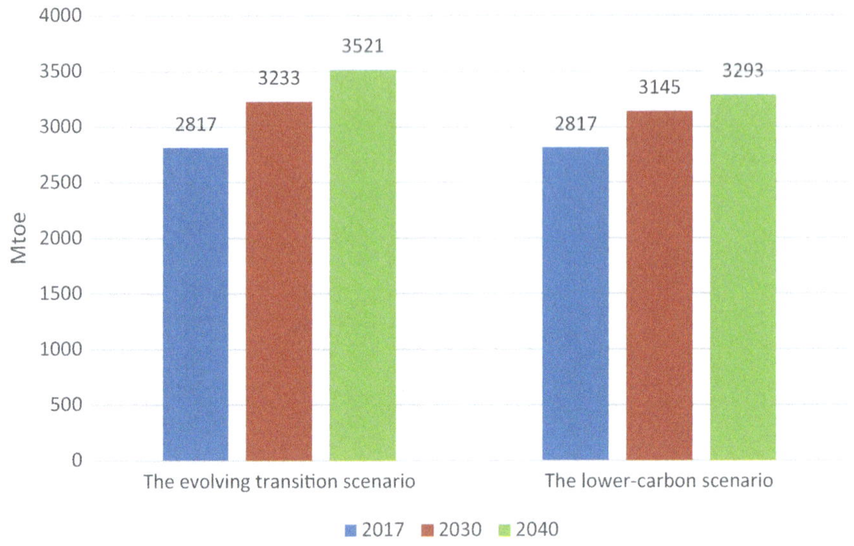

Fig. 10.15 Prediction of global fossil energy consumption of the transport sector in the two policy scenarios. *Data source* The authors, edited from BP World Energy Outlook 2019 (BP, 2019)

10.1.4.2 Prediction of Global CO_2 Emissions

IEA: CO_2 Emissions of the Transport Sector Can Be Reduced by 38% from 2017 to 2040 in the Sustainable Development Scenario, but It Will Become the Largest CO_2 Emitter

According to the IEA World Energy Outlook 2018, the transport sector will become the largest CO_2 emitter by 2040. According to the report, CO_2 emissions of the transport sector will be 7932 Mtoe in 2025, 7326 Mtoe in 2030, 6373 Mtoe in 2035 and 5563 Mtoe in 2040 in the sustainable development scenario, which will be reduced by 54 Mtoe (0.7%) in 2025, 660 Mtoe (8.3%) in 2030, 1613 Mtoe (22.0%) in 2035, and 2423 Mtoe (38.0%) in 2040 compared with 2017 (Fig. 10.16).

BP: In the Lower-Carbon Transport Scenario, CO_2 Emissions of Transport Will Be 8.1 Gt in 2040, 2% (0.2 Gt) Lower Than 2017

BP (2019) believes that CO_2 emissions of the transport sector will continue to grow in the evolving transition scenario, despite the significant improvement of vehicle efficiency and automobile electrification level. BP (2019) forecast that CO_2 emissions of transport will increase by 1.1 Gt in 2040 compared with 2017 in the evolving transition (ET) scenario, reaching 9.4 Gt.

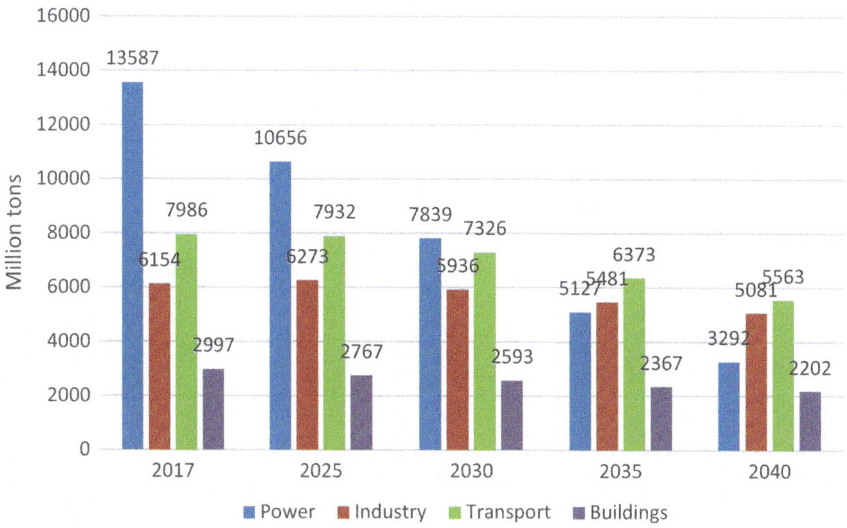

Fig. 10.16 Prediction of global CO_2 emissions from fossil energy consumption of different sectors in the sustainable development scenario. *Data source* The authors, edited from World Energy Outlook 2018 (IEA, 2018)

The alternative 'lower-carbon transport' (LCT) scenario includes a large number of measures designed to reduce carbon emissions in the transport sector, including:

(1) Further tightening vehicle efficiency standards, so that the average energy efficiency of internal-combustion-engine cars in 2040 will be around 55% more efficient than it is today; energy efficiency improvements in new trucks and ships also accelerate.
(2) Promoting transport electrification, including bans on the sales of all internal-combustion engine cars in most parts of the OECD and China by 2040 or before 2040; half of global sales of new trucks and buses are driven by electrical or hydrogen energy by 2040.
(3) Increasing penetration of shared mobility services, including more consumer-friendly 'mini-buses', increasing passenger transport driven by electrical energy.

Increasing the share of biofuels in road transport in the OECD and China to 20% by 2040, while that in the rest of the world will be 10%; similarly in aviation, increasing the share of biofuels in jet fuel to 20% in the developed world by 2040; car scrappage schemes which reduce the typical lifespan of a car from around 12 years to 8 years by 2040, improving the average efficiency of the global car parc and the pace of electrification.

BP (2019) believes that with these measures, the total CO_2 emissions in the transport sector will be 8.1 Gt in the low carbon transport scenario, which is 2% (0.2 Gt) lower than the level in 2017 (Fig. 10.17).

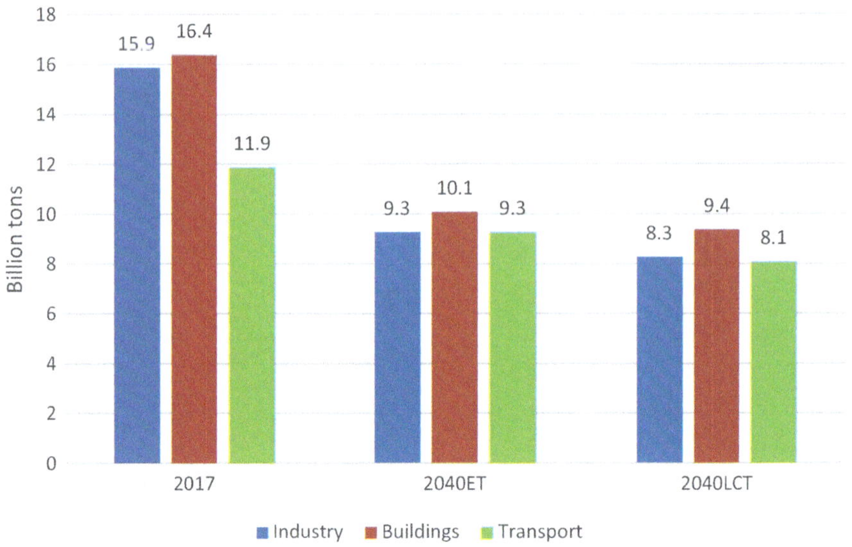

Fig. 10.17 Prediction of global CO_2 emissions from fossil energy consumption of different sectors in the two scenarios. *Data source* The authors, edited from BP World Energy Outlook 2018 (BP, 2018)

10.2 Transport Energy Consumption and CO_2 Emissions in China

10.2.1 Transport Energy Consumption in China

As a service industry, the transport sector is the carrier of freight and passenger flows in social and economic activities, and plays an important role in national economic and social development. The transport sector is the second largest energy consumption sector in China after the industry sector, and is one of the important sources of CO_2 emissions. At present, there is a lack of statistical data relating to energy consumption of motorcycles and low-speed vehicles (agricultural transport vehicles) in the statistical work on transport energy in China. These vehicles' energy consumption is large and grows fast. Therefore, the actual energy consumption of transport activities is larger than the data in China Energy Statistical Yearbooks. With the increasing demand for energy in the transport sector and the severe energy situation in China, effective measures must be taken to strengthen energy conservation in order to maintain the sustainable development of the economy and environment in China. In the statistical yearbook of China, transport, warehousing and postal services are taken as a whole when collecting statistics; therefore, transport, warehousing and postal services are regarded as the transport sector in this study.

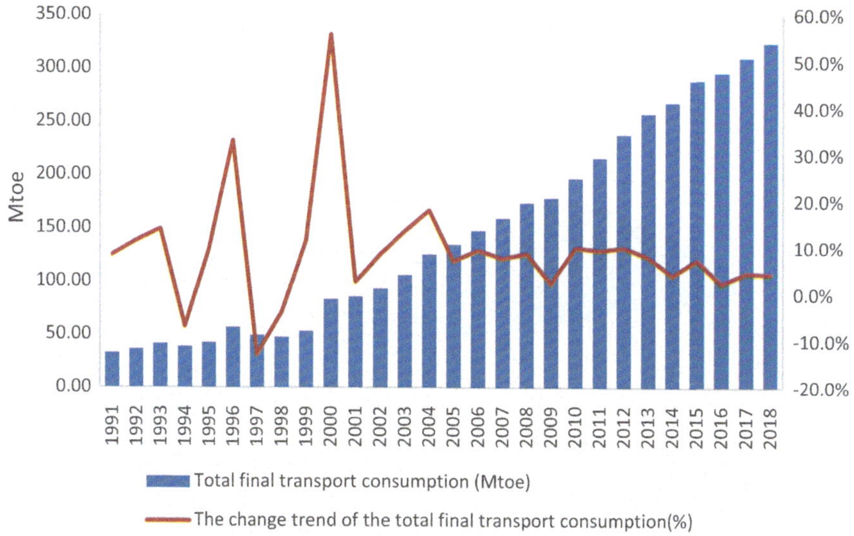

Fig. 10.18 The change trend of the total final transport consumption of China from 1990 to 2018. *Data source* The authors, edited from World Energy Balances Highlights 2020edn (IEA, 2020b)

10.2.1.1 Overall Characteristics of Transport Energy Consumption

The Growth Rate of Transport Energy Consumption Fluctuated Greatly at the End of 1990, and It Showed a Downward Trend After Entering the New Century

The end-use energy consumption of the transport sector in China is shown in Fig. 10.18. According to the IEA, the end-use energy consumption increased from 30.2 Ktoe to 83.64 Mtoe from 1990 to 2016, an average annual growth of 2044.4 Ktoe. The average growth rate was about 4% from 1990 to 2016. There was a large fluctuation in the growth rate of transport energy from 1991 to 2013, which reached 56% in 2000. The growth rate showed a slight fluctuating downward trend after 2003, especially in 2009, which rapidly shrunk to 2.50% due to the financial crisis. Following the macroeconomic adjustment in mainland China after 2009, it showed robust growth. During the 12th Five-Year Plan period (2010–2015), the average annual growth rate was more than 8%, but the growth rate decreased year by year after 2015.

Gasoline, Kerosene, and Diesel Oil Are the Main Types of Transport Energy

There are many types of energy, but end-use energy consumption varies greatly. According to the China Energy Statistical Yearbook (2017), the end-use energy consumption of China's transport sector in 2016 was still dominated by coal and oil products. More specifically, the end-use consumption of gasoline, kerosene and

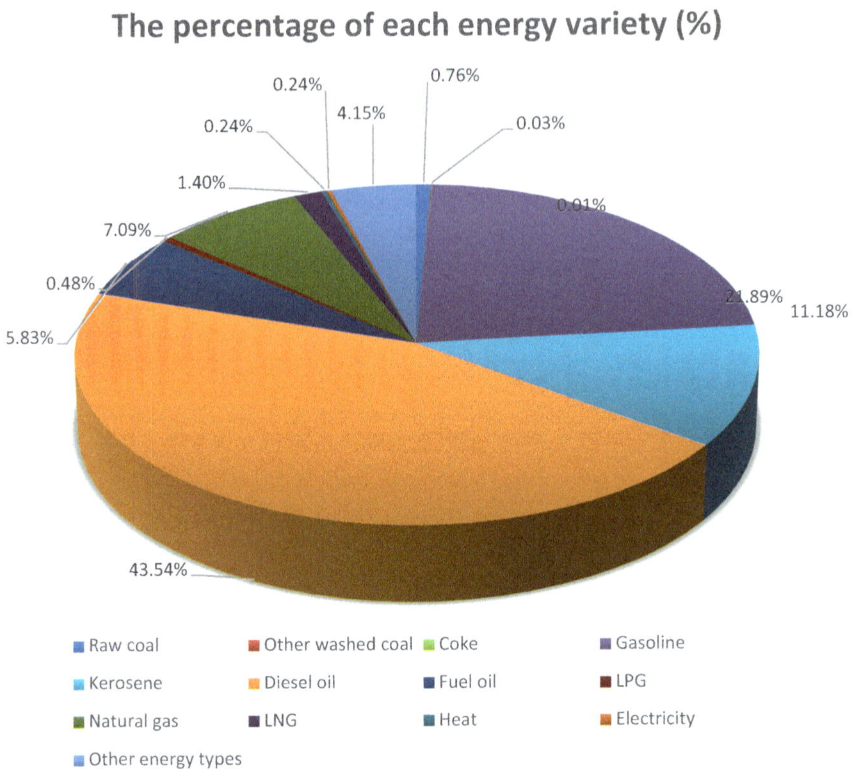

Fig. 10.19 The percentage of each energy variety in transport energy consumption in China in 2016. *Data source* The authors, edited from the China Energy Statistical Yearbook (2017)

diesel oil was relatively large. The proportion of these three types in the total energy consumption of the transport sector is 22%, 11% and 44%, respectively (Table 10.3; Fig. 10.19). They all cause serious pollution. In fact, the vigorous development of the automobile industry has brought about the influx of various means of transport into the market, and national economic development and rising incomes have led to the improvement of residents' purchasing power, resulting in an increase in the energy demand of private cars and other means of transport (Fig. 10.20 and Table 10.3).

The Proportion of Transport Energy Consumption in Total Energy Consumption Rose Rapidly in the Late 1990s and Went Steadily up in the Twenty-First Century

The experience of developed countries shows that there is a corresponding relationship between the level of economic development and the proportion of transport energy consumption in total energy consumption. Generally speaking, the more

Table 10.3 Specific types of transport energy in China in 2016

Energy varieties	Total final consumption (10,000 ktoe)	The percentage of each energy variety (%)
Raw coal	282.91	0.76
Other washed coal	12.34	0.03
Coke	3.12	0.01
Gasoline	8109.1	21.89
Kerosene	4141.9	11.18
Diesel oil	16,127.89	43.54
Fuel oil	2159.16	5.83
LPG	176.53	0.48
Natural gas	2625.62	7.09
LNG	518.74	1.40
Heat	90.06	0.24
Electricity	1538.08	0.24
Other energy types	1253.88	4.15

Data source The authors, edited from China Energy Statistical Yearbook (2017)

developed the economy, the bigger the share of transport energy consumption to total energy consumption. IEA data show that the total transport energy consumption in China slowly increased with fluctuations in the 1990s, but its proportion in total energy consumption increased rapidly. In 1990, the proportion of transport energy consumption in China's total energy consumption was 4.6%, while that increased to 10.7% in 2018, reaching 83,640.8 ktoe (Fig. 10.20).

In the new century, the total transport energy consumption in China shows a rapid growth trend, and its proportion in total energy consumption also shows a steady upward trend. In 2018, its proportion in total energy consumption reached 15.8%, a historic high.

China today is in a period of rapid economic growth, industrialization, and urbanization, with huge demand for transport, which makes transport energy consumption increase year by year, especially oil consumption. The transport sector will inevitably become one of the fastest-growing sectors of energy consumption in China, and the proportion of transport energy consumption in total energy consumption will continue to rise.

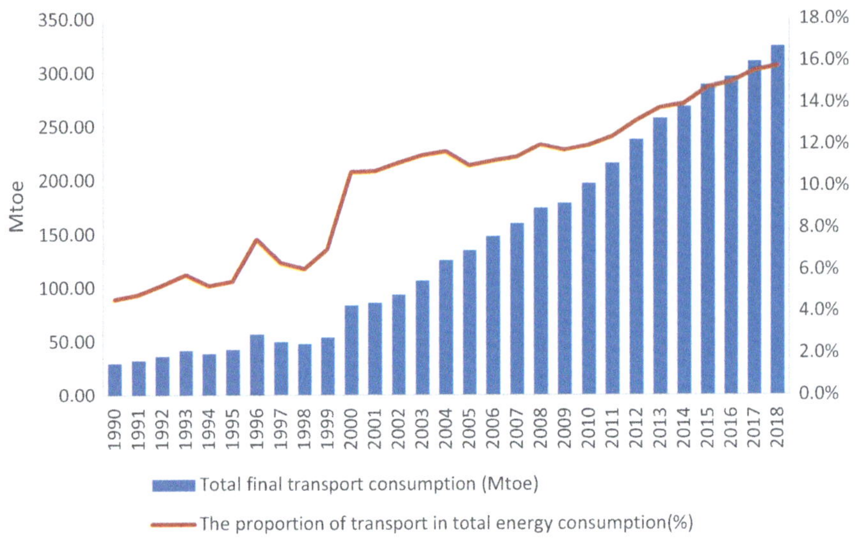

Fig. 10.20 The proportion of transport energy consumption in total energy consumption in China from 1990 to 2018. *Data source* The authors, edited World Energy Balances Highlights 2020edn (IEA, 2020b)

The Growth Rate of Per Capita GDP is Slightly Higher Than that of Per Capita Transport Energy Consumption, and the Utilization Efficiency of Transport Energy Shows an Improvement Trend

Per capita GDP can reflect the level of economic development, and per capita transport energy consumption can reflect the energy consumption intensity of a country's transport sector. Generally speaking, the higher the level of per capita GDP, the greater the per capita transport energy consumption. There is a positive correlation between per capita GDP and per capita transport energy consumption. From 2000 to 2018, China's socio-economic development was in a phase of rapid growth, and the real per capita GDP in China grew rapidly, rising from RMB 7910 in 2000 to RMB 34,930 in 2018. According to the data of China's transport energy provided by IEA and the Chinese population provided by National Bureau of Statistics of China, the per capita transport energy consumption increased from 0.092 TSC to 0.66 TSC, an increase of 3.5 times; clearly, the growth rate of per capita GDP is slightly higher than that of per capita transport energy consumption, which indicates that transport energy efficiency has been improved to a certain extent in China (Fig. 10.21).

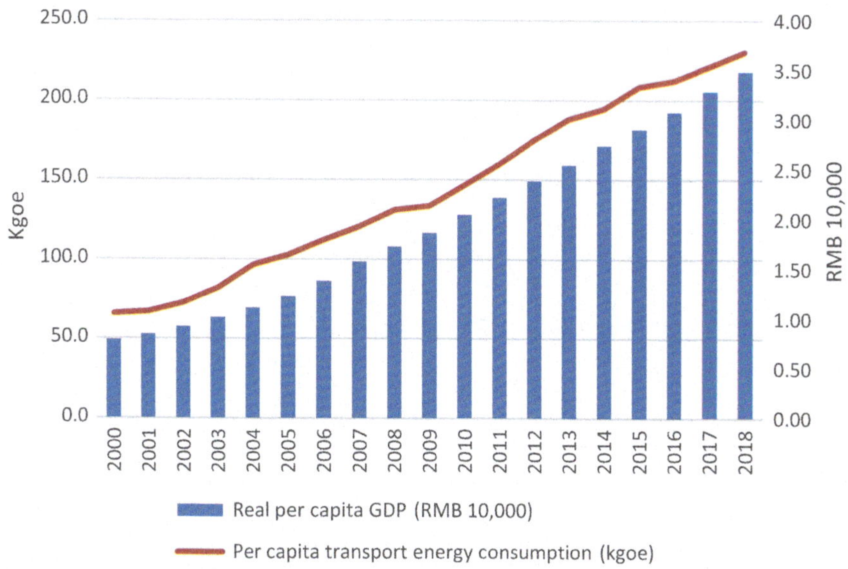

Fig. 10.21 Per capita transport energy consumption from 2009 to 2016. *Data source* The authors, edited from World Energy Balances Highlights 2020edn (IEA, 2020b); China Statistical Yearbook (NBSC, 2021)

10.2.1.2 Provincial Characteristics of Transport Energy Consumption

The Transport Energy Consumption in Eastern and Central Areas is Higher Than that of Western Areas

According to energy consumption characteristics, China's 30 provinces can be basically divided into three groups, namely provinces of high energy consumption, provinces of medium energy consumption and provinces of low energy consumption. The provinces with high energy consumption include Guangdong, Shanghai, Jiangsu, Shandong, Liaoning, Hubei, Sichuan, Henan and Hunan, whose annual transport energy consumption is 16 Mtoe. Except Sichuan, all other provinces are located in eastern coastal or central regions. Beijing, Hebei, Fujian, Yunnan, Shanxi, Heilongjiang, Anhui, Xinjiang, Guangxi and Shaanxi are representative provinces of medium energy consumption. The energy consumption of the transport sector in these provinces is above 9.1 Mtoe (Figs. 10.23 and 10.24). Except Beijing and Fujian, these provinces are located in central or western regions. The energy consumption of the transport sector in other provinces is at a low level. Moreover, most of these provinces are located in the western region, with a small population base, resulting in a low demand for transport energy.

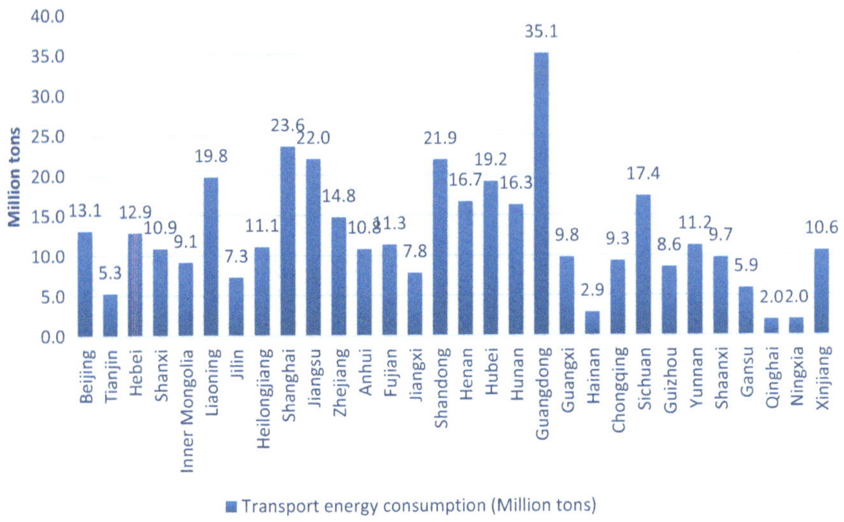

Fig. 10.22 Transport energy consumption in China in 2016. *Data source* The authors, edited from provincial statistical yearbooks of China

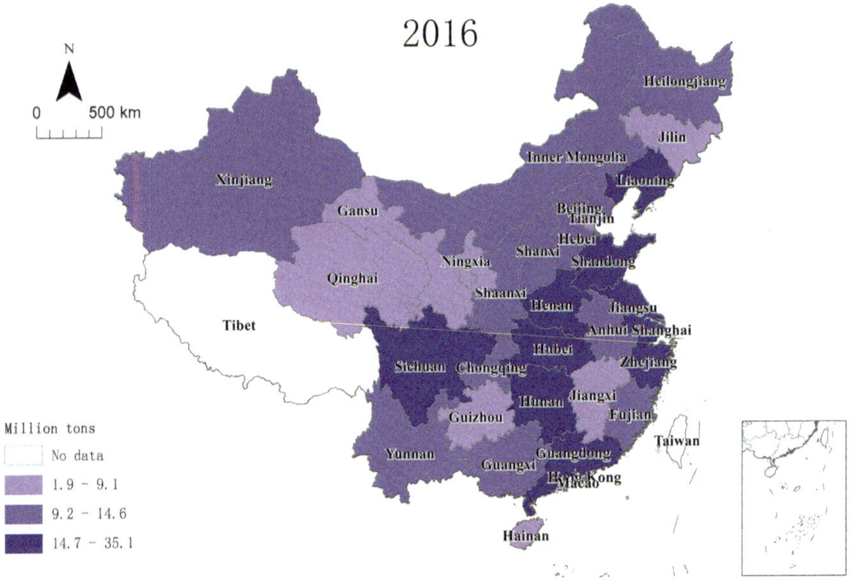

Fig. 10.23 Transport energy consumption in China in 2016 *Data source* Provincial statistical yearbooks of China

Table 10.4 The Gini coefficient of per capita transport energy consumption in Chinese provinces from 2009 to 2016

Year	Standard deviation	Gini coefficient
2009	160.21	0.253
2010	163.08	0.282
2011	156.46	0.287
2012	160.48	0.305
2013	139.51	0.346
2014	138.47	0.342
2015	144.64	0.358
2016	153.29	0.372

Data source The authors, edited from provincial statistical yearbooks of China

The Difference in Per Capita Transport Energy Consumption in Different Provinces is Widening

The Gini coefficient is an international index to measure the income gap of residents in a country or region, and was first proposed by Corrado Gini, an Italian statistician and sociologist, in 1912. We use the Gini coefficient to analyze the per capita energy consumption gap between different provinces. As shown in Table 10.4, the Gini coefficient of per capita transport energy consumption in China showed an upward trend from 2009 to 2016, rising from 0.253 in 2009 to 0.372 in 2016, an increase of 47%. It shows that the difference in per capita transport energy consumption in Chinese provinces increased significantly during the study period. The astonishing economic development in China has led to the development of the transport sector. Due to the differences in economic level, geographical location and resource endowment among provinces, the transport sector presents the characteristics of unbalanced development. At this stage, the extensive development of the transport sector in China is characterized by high energy consumption, and the difference in per capita transport energy consumption is gradually increasing.

Per Capita Transport Energy Consumption in Central and Western Regions is Increasing Comparatively Fast

Heilongjiang and Anhui had the highest growth rate of per capita transport energy consumption, which reached 144.5% and 132.1% from 2009 to 2019, respectively. The growth rate in Jiangxi, Henan, Hunan, Chongqing, Guizhou and Qinghai was more than 70% from 2009 to 2019. These provinces are all located in central and western regions. Due to the backward economic development level, transport development level and technological innovation level, the per capita transport energy consumption in these provinces has increased relatively fast (Fig. 10.25).

The growth rate of transport energy consumption in Beijing, Tianjin, Inner Mongolia, Shanghai and Shaanxi was relatively low, which was below 10% from

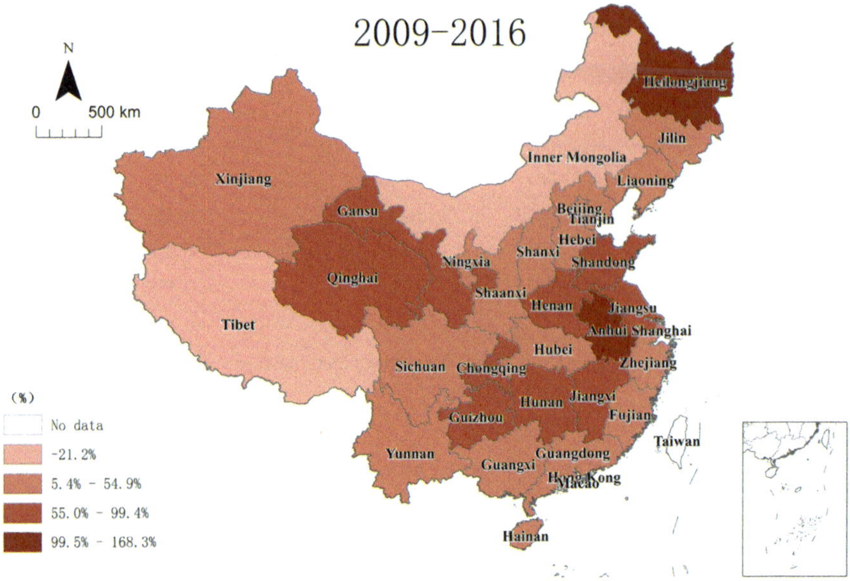

Fig. 10.24 The growth rate of per capita transport energy consumption in China from 2009 to 2016. *Data source* The authors, edited from provincial statistical yearbooks of China

2009 to 2019. In particular, the per capita transport energy consumption in Tianjin and Inner Mongolia showed a negative growth rate, which was 2.6% and 23.9%, respectively. Beijing, Tianjin and Shanghai are the three major municipalities. With their strong economic foundation, advanced scientific and technological level and the large energy consumption base of the transport sector, the growth rate of per capita transport energy consumption in Beijing, Tianjin and Shanghai was at a low level. The growth rate of per capita transport energy consumption in Inner Mongolia decreased significantly during the study period. Contrary to Inner Mongolia, per capita transport energy consumption in Shaanxi maintained a low growth rate during the study period (Fig. 10.24).

10.2.2 Measurement and Analysis of Transport CO_2 Emissions in China

10.2.2.1 Measurement of Transport CO_2 Emissions

The main GHG of the transport sector is CO_2. CO_2 emissions of the transport sector in this study come from fossil fuel combustion, electricity production, and heat production. The Chinese government has not published data on CO_2 emissions of the transport sector, which need to be calculated.

According to the method in IPCC Guidelines, the formula for calculating CO_2 emissions from transport energy consumption is as follows (IPCC, 2006):

$$CC_{i,t} = \sum_{j=1}^{n} \left(E_{i,t,j} \times ALCV_j \times CCF_j \times COF_j \times \frac{44}{12} \right) \tag{10.1}$$

In Eq. (10.1), $CC_{i,t}$ stands for the total CO_2 emissions from transport energy consumption of province i in year t, j represents different fossil fuel types, $E_{i,t,j}$ is the total consumption of fuel type j in province i in year t, $ALCV_j$ identifies the average low calorific value (ALCV) of fuel type j, CCF_j denotes the carbon content factor (CCF) of fuel type j, and COF_j stands for the carbon oxidation factor (COF) of the carbonaceous j. The data on ALCV, CCF and COF were collected from Shan et al. (2018a, 2018b). The data on fossil fuel consumption were obtained from China Energy Statistical Yearbook (CESY) (2011–2017).

The IPCC does not give specific carbon emissions factors for heat and electricity, and scholars have not yet agreed upon uniform or normative calculations. We first calculate the total carbon emissions from energy consumption for generating heat, subsequently assigning that to the transport sector based on its share of total heat energy consumption. Based on Eq. (10.1), its calculation formula is as follows:

$$HC_{i,t} = \frac{\sum_{j=1}^{n} \left(E_{i,t,j,H} \times ALCV_j \times CCF_j \times COF_j \times \frac{44}{12} \right)}{H_{i,t,total}} H_{i,t,transport} \tag{10.2}$$

where $HCE_{i,t}$ represents the total CO_2 emissions from heat generation of the transport industry of province i in year t, $E_{i,t,j,H}$ is the consumption of energy j for the production of hot power, $H_{i,t,total}$ is the total heat energy consumption of province i in year t, and $H_{i,t,transport}$ is the total heat energy consumption in the transport industry of province i in year t.

This study applies the carbon emissions coefficient of electrical energy published by the National Development and Reform Commission (NDRC) for calculating electricity carbon emissions; its estimation method is written as:

$$EC_{i,t} = E_{i,t} \times EF_{grid,OM,y,i,t} \tag{10.3}$$

In Eq. (10.3), $EC_{i,t}$ refers to the total electrical CO_2 emissions from the transport sector of province i in year t. $EF_{grid,OM,y,i,t,}$ denotes the total electric power consumption by the transport industry in province i in year t, and the data are obtained from CESY (2011–2017). $EF_{i,t}$ is the baseline emissions factors for regional power grids, which can be collected from NDRC (2010–2016) (Table 10.5).

Table 10.5 Baseline emission factors for regional power grids in China

Provinces (EF_{grid}, OM, y, tCO_2/MWh)	2009	2010	2011	2012	2013	2014	2015	2016
Beijing, Tianjin, Hebei, Shanxi, Shandong, Inner Mongolia	0.9914	0.9914	0.9803	1.0021	1.0302	1.058	1.0416	1
Liaoning, Jilin, Heilongjiang	1.1109	1.1109	1.0852	1.0935	1.112	1.1281	1.1291	1.1171
Shanghai, Jiangsu, Zhejiang, Anhui, Fujian	0.8592	0.8592	0.8367	0.8244	0.81	0.8095	0.8112	0.8086
Henan, Hubei, Hunan, Jiangxi, Sichuan, Chongqing	1.0871	1.0871	1.0297	0.9944	0.9779	0.9724	0.9515	0.9229
Shaanxi, Gansu, Qinghai, Ningxia, Xinjiang	0.9947	0.9947	1.0001	0.9913	0.972	0.9578	0.9457	0.9316
Guangdong, Guangxi, Guizhou, Yunnan	0.9762	0.9762	0.9489	0.9344	0.9223	0.9183	0.8959	0.8676
Hainan	0.7972	0.7972	0	0	0	0	0	0

Data source National Development and Reform Commission (2010–2017)

10.2.2.2 Characteristics of CO_2 Emissions from Transport in China

The Total CO_2 Emissions and Per Capita CO_2 Emissions from Transport Increased Significantly

Figure 10.26 shows that CO_2 emissions of the transport sector in China increased significantly in 2016 compared with 2009, regardless of the calculation results of this study or the data published by IEA.

Unlike IEA, CO_2 emissions of the transport sector estimated in this study showed a downward trend in 2013, and an upward trend in the rest years. In 2012, the 18th National Congress of the Communist Party of China put forward the idea of building ecological civilization and changing the economic growth mode. In order to implement these policies, the demand for the transport of bulk energy products was reduced, and the energy conservation and emission reduction efforts were increased in the transport sector. In 2013, transport energy and CO_2 emissions decreased significantly. However, the robust economic growth and the sustained growth of total transport demand resulted in a strong rebound in the growth of CO_2 emissions of the transport sector.

China is a developing country, and the transport sector will continue to face great pressure to control its carbon emissions. The main reasons are as follows: Firstly, continued economic growth will stimulate growth in transport demand across sectors. For example, the supply of raw materials and the transport of products will continue to

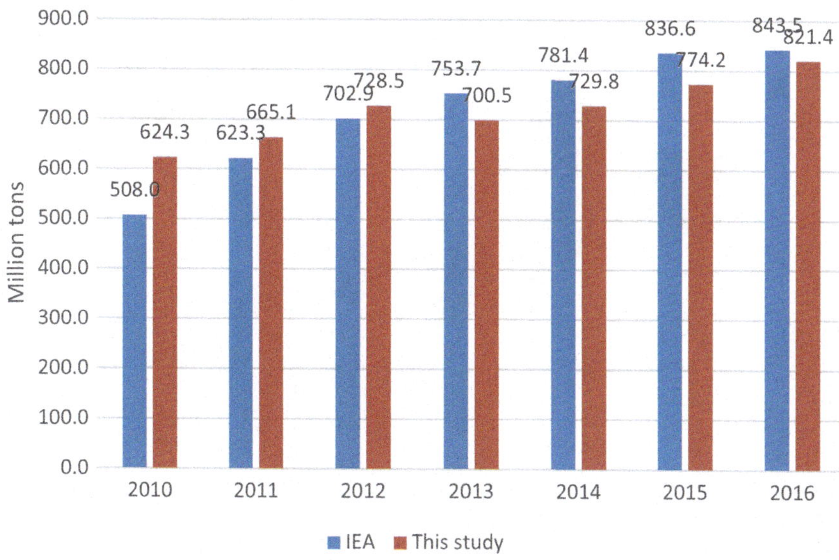

Fig. 10.25 The comparison of CO_2 emissions in China computed by IEA and this study. *Data source* The authors, edited from CO_2 Emissions From Fuel Combustion 2012-2018edn (IEA, 2012–2018)

increase. Secondly, urbanization is accelerating in China, and further urbanization in the future will inevitably lead to an increase in urban transport demand. Furthermore, the energy consumption structure of China's transport sector is dominated by fossil fuels (such as diesel, gasoline, kerosene and fuel oil), which is difficult to change significantly in the short term. Based on the above three points, it can be concluded that immediate emission reduction is required in the transport sector.

The Growth Rate of Transport CO_2 Emissions in Provinces Presents Spatial Agglomeration

In terms of provincial differences, CO_2 emissions of the transport sector in Inner Mongolia, Shandong and Hainan decreased by 21.3%, 5.5% and 1.4% from 2009 to 2016, respectively. Under the implementation of shifting road freight to rail freight, the energy consumption and CO_2 emissions of these three provinces showed different degrees of decline.

Based on the natural breaks classification method (Jenks, 1967), the provinces with growing transport CO_2 emissions can be divided into three classes: high, medium and low. Anhui, Xinjiang, Heilongjiang and Jiangxi belong to the high class, the growth rates of which were between 104.1% and 132.2% between 2009 and 2016. Chongqing, Qinghai, Hunan, Henan, Guizhou and Jiangsu belong to the medium class, the growth rates of which ranged from 75.4% to 101.1% from 2009 to 2016.

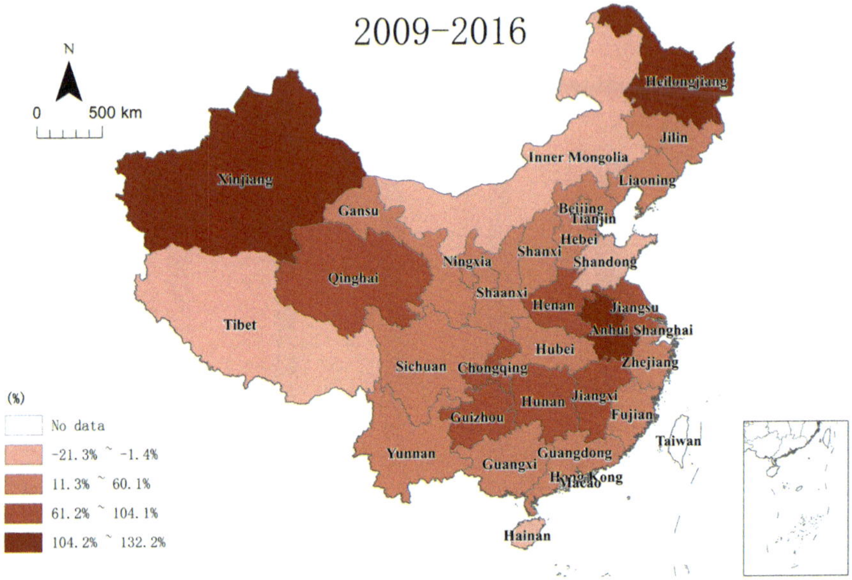

Fig. 10.26 The growth rate of transport CO_2 emissions in China from 2009 to 2016. *Data source* Edited by the authors

Most of these provinces are in central regions. The other provinces belong to the low class, the growth rates of which were between 11.3% and 60.1% from 2009 to 2016 (Fig. 10.26).

The Spatial Characteristics of Transport CO_2 Emissions Present Transverse Distribution

Using ArcGIS 10.4, this study makes the interval diagram of cumulative transport CO_2 emissions in each province from 2009 to 2016, as shown in Figs. 10.27, 10.28, and 10. 29 showing the characteristics of horizontal spatial differences. The cumulative transport CO_2 emissions in each province are divided into four levels from high to low using the Jenks method.

The first class includes Guangdong, Shandong, Shanghai and Liaoning. The industrial base in central and southern Liaoning, industrial bases in the Pearl River Delta, some of the industrial bases in the Yangtze River Delta and emerging industrial bases in the Shandong Peninsula are located in Liaoning, Guangdong, Shanghai and Shandong respectively. These regions are densely populated with a prosperous economy and huge demand for passenger and freight transport, which brings huge transport energy demand and CO_2 emissions.

The second class includes Jiangsu, Hubei, Zhejiang, Henan, Inner Mongolia, Sichuan, Hunan, Beijing and Hebei. These provinces had cumulative CO_2 emissions

Fig. 10.27 The total amount of transport CO_2 emissions in China from 2009 to 2016. *Data source* edited by the author

of more than 1.9 million tons from 2009 to 2016. Jiangsu and Zhejiang are home to the industrial bases of the Yangtze River Delta, while Beijing and Hebei are home to Beijing-Tianjin-Tangshan industrial bases. Hubei, Hunan and Henan have brought huge passenger and freight demand under the Rise of Central China Plan. Sichuan has the second largest population in China, and its demand for transport energy is in the forefront. Inner Mongolia is an important natural resource base and tourism destination in China, with large volumes of commodities and tourists.

Shanxi, Shaanxi, Yunnan, Heilongjiang, Fujian, Guangxi, Anhui, Chongqing, Xinjiang, Guizhou, Jilin and Jiangxi are in the third class. Most of these provinces are located in central and western regions, and the passenger and freight volumes are much lower than those in eastern regions.

Gansu, Tianjin, Hainan, Ningxia and Qinghai are in the fourth class, and they are the provinces with the least CO_2 emissions. Gansu and Ningxia are economically underdeveloped, characterized by small passenger and freight flows and a low level of production technology of various sectors including the transport sector, resulting in higher transport CO_2 emissions. Although Tianjin has a relatively developed economy and large passenger and freight flows, it has a high level of energy saving in the transport sector, resulting in low CO_2 emissions. Hainan has a small population and is surrounded by the sea, so its transport demand and CO_2 emissions are low (Figs. 10.28, 10.29 and 10.30).

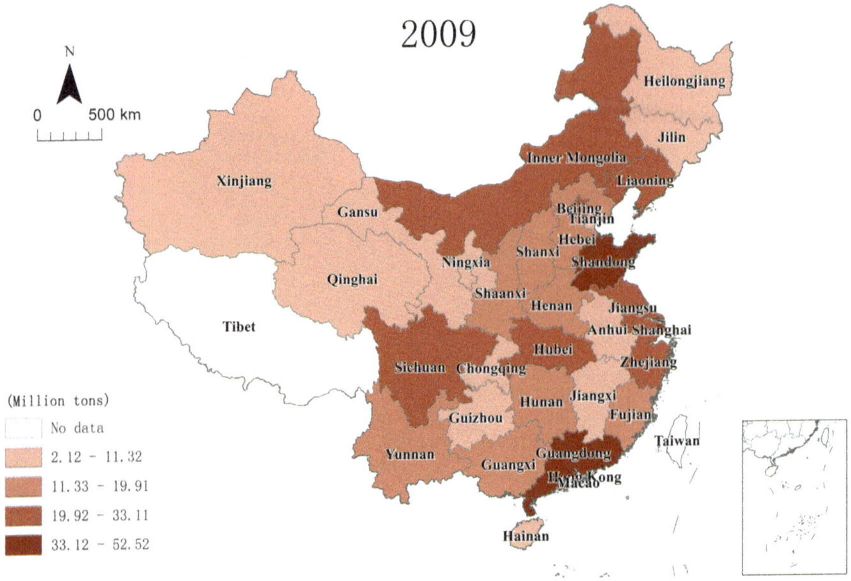

Fig. 10.28 Transport CO_2 emissions in China in 2009. *Data source* Edited by the authors

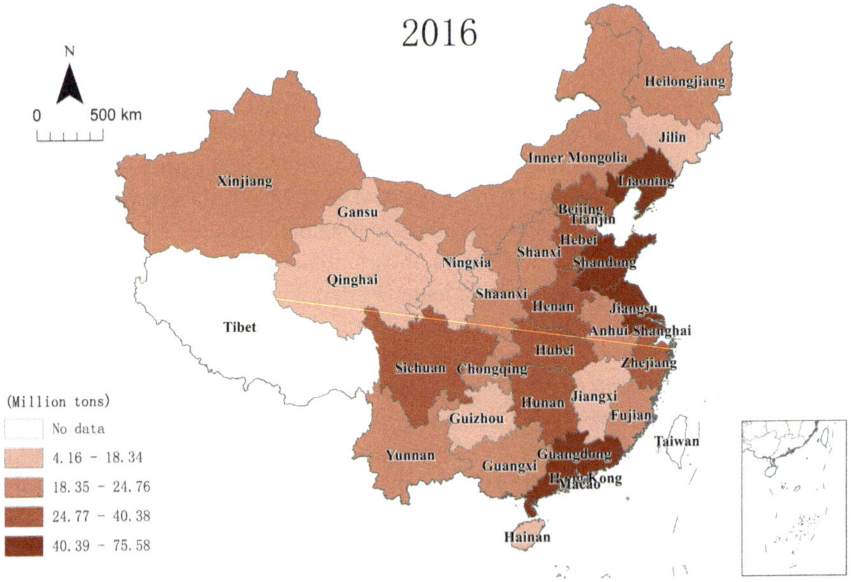

Fig. 10.29 Transport CO_2 emissions in China in 2016. *Data source* Edited by the authors

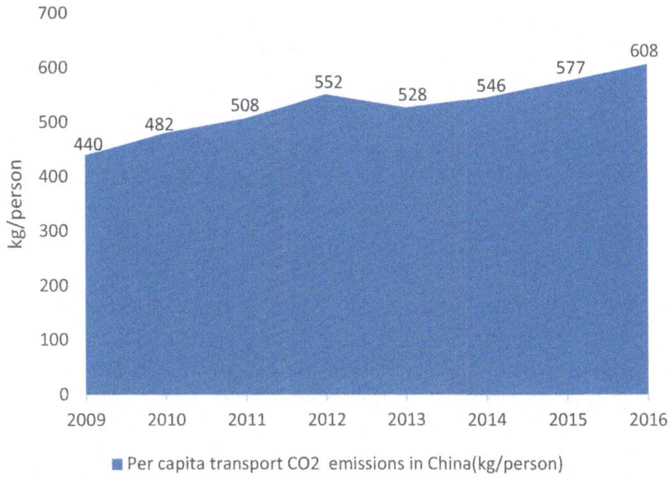

Fig. 10.30 Per capita transport CO_2 emissions in China from 2009 to 2016. *Data source* Edited by the authors

Per Capita Transport CO_2 Emissions in China Showed an Overall Upward Trend

The per capita transport CO_2 emissions in China increased from 423.3 kg/person in 2009 to 594.0 kg/person in 2016, with an average annual growth rate of about 5% (Fig. 10.30). In order to save energy and reduce emissions, the Chinese government implemented a series of financial subsidies and management policies during the 12th Five-Year Plan period. For example, in June 2011, the Ministry of Finance and the Ministry of Transport issued "Interim Measures for the Management of Special Funds for Energy Conservation and Emission Reduction in transport". Although it can improve the utilization efficiency of transport energy to a certain extent, under the background of the rapid economic growth in China, the growth of transport energy and CO_2 emissions brought by the growth of transport demand offset the emissions reduction from the improvement of transport energy utilization efficiency. In the same period, the population growth rate in China continued to slow down, so the per capita transport CO_2 emissions continued to grow (Fig. 10.31).

10.2.2.3 Analysis of CO_2 Emission Intensity of the Transport Sector in China

The Concept of CO_2 Emission Intensity of the Transport Sector

CO_2 emission intensity is an indicator used to describe the relationship between GDP and CO_2 emissions. It refers to the volume of CO_2 emissions per unit of GDP, which is an important index to measure the environmental cost per unit of GDP. According to this concept, the CO_2 emission intensity of the transport sector is summarized as the volume of CO_2 emissions per unit of GDP of the transport sector. If the transport sector keeps its products increasing and the volume of CO_2 emissions per unit of economic output value decreasing, it indicates that the transport sector is witnessing a transition to a low-carbon mode. The reduction in the index indicates that the environmental cost for the growth of the transport sector is decreasing.

Characteristics of CO_2 Emission Intensity of the Transport Sector in China

In this study, the ratio of CO_2 emissions to the added value of the transport sector is taken as the CO_2 emission intensity of the transport sector, and the unit is ton/10^4 RMB. The calculation results are shown in Table 10.6.

As Fig. 10.32 shows, there are certain spatial differences in the CEITS among provinces in China. The provinces with high CEITS are mainly concentrated in the northeast, southwest and northwest regions of China. Due to relatively slow economic development and geographical location, transport in these regions is underdeveloped. The eastern coastal provinces occupy a superior geographical position, and can rely on convenient transport to produce economic benefits. Although carbon emissions of the transport sector are increasing, the CEITS is significantly decreasing. Among them, Beijing and Shanghai, two metropolitan cities in China, have been at the forefront of economic development, but the growth rate of transport products is lower than that of CO_2 emissions, and CO_2 emissions of the transport sector have not been well controlled. They still face great pressure to reduce CO_2 emissions.

National CO_2 emissions of the transport sector generally showed a downward trend, while CO_2 emissions per 10,000 RMB of value added in the transport sector decreased from 3.43 tons in 2009 to 2.49 tons in 2016. It indicates that there was a decrease in CO_2 emissions per unit of value added in the transport sector, and that the growth rate of the transport sector was greater than that of CO_2 emissions in the transport sector (Fig. 10.32). In a word, the environmental cost per unit of transport development has been reduced, the economic benefit of the transport sector has increased significantly, and the development mode of the transport sector is shifting to a low-carbon one.

In fact, the Chinese government has been encouraging the low-carbon economy and low-carbon life. In the field of transport, a series of measures have been taken to achieve "carbon reduction and carbon neutrality", some policy objectives have been achieved, and CEITS showed a downward trend. However, with further economic development and motorization, transport demand will continue to grow. It is estimated that the growth of energy consumption of the transport sector is much faster than that of all sectors, and that the share of transport energy consumption in the total energy consumption of all sectors will continue to increase (Liu, 2013). Therefore, the per capita transport CO_2 emissions show an upward trend (Table 10.7 and Fig. 10.34).

Table 10.6 The CO_2 emission intensity of the transport sector (CEITE) in 31 provinces of China from 2009 to 2016

Province	2009	2010	2011	2012	2013	2014	2015	2016	Mean value
Beijing	3.96	3.30	2.28	3.30	2.93	2.84	2.83	2.74	3.02
Tianjin	2.00	1.80	1.46	1.80	1.54	1.55	1.66	1.71	1.69
Hebei	1.24	1.26	0.82	1.13	1.13	1.08	1.11	1.25	1.13
Shanxi	3.67	3.02	2.75	2.58	2.94	2.91	2.66	2.63	2.90
Inner Mongolia	3.20	3.25	2.97	2.97	1.98	1.97	2.41	1.71	2.56
Liaoning	4.19	3.96	3.49	3.33	2.87	2.89	2.61	3.69	3.38
Jilin	3.21	3.12	2.86	2.75	3.21	3.21	3.25	3.04	3.08
Heilongjiang	2.61	2.38	3.30	3.43	3.71	3.48	3.41	3.27	3.20
Shanghai	6.68	5.41	5.04	4.97	4.74	4.24	4.09	4.12	4.91
Jiangsu	1.84	1.73	1.54	1.52	1.59	1.63	1.64	1.62	1.64
Zhejiang	2.74	2.30	2.15	2.22	2.10	2.02	2.03	1.91	2.18
Anhui	2.08	2.09	2.37	2.73	2.67	2.74	2.77	2.74	2.53
Fujian	2.00	1.94	1.90	1.74	1.66	1.61	1.44	1.42	1.71
Jiangxi	2.13	2.28	2.23	1.87	2.12	2.08	2.24	2.15	2.14
Shandong	2.96	2.78	2.63	2.72	2.09	1.94	1.86	1.79	2.35
Henan	2.42	2.58	2.67	2.44	2.21	1.35	1.94	1.91	2.19
Hubei	4.04	3.73	3.84	3.57	3.01	2.78	2.72	3.11	3.35
Hunan	2.52	2.52	2.41	1.95	2.23	2.26	2.46	2.46	2.35
Guangdong	3.29	3.21	2.91	2.67	2.52	2.36	2.30	2.35	2.70
Guangxi	3.88	3.36	2.96	3.01	2.23	2.58	2.49	2.45	2.87
Hainan	6.50	5.86	5.38	4.89	4.03	3.06	3.15	2.84	4.46
Chongqing	2.94	3.15	2.92	3.03	2.62	2.33	2.55	2.43	2.75
Sichuan	4.57	4.44	4.27	3.87	2.41	2.31	2.01	2.30	3.27
Guizhou	2.58	2.46	2.23	2.24	1.86	1.85	1.81	1.86	2.11
Yunnan	8.44	9.68	9.28	8.70	7.48	8.06	7.42	7.26	8.29
Shaanxi	4.51	4.35	4.06	3.78	3.22	3.08	2.92	2.75	3.58
Gansu	3.96	4.08	3.43	3.30	5.16	5.05	4.98	4.94	4.36
Qinghai	4.30	3.41	4.01	4.12	4.15	4.03	4.06	4.38	4.06
Ningxia	2.86	2.47	2.17	2.07	2.11	2.17	2.21	2.26	2.29
Xinjiang	4.78	4.77	4.58	3.79	4.26	3.72	3.98	3.89	4.22
China	3.43	3.32	3.04	3.07	2.69	2.56	2.54	2.49	2.89

Data source Edited by the authors

Fig. 10.31 The average value of CO_2 emission intensity of the transport sector in 30 provinces of China from 2009 to 2016. *Data source* Edited by the authors

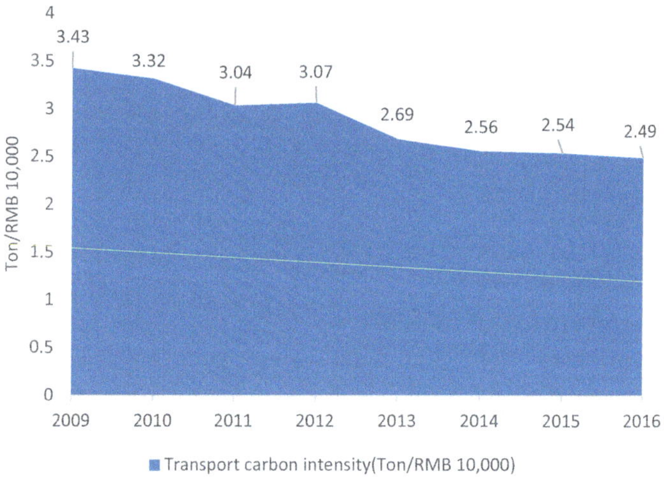

Fig. 10.32 CO_2 emission intensity of the transport sector in China from 2009 to 2016. *Data source* Edited by the authors

Table 10.7 CO_2 emission intensity gap of the transport sector in China from 2009 to 2016

	2009	2010	2011	2012	2013	2014	2015	2016
Standard deviation	1.540	1.586	1.534	1.387	1.286	1.319	1.211	1.213
Coefficient of variation	0.435	0.472	0.485	0.450	0.445	0.476	0.438	0.439
Gini coefficient	0.137	0.164	0.178	0.143	0.145	0.145	0.142	0.139

Data source Edited by the authors

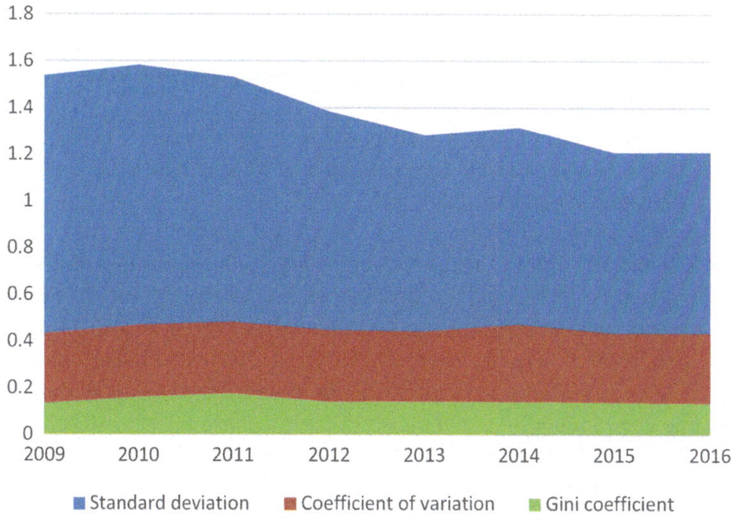

Fig. 10.33 The trend of CO_2 emission intensity gap of the transport sector in China from 2009 to 2016. *Data source* Edited by the authors

As can be seen from the three indicators, the difference in CEITS among provinces generally increased first and then decreased. The three indicators showed similar tendencies (Table 10.7; Fig. 10.33). The value of the standard deviation was relatively high, which indicates that the absolute difference in CEITS among provinces was relatively large. It can be attributed to differences in economic level, natural resource endowments and other aspects among provinces. The coefficient of variation was lower than the standard deviation, and the Gini coefficient was at a low level.

It is found that the standard deviation in 2010 was the largest, and the standard deviation in 2016 was 23.5% lower than that in 2010 (the highest point). The largest values of the variation coefficient and the Gini coefficient were both in 2011, which increased by 11.4% and 29.7% respectively compared with 2009. Compared with the highest point in 2011, the variation coefficient and the Gini coefficient decreased by 9.5% and 22% respectively in 2016.

After entering the 2010s, the rapid economic development drives the development of the transport sector, the development model of high energy consumption and emissions of the transport sector has been gradually transformed, and the difference in provincial CEITS is gradually narrowing. Especially in the 12th Five-Year Plan period, the Chinese government took a series of energy saving and emission reduction measures in the transport sector. Therefore, CEITS has been effectively controlled, and the provincial difference has slowly narrowed.

10.2.3 Prediction of Transport Energy Consumption and Carbon Emissions in China

10.2.3.1 Prediction of Transport Energy Consumption in China

IEA: The Growth Rate of China's Transport Energy Consumption is Much Higher Than the Global Average Level, and China Plays a Key Role in Global Transport Emission Reduction

According to the IEA World Energy Outlook 2019, in the current scenario (CS), the new policy scenario (NPS) and the sustainable development scenario (SDS) (IEA, 2019b), transport energy consumption in China will be 503 Mtoe, 468 Mtoe and 436 Mtoe in 2030, respectively, increasing by 53.8%, 43.1% and 33.3% compared with 2018 (327 Mtoe). While in the current scenario and the sustainable development scenario, the growth rates of global transport energy consumption will be 16.2% and 3.2% from 2018 to 2030, respectively.

In the current scenario, the new policy scenario and the sustainable development scenario, transport energy consumption will be 583 Mtoe, 517 Mtoe and 399 Mtoe in 2040, respectively, increasing by 78.3%, 58.1% and 22% compared with 2018 (327 Mtoe). While in the current scenario and the sustainable development scenario, the growth rates of global transport energy consumption will be 25.9% and 8.7% from 2018 to 2040, respectively (Fig. 10.34).

The IEA report shows that the growth rate of transport energy consumption in China is much higher than that in the world whether it is in 2030 or 2040, and China plays a key role in global transport emission reduction.

BP: In the Evolving Transition Scenario, Transport Energy Consumption in China Will Grow by 50% in 2040 Compared with 2017, Reaching 544 Mtoe, Which is Second to Industrial Energy Consumption

According to the BP World Energy Outlook 2019, in the evolving transition scenario, transport energy consumption will reach 544 Mtoe, 50% higher than that in 2017 (363 Mtoe), accounting for 13.5% of the total energy consumption of all sectors, while it only accounted for 11.5% in 2017. However, the proportion of transport energy consumption in the total energy consumption in China is lower than the global average. BP (2019) believes that the proportion of global transport energy

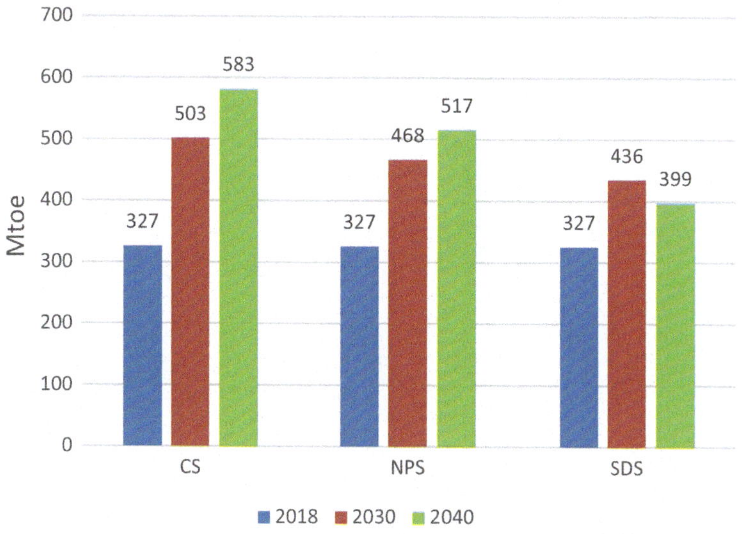

Fig. 10.34 Prediction of final transport energy consumption in China in different policy scenarios. *Data source* The authors, edited from IEA World Energy Outlook 2019edn (IEA, 2019d)

consumption in the total energy consumption of all sectors will be 20% by 2040 in the evolving transition scenario, which is much higher than that in China. In the evolving transition scenario, the proportion of transport energy consumption in China to that in the world in 2040 is 7.3%, significantly higher than that in 2017 (6.2%). This is consistent with the conclusion of the IEA World Energy Outlook 2019edn: transport energy emission reduction in China is the key to global emission reduction (IEA, 2019b).

10.2.3.2 Prediction of Transport CO_2 Emissions from Oil Energy Consumption in China

IEA: In the sustainable development scenario, transport CO_2 emissions from oil energy consumption in China in 2040 can be reduced by 63% compared with 2017

According to the IEA World Energy Outlook 2018edn, transport CO_2 emissions from oil energy consumption in China will show a decreasing trend in the current policy scenario, the new policy scenario and the sustainable development scenario (IEA, 2018). transport carbon emissions from oil energy consumption will be reduced by 63% in 2040 compared with 2017. The use of electric vehicles will reduce the consumption of oil energy. The improvement of transport oil utilization efficiency will reduce transport CO_2 emissions from oil energy consumption. IEA (2018) believes that the consumption of transport oil will be reduced from 284 Mtoe in

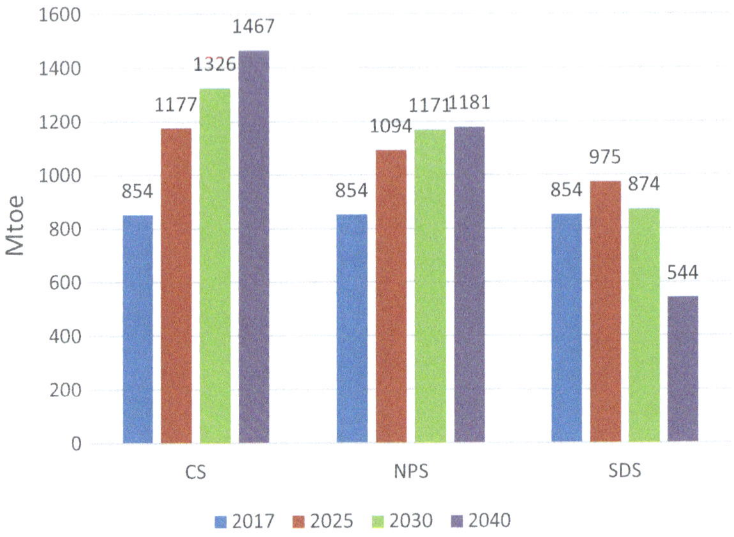

Fig. 10.35 Prediction of oil-related CO_2 emissions in the transport sector in different policy scenarios. *Data source* The authors, edited from IEA World Energy Outlook 2018edn (IEA, 2018)

2017 to 180 Mtoe in 2040 in the sustainable development scenario, a reduction of 36.7% (Fig. 10.35).

10.3 Conclusions

10.3.1 The Transport Sector Has Become the Largest Energy Consuming Sector in the World

Worldwide, with the sustained and rapid development of the transport sector, the proportion of transport energy consumption in end-use energy consumption is increasing. In the two periods 1996–2004 and 2017–2018, the transport sector exceeded the industry sector and became the largest energy consumer. According to the World Energy Outlook 2019edn (IEA, 2019d), the global transport energy demand is expected to grow by more than 25% from 2017 to 2040. With the improvement of residential purchasing power, personal car ownership will continue to grow. Economic development and the increase of personal purchasing power will drive the growth of global trade in goods and services, which will in turn drive the energy demand of commercial transport (heavy-load, air, maritime and railway transport). In addition, the rapid growth of various economic activities and the size of the middle class, especially the middle class in emerging economies, will lead to the highest annual growth rate of aviation demand, which will reach 2.2% from 2017 to 2040 (ExxonMobil Corporation, 2019).

10.3.2 The Transport Sector Has Become the Second Largest CO_2 Emission Source Sector

The transport sector is an important source of CO_2 emissions from fossil energy consumption. There is a clear correlation between transport CO_2 emissions and global warming. Based on the data of Greenhouse Gas Emissions from Energy Highlights 2020edn (IEA, 2020c), after 1989, the transport sector surpassed the manufacturing and construction industry and became the second largest CO_2 emission source sector after the power and heat industry. Based on the data of CO_2 Emissions From Fuel Combustion (IEA, 2019a), in 2017, CO_2 emissions from transport fossil energy consumption reached 8.039 billion tons, an increase of 33.2% compared with 6.6325 billion tons in 2007, accounting for 24.5% of the global fossil energy carbon emissions.

10.3.3 The Energy Consumption of the Transport Sector in China Rose Steadily in the Twenty-First Century

Due to the particularity of the energy consumption statistical system in China, the proportion of transport energy consumption in total energy consumption is far lower than the world average. The transport sector is the second largest energy consumption sector in China. However, the growth of transport energy consumption in China has certain particularities. The proportion of transport energy consumption in total energy consumption rose rapidly in the late 1990s and went steadily up in the twenty-first century. The transport energy consumption in eastern and central areas is higher than that of western areas. Per capita transport energy consumption in central and western regions is increasing comparatively fast.

10.3.4 The Total CO_2 Emissions and Per Capita CO_2 Emissions of the Transport Sector in China Have Increased Significantly in Recent years

Although the Chinese government implemented a series of financial subsidies and management policies, which improved the utilization efficiency of transport energy to a certain extent, under the background of the rapid economic growth in China, the growth of transport energy and CO_2 emissions brought by the growth of transport demand offset the emissions reduction from the improvement of transport energy utilization efficiency. In the same period, the population growth rate in China continued to slow down. The transport sector in China will continue to face great pressure to control its carbon emissions.

References

BP. (2018). BP Statistical Review of World Energy, 2018 edn. https://www.docin.com/p-1933943012.html

BP. (2019). World Energy Outlook, 2019 edn. https://www.bp.com/en/global/corporate/energy-economics/energy-outlook.html

China Energy Statistical Yearbook (CESY). (2017). China Statistical Publishing House, Beijing. http://tongji.oversea.cnki.net/oversea/engnavi/YearBook.aspx?id=N2018070147&floor=1. Accessed April 23, 2020.

China Petroleum Group Economic and Technological Research Institute. (2019). Energy outlook of the world and China in 2050. https://www.sohu.com/a/358633090_680938

Exxon Mobil. (2019). Outlook for energy: A Perspective to 2040. https://www.sohu.com/a/343214502_810912

International Energy Agency (IEA). (2012–2018). CO_2 emissions from fuel combustion: Highlights, Paris. https://www.oecd-ilibrary.org/fr/energy/co2-emissions-from-fuel-combustion-2019_2a701673-en

International Energy Agency (IEA). (2018). World Energy Outlook 2018. https://www.iea.org/reports/world-energy-outlook-2018

International Energy Agency (IEA). (2019a). CO_2 emissions from fuel combustion: Highlights, Paris. https://www.oecd-ilibrary.org/fr/energy/co2-emissions-from-fuel-combustion-2019_2a701673-en. Accessed December 7, 2019.

International Energy Agency (IEA). (2019b). World Energy Outlook 2019. https://www.iea.org/reports/world-energy-outlook-2019

International Energy Agency (IEA). (2020a). Key world energy statistics. https://www.iea.org/reports/key-world-energy-statistics-2020

International Energy Agency (IEA).(2020b). World Energy Balances 2020. https://www.iea.org/data-and-statistics/data-product/world-energy-balances#data-sets

International Energy Agency(IEA). (2020c). Greenhouse Gas Emissions from Energy Highlights 2020edn. https://www.iea.org/data-and-statistics/data-product/greenhouse-gasemissions-from-energy-highlights

IPCC. (2006). Guidelinesfornational GHG inventories. Report of the Intergovernmental Panel on Climate Change, vol 2 (Chapter 6). https://www.ipcc-nggip.iges.or.jp/public/2006gl/vol3.html. Accessed December 7, 2019.

IPCC. (2007). *Climate change2007: The physical science basis, working group I contribution to the fourth assessment report of the intergovernmental panel on climate change*. Cambridge University Press. https://xueshu.baidu.com/usercenter/paper/show?paperid=1x6w0tp0ya660vp0nm550c90jc144721&site=xueshu_se

Jenks, G. F. (1967). The data model concept in statistical mapping. *International Yearb. Cartography 7*, 186–190. https://xueshu.baidu.com/usercenter/paper/show?paperid=3a917c675547e8306d705d50ee3c73a6

Liu, J. C. (2013). Energy saving potential and carbon emissions prediction for the transportation sector in China. *Resources Science, 33*(04):640–646. (In Chinese) https://xueshu.baidu.com/usercenter/paper/show?paperid=72abf505ae5bfa497156712127b881f8

National Bureau of Statistics of China (NBSC). (2021). http://data.stats.gov.cn/easyquery.htm?cn=E0103

National Development and Reform Commission (NDRC). (2010–2017). Baseline emissions factors for regional power grids in China. Beijing. http://m.tanpaifang.com/article/62660.html

Shan, Y. L., Guan, D. B., Zheng, H. R., Ou, J. M., Li, Y., Meng, J., Mi, Z. F., Liu, Z., & Zhang, Q. (2018a). China CO_2 emission accounts 1997–2015. *Scientific Data, 5*, 170201. https://doi.org/10.1038/sdata.2017.201

Shan, Y., Guan, D., Zheng, H., Ou, J., Li, Y., Meng, J., Mi, Z., Liu, Z. & Zhang, Q. (2018b). China CO_2 emission accounts 1997–2015. *Scientific Data. 5*, 170201. https://www.x-mol.com/paper/527136

The World Bank. (2021). World Development Indicators. https://data.worldbank.org.cn/indicator

The World Bank. (2021). World Development Indicators 2021edn. https://data.worldbank.org.cn/indicator

World Meteorological Organization (WMO). (2020). Global climate in 2015–2019. https://www.vzkoo.com/doc/10370.html?a=4

Chapter 11
Transport Environmental Efficiency in China

11.1 Research Progress in TEE

11.1.1 Existing Methods

Transport sustainability assessment has become a hot issue for scholars since the 1990s. As research progresses, the evaluation of transport sustainability is mainly based on the DEA method.

There are two total-factor research approaches operating within this context. The first uses carbon emissions as input by employing a radial DEA model, which includes the CCR DEA or BCC DEA model. The CCR DEA model assumes constant returns to scale (CRS), and is the earliest DEA model introduced by Charnes et al. in 1978. Subsequently, Banker et al. (1984) proposed the BCC DEA model, which is based on the assumption of variable returns to scale (VRS). Both have the characteristic that inputs and outputs are assumed to increase or decrease proportionally when estimating efficiency. Lan and Zhang (2014) applied the CCR DEA model and treated capital stock, labor force, and carbon emissions in the transport industry as three inputs. The added value of the transport industry was treated as an output to evaluate the transport carbon emission efficiency in 30 Chinese provinces between 2006 and 2010. Chen et al. (2019) regarded carbon emissions from the transport industry as a key input indicator when using both CCR DEA and BCC DEA methods to calculate the transport carbon emission efficiency in Beijing from 2000 to 2017. However, the radial DEA model is unable to consider the effect of non-radial slacks when calculating efficiency, which does not square with the facts (Yang et al., 2018).

The second approach selects carbon emissions as an undesirable output and is based on a non-radial DEA model with undesirable outputs. In dealing with the problem of the radial DEA model, Tone (2001) proposed a non-radial SBM model which takes input and output slacks into account when calculating efficiency, but the stranded SBM DEA model cannot consider undesirable outputs. Tone (2004) then proposed the SBM DEA model with undesirable outputs that can accommodate undesirable output factors. Chang et al. (2013) analyzed TEE in China based on

P. Zhao and L. Zeng, *Transport Efficiency and Safety in China*, Population, Regional Development and Transport, https://doi.org/10.1007/978-981-99-1055-7_11

CO_2 emissions. They used the SBM DEA model with undesirable outputs, and they found that most Chinese provinces did not have an eco-efficient transport industry. In order to realize further improvements in evaluating and comparing the DMUs, which are concurrently on the production frontier, Ma et al. (2018a) measured the sustainable transport efficiency in different Chinese provinces by employing a Super-SBM DEA model with undesirable outputs. They found distinguishable levels of sustainable transport efficiency among different regions, with eastern regions exhibiting the highest level, followed by western and central regions, and western regions exhibiting the lowest level. Ma et al. (2018b) applied a Super-SBM DEA model with undesirable outputs and evaluated the integrated transport efficiency of 31 Chinese provinces from 2009 to 2016. They found that the mean value of integrated transport efficiency demonstrated a linear growth.

11.1.2 Influencing Factors of TEE

The influencing factors of TEE can be summarized into mainly four categories: economic factors, technical factors, policy factors and social factors.

11.1.2.1 Economic Factors

The main economic factors are the levels of economic development, urbanization and infrastructure.

The higher the level of economic development, the higher the living standards of residents, the higher the requirements of residents for environmental quality, the more abundant the funds for pollution control. The improvement of residents' living standards also affects the choice of the mode of transport, thus affecting TEE. Yuan et al. (2017) believed that the higher the level of economic development, the higher the efficiency of transport carbon emissions. However, some scholars believe that the higher the level of economic development, the greater the transport demand, which means more transport pollutant emissions, resulting in the rebound effect. Wu et al. (2015), Li et al. (2019), and Wang et al. (2018) believed that the improvement of economic development level is an important driving factor of transport carbon emissions.

The increase of transport demand caused by the improvement of urbanization level is an important reason for transport pollutant emissions. Wu et al. (2015), Xie et al. (2017) and Lv et al. (2019) all believed that the improvement of urbanization level leads to the sharp increase of transport carbon emissions. Zhang (2012) used the vector autoregressive model to analyze the driving factors of the carbon emissions of urban road transport. Their research showed that if the urbanization level increases by 1%, the carbon emissions of urban road transport will increase by 0.928% over the long term. Bian and Ji (2019) found that the transport carbon emissions in Qinghai would continue to increase with the continuous improvement of urbanization level.

The more perfect the transport infrastructure and the more reasonable the construction, the higher the utilization rate and the better it is for improving TEE. The lack of transport infrastructure and insufficient transport supply may lead to traffic congestion (Oduyemi & Davidson, 1998). Yang et al. (2017) believed that traffic congestion increases the emissions of motor vehicle exhaust in Beijing. Zhao et al. (2022) believed that the higher the level of per capita transport infrastructure, the higher the transport carbon emission efficiency.

11.1.2.2 Technical Factors

Technical factors mainly include the degree of informatization and the level of energy-saving technology in the transport sector. With the application of information technology in the transport sector, residential transport is becoming more and more comfortable and convenient. The higher the degree of informatization, the better it is for solving traffic congestion and environmental problems and improving TEE. Cui and Li (2015) estimated the carbon emission efficiency of the transport sector in 15 countries from 2003 to 2010, and found that the higher the R&D funds, the higher the carbon emission efficiency of transport. With the development of technology, the regional land use rate has improved, and the density of the road network has also increased. The higher the density of the road network, the better it is for improving TEE. Ji et al. (2022) believed that urban transport carbon emission can be controlled by technical means.

11.1.2.3 Social Factors

Transport structure, population density, transport management level and population quality are social factors. The energy consumption and pollutant emissions per unit of highway freight are 7 times and 13 times higher than that of railway freight respectively (Ministry of Ecology and Environment of China, 2018). Wei et al. (2013), Cui and Li (2015), Yuan et al. (2017), and Wang et al. (2018) believed that reasonable transport structure can inhibit the emissions of transport pollutants.

There is still controversy about the relationship between population density and transport pollution emissions. Most scholars believe that the higher the population density is, the more problems are likely to appear, which is not conducive to the improvement of regional ecology. For example, Yang et al. (2015) also found that population density has a positive effect on transport carbon emissions. Darido et al. (2014) found that urban expansion, compact population density, and diverse travel modes can result in the increase of urban transport carbon emissions. However, Lawrence et al. (2000) found that population density is negatively correlated with the carbon emissions of the transport sector. Lee and Lee (2014) made an analysis on PUMS data, and they found that if the urban population density doubled, transport carbon emissions would decrease by 48%. Yang et al. (2019) found that population density had an inhibitory effect on the carbon emissions of the transport sector in China from 2000 to 2015.

Both the level of transport management and the quality of residents are human factors. The level of transport management is mainly determined by the action of the personnel of the transport management department, and the quality of residents is mainly the embodiment of the action of traveling residents. The higher the level of transport management, the better it is for improving the ecology. The quality of residents determines the travel mode of residents. High-quality residents generally have a stronger awareness of environmental protection, and are more likely to choose public transport as the travel mode, which has a certain impact on TEE.

11.1.2.4 Policy Factors

The policy factors are mainly related to the policy system of transport. Timilsina and Shrestha (2009) believed that fiscal policy, fuel economy, and the policies of developing and promoting clean energy and energy-efficient vehicles also play a positive role in curbing the carbon emissions of transport. Cui and Li (2015) found that tax relief for transport technology enterprises can significantly improve the carbon emissions efficiency of the transport sector. From 2003 to 2006, the London government began to levy congestion charges, and the policy reduced the concentrations of NO_x, PM10 and SO_2 emitted by transport, which decreased by 17%, 24% and 3%, respectively. In 2008, the London government implemented the Low Emission Zone policy; the policy stipulates that vehicles driving in low emission zones must meet certain emission standards, otherwise they will be charged. The results show that PM10 concentration has dropped by between 2.46% and 3.07% compared with zones outside low emission zones.

11.2 Methods

The area of study covers 30 Chinese mainland provinces (except Tibet) from 2009 to 2017. As shown in Fig. 11.1, these provinces are divided into eight regions in accordance with Zhao et al. (2022) and Zeng et al. (2019).

Following the studies of Chang et al. (2013) and Wang and He (2017), the transport sector in this study consists of transport, warehousing, and postal services industries. Based on existing research results, such as those of Song et al. (2016), Wang and He (2017), and Feng and Wang (2018), this study treats the capital stock, labor force, and energy consumption of the transport sector as three inputs. The value added by the transport sector is a desirable output, while the CO_2 emissions of the transport sector constitute an undesirable output (Table 11.1).

We use the total number of employees in the transport sector as the labor force, and relevant data are taken from China Statistical Yearbooks (2011–2017). Data on

Fig. 11.1 Eight economic zones of China. *Data source* Edited by the authors

Table 11.1 TEE measurement index system

Primary indexes	Secondary indexes	Tertiary indexes
Inputs	Capital	Capital stock of the transport sector (unit: 100 million RMB)
	Labor	Total number of employees in the transport sector (unit: 10,000 persons)
	Energy	Total energy consumption of the transport sector (unit: 10,000 tons of standard coal)
Outputs	Desired outcomes	Added value of the transport sector (unit: 100 million RMB)
		CO_2 emissions from the transport sector (unit: 10^4 tons)

Data source China Statistical Yearbook (2009–2017), China Fixed Assets Statistical Yearbook (2009–2013, 2014–2017), and statistical yearbooks of provinces over the years

energy consumption are gathered from all provincial statistical yearbooks (2011–2017). The capital stock of the transport sector is generally measured by social capital stock with the perpetual inventory method (Goldsmith, 1951) as follows: $K_{i,t} = I_{i,t} + (1\text{-}\delta)\ K_{it\text{-}1}$. Here K represents the social capital stock; I is the fixed capital investment in the transport sector; δ indicates the depreciation rate of capital stock; i represents the province, municipality, or autonomous region; and t stands for the year. Learning from Li and Zhang (2016), we set the value of δ to 8.76%; Consistent with the study conducted by Zhang et al. (2004), the capital stock of the transport sector in 2009 is equal to the value of the fixed capital investment in the transport sector in 2010 divided by 10%. The data on fixed capital investments in the transport sector are taken from the National Bureau of Statistics of China (NBSC, 2020). The calculation process of CO_2 emissions of the transport sector is shown in Chap. 10, which is not repeated here.

11.3 Measurement Results

According to the EBM model and variable selection, we apply MaxDEA 7.9 ultra software to calculate the TEE of 30 provinces in China from 2009 to 2016. The results are listed in Table 11.2. Using ArcGIS 10.1 software, we draw the spatial distribution map of TEE based on the average value of provincial TEE from 2009 to 2016 (Fig. 11.2)

11.3.1 The Overall Characteristics of TEE in China

See the Fig. 11.2.

11.3.1.1 The Spatial Difference in TEE Was Very Large. The Huang-Huai-Hai Region Had the Highest Level of TEE, While Western Provinces Had the Lowest Level of TEE

Using ArcGIS 10.1 software, we draw the spatial distribution map of TEE based on the average value of provincial TEE from 2009 to 2016. The national TEE was relatively low, and the average value of TEE of 30 provinces from 2009 to 2016 was 0.611—there is certain room for improvement. In addition, the spatial difference in TEE was very large.

The Huang-Huai-Hai Region, which mainly includes Henan, Shandong and Jiangsu, had the highest level of TEE, while western provinces had the lowest level of TEE. The provinces in the Huang-Huai-Hai Region had high transport capital stock and transport output, as well as developed transport and production technology. Therefore, despite high transport pollutant emissions, the provinces in the Huang-Huai-Hai Region still maintained a high level of TEE. In western provinces (except Inner Mongolia, Guizhou and Ningxia), the impact of transport pollutant emissions on the ecological environment was relatively low. Under the background of the Western Development Program, the capital stock of the transport sector increased rapidly, but the transport output remained at a low level, resulting in a low level of regional TEE (Fig. 11.3).

Table 11.2 TEE of 30 provinces in China from 2009 to 2016

Province	2009	2010	2011	2012	2013	2014	2015	2016	Mean
Beijing	0.526	0.587	0.56	0.533	0.593	0.648	0.684	0.746	0.610
Tianjin	0.682	0.723	0.7	0.666	0.652	0.717	0.738	0.75	0.704
Hebei	1	1	1	1	1	1	1	1	1
Shanxi	0.471	0.516	0.501	0.521	0.476	0.504	0.58	0.615	0.523
Inner Mongolia	0.769	0.709	0.7	0.695	0.708	0.742	0.634	0.674	0.704
Liaoning	0.615	0.623	0.639	0.67	0.694	0.71	0.818	0.626	0.674
Jilin	0.527	0.498	0.472	0.477	0.506	0.523	0.518	0.518	0.505
Heilongjiang	0.515	0.496	0.531	0.43	0.446	0.508	0.522	0.537	0.498
Shanghai	0.425	0.51	0.463	0.475	0.523	0.616	0.686	0.739	0.555
Jiangsu	0.898	1	0.968	1	1	1	1	1	0.983
Zhejiang	0.616	0.632	0.606	0.587	0.635	0.68	0.708	0.738	0.650
Anhui	0.668	0.651	0.591	0.607	0.65	0.658	0.619	0.593	0.630
Fujian	0.761	0.714	0.621	0.616	0.604	0.647	0.751	0.837	0.694
Jiangxi	0.672	0.659	0.621	0.723	0.743	0.77	0.764	0.784	0.717
Shandong	1	1	1	1	0.764	0.859	0.881	0.906	0.926
Henan	0.872	0.791	0.695	0.765	0.919	1	1	1	0.880
Hubei	0.508	0.532	0.5	0.5	0.545	0.568	0.569	0.543	0.533
Hunan	0.539	0.552	0.48	0.507	0.541	0.589	0.616	0.642	0.558
Guangdong	0.625	0.627	0.595	0.644	0.644	0.715	0.748	0.791	0.674
Guangxi	0.409	0.446	0.471	0.453	0.467	0.497	0.53	0.54	0.477
Hainan	0.338	0.326	0.336	0.362	0.391	0.476	0.452	0.457	0.392
Chongqing	0.436	0.458	0.418	0.416	0.462	0.485	0.517	0.548	0.468
Sichuan	0.359	0.339	0.366	0.315	0.265	0.358	0.4	0.461	0.358
Guizhou	0.724	0.691	0.708	0.754	0.737	0.826	0.896	0.899	0.779
Yunnan	0.21	0.196	0.181	0.191	0.195	0.197	0.211	0.22	0.200
Shaanxi	0.449	0.44	0.433	0.452	0.432	0.487	0.494	0.514	0.463
Gansu	0.815	0.734	0.703	0.709	0.556	0.495	0.431	0.371	0.602
Qinghai	0.271	0.299	0.282	0.266	0.249	0.248	0.26	0.265	0.268
Ningxia	0.797	0.877	0.862	0.874	0.885	0.854	0.818	0.763	0.841
Xinjiang	0.386	0.356	0.336	0.425	0.46	0.537	0.554	0.561	0.452
China	0.596	0.599	0.578	0.588	0.591	0.630	0.647	0.655	0.611

Data source Edited by the authors

Fig. 11.2 The mean value of TEE intervals in 30 provinces of China from 2009 to 2016. *Data source* Edited by the authors

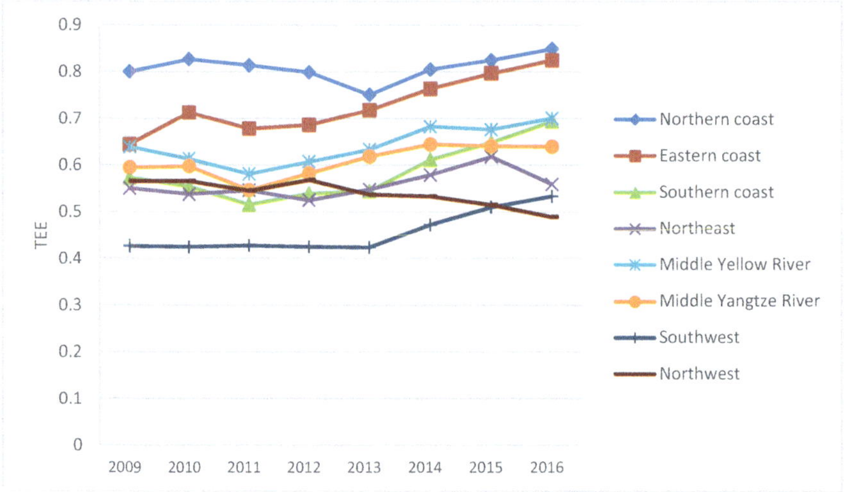

Fig. 11.3 The trend of TEE in China and eight economic zones from 2009 to 2016. *Data source* Edited by the authors

11.3.1.2 The National TEE First Decreased and then Increased, and the National Policy Plays a Significant Role in TEE

The national average TEE showed different trends in two phases. Phase one lasted from 2009 to 2011, and the national average TEE exhibited a general downward trend and fell from 0.596 in 2009 to 0.578 in 2011. In the second phase, lasting from 2011 to 2016, the national average TEE showed an upward trend, reaching a value of 0.655 in 2016 and thereby exceeding the initial TEE level.

In order to confront the global economic crisis of 2008, China executed a positive financial policy and permissive monetary policy. This allowed various provinces to increase their investments in transport infrastructure, resulting in over-investment in transport (Li et al., 2015; Lu, 2012; Zhang & Zhou, 2020). Therefore, the national TEE exhibited a downward trend from 2010 to 2012. In 2011, the Ministry of Transport (2011) developed many policies intended to curb the release of transport pollutants. Notably, in 2013, the Communist Party of China (CPC) (2013) promoted the building of ecological civilization, and various provinces in China increased efforts to achieve energy conservation and emissions reduction. Many new energy-saving technologies were popularized and applied in the transport sector. These were conducive to improving TEE and can explain the upward trend in national TEE after 2011 (Fig. 11.3).

11.3.1.3 Coastal Regions Had the Highest Level of TEE, Followed by Central and Western Regions

A comparison of regions indicates that the Northern coast had the highest level of TEE, followed by the Eastern coast, Middle reaches of the Yellow River, Middle reaches of the Yangtze River, Southern coast, Northeast, Northwest, and Southwest.

Coastal provinces enjoy geographical advantages and high levels of economic development. They boast relatively highly developed transport industries, advanced education systems, and large populations of professional and technical personnel, and they have successfully implemented and promoted advanced foreign technologies and business ideas. These provinces have been at the forefront in implementing a number of energy-saving and emissions reduction measures, including many electrification improvement projects for logistics parks, docks, and integrated transport hubs. As a result, they may utilize transport energy more efficiently and realize higher TEE than other regions do (Feng & Wang, 2018).

Middle reaches of the Yellow River and Middle reaches of the Yangtze River closely follow coastal areas and have been opened up to the outside world. The educational systems in these regions are also relatively developed, energy-saving technologies are fully applied, internal infrastructure is continuously improved, and TEE levels closely follow those of coastal provinces.

As a result of geographical factors, Northeast, Northwest and Southwest regions are less able to acquire and introduce advanced technologies. After the reform and opening-up, a large number of talented workers moved from the Northeast, Northwest, and Southwest to coastal areas (Cheng & Lu, 2017). The resulting loss of talent has made it difficult to improve production technologies in various industries, especially in transport, so TEE levels remain low in these regions (Table 11.3).

11.3.1.4 Eastern Coast and Southwest Showed an Obvious Upward Trend for TEE, While Northwest Areas Showed a Downward Trend, and TEE in Other Regions was Generally Consistent with the National Trend

Due to the high level of emission reduction technology in eastern regions, the TEE in Eastern coast continued to improve. Southwest lags behind coastal regions when it comes to emission reduction technology, but with the support of the Western Development Program, the regional resources have been developed, resulting in the increase of transport demand, the rapid growth in transport output, and thus continuous improvement of TEE in Southwest. In addition, with the implementation of the Western Development Program, the country continues to increase the investment in transport infrastructure in northwest regions, while the output level of the transport sector is low in the short term, which inhibits TEE (Table 11.3).

Table 11.3 TEE of eight economic zones from 2009 to 2016

Regions	2009	2010	2011	2012	2013	2014	2015	2016	Mean
Northern coast	0.802	0.828	0.815	0.800	0.752	0.806	0.826	0.850	0.810
Eastern coast	0.646	0.714	0.679	0.687	0.719	0.765	0.798	0.826	0.729
Southern coast	0.575	0.556	0.517	0.541	0.546	0.613	0.650	0.695	0.587
Northeast	0.552	0.539	0.547	0.526	0.549	0.580	0.619	0.560	0.559
Middle reaches of the Yellow River	0.640	0.614	0.582	0.608	0.634	0.683	0.677	0.701	0.643
Middle reaches of the Yangtze River	0.597	0.599	0.548	0.584	0.620	0.646	0.642	0.641	0.610
Southwest	0.428	0.426	0.429	0.426	0.425	0.473	0.511	0.534	0.456
Northwest	0.567	0.567	0.546	0.569	0.538	0.534	0.516	0.490	0.541
China	0.596	0.599	0.578	0.588	0.591	0.630	0.647	0.655	0.611

Data source Edited by the authors

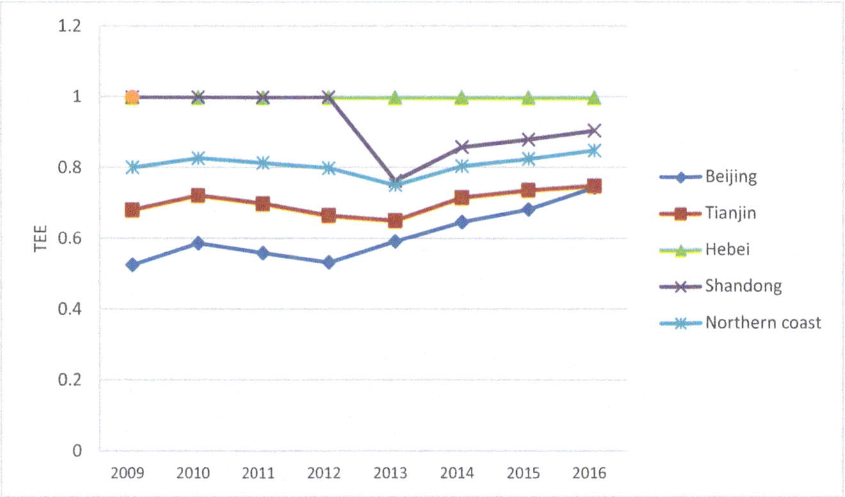

Fig. 11.4 The trend of TEE in Northern coast from 2009 to 2016. *Data source* Edited by the authors

11.3.2 The Regional Characteristics of TEE in China

11.3.2.1 Northern Coast

During the study period, the provincial average value of TEE in Northern coast was 0.810, which is the highest in the eight regions. It declined first and then rose, with the lowest value in 2013.

In terms of provincial differences, the overall level of TEE in Hebei was higher, which was at the forefront of production during the study period. The overall level of TEE in Shandong was in the medium level, and the value of TEE in Shandong was at the forefront of production from 2009 to 2012; in 2013, it failed to be at the forefront of production and was at the lowest point during the study period, but then it showed an upward trend after 2013. The overall levels of TEE in Beijing and Tianjin were relatively low, with annual average values of 0.610 and 0.704 respectively, the trend of which was roughly consistent with the overall change trend of the national TEE; but the lowest point in Beijing is in 2012, and that in Tianjin is in 2013 (Figs. 11.4 and 11.5).

11.3.2.2 Eastern Coast

During the study period, the average annual TEE in Eastern coast was 0.729, which was at a high level as a whole. The average TEE in the region showed an upward trend, except for 2011, rising from 0.646 in 2009 to 0.826 in 2016.

Fig. 11.5 The mean value of TEE in Northern coast from 2009 to 2016. *Data source* Edited by the authors

In terms of provincial differences, Jiangsu always maintained a high level of TEE, which was kept at the forefront of production in the study years, except 2009 and 2012. The level of TEE in Zhejiang was lower than that in Shanghai, and the change trend of TEE in Zhejiang was roughly consistent with that of the whole country. The TEE of Zhejiang was lowest in 2012. Shanghai had the lowest level of TEE, and the change trend of TEE in Shanghai was consistent with the region. The TEE improved during the study period, except for 2011 (Figs. 11.6 and 11.7).

11.3.2.3 Southern Coast

The annual average value of TEE in Southern coast during the study period was 0.587, slightly lower than the national average value. It showed a downward trend in 2008–2011, an upward trend in 2011–2015, and a significant decline trend in 2016. In terms of provincial differences, the levels of TEE in Fujian and Guangdong were relatively high, slightly higher than the national average level, and the change trend in the two provinces was roughly consistent with the regional change trend. The TEE in Hainan was relatively low, which first rose and then declined, with the highest point in 2014 (Figs. 11.8 and 11.9).

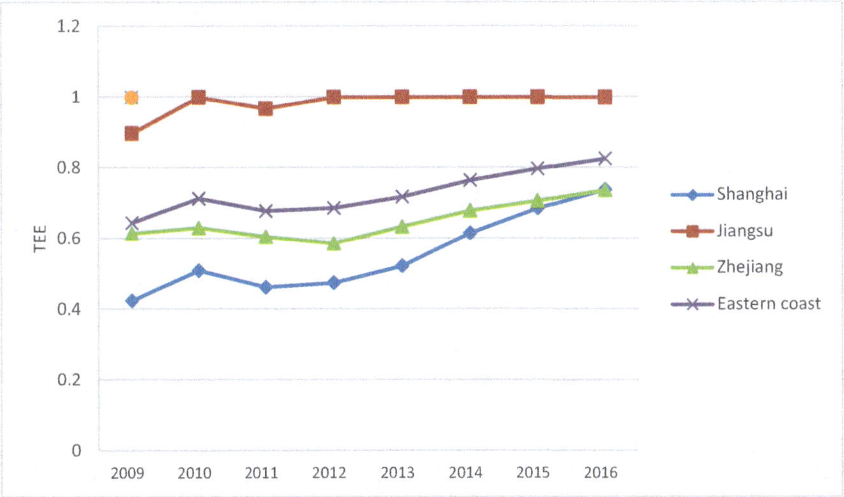

Fig. 11.6 The trend of TEE in Eastern coast from 2009 to 2016. *Data source* Edited by the authors

Fig. 11.7 The mean value of TEE in Eastern coast from 2009 to 2016. *Data source* Edited by the authors

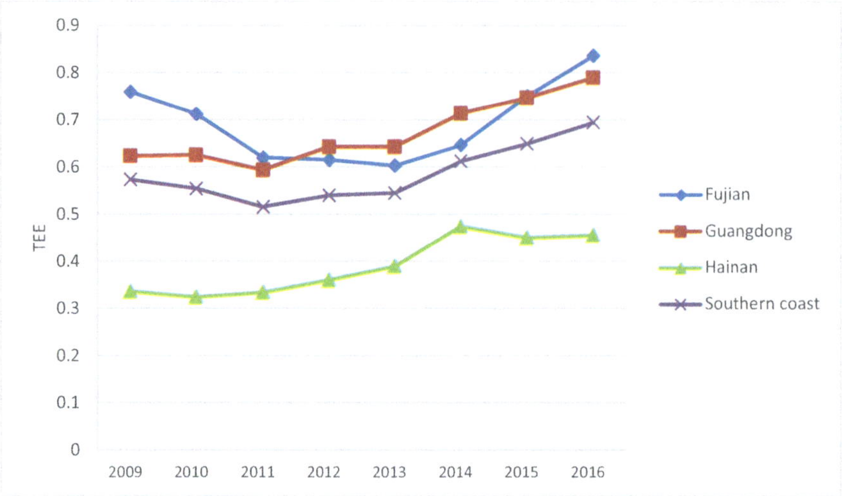

Fig. 11.8 The trend of TEE in Southern coast from 2009 to 2016. *Data source* Edited by the authors

Fig. 11.9 The mean value of TEE in Southern coast from 2009 to 2016. *Data source* Edited by the authors

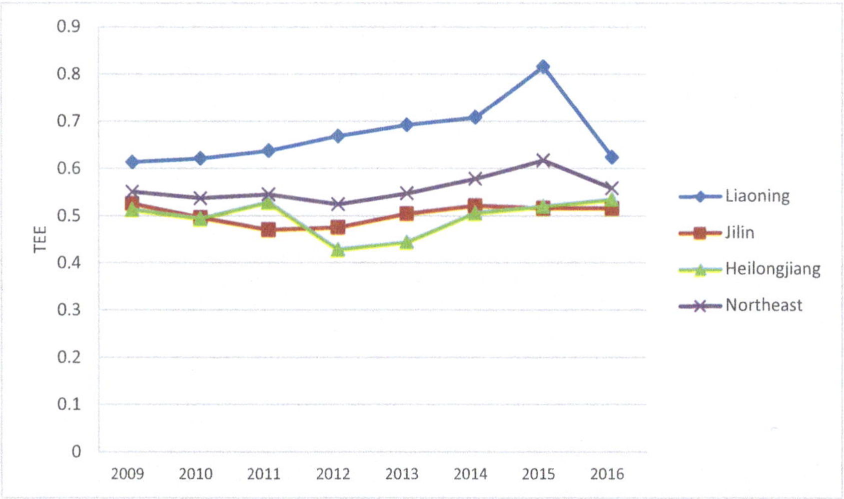

Fig. 11.10 The trend of TEE in Northeast from 2009 to 2016. *Data source* Edited by the authors

11.3.2.4 Northeast

During the study period, the annual average value of TEE in Northeast China was 0.559, which was slightly lower than that in the whole country. The TEE fluctuated at a low level between 0.526 and 0.549 from 2009 to 2013; it showed a significant improvement in 2014 and 2015, but declined in 2016.

In terms of provincial differences, the levels of TEE in Jilin and Heilongjiang were low, lower than the national average level; they generally first declined and then rose. The lowest level of TEE in Jilin was in 2011 and that in Heilongjiang was in 2012. Liaoning had the highest level of TEE in the region; from 2009 to 2015, the TEE in Liaoning showed an upward trend from 0.818 to 0.626, and presented a significant downward trend in 2016, with a decline rate of 30.7%. From 2015 to 2016, the TEE of Jilin remained unchanged, while that of Heilongjiang showed a low level of improvement. This shows that the decline of TEE in Liaoning in 2016 is the root cause of the sharp decline of TEE in the region in 2016 (Figs. 11.10 and 11.11).

11.3.2.5 Middle Reaches of the Yellow River

During the study period, the annual mean value of TEE in Middle reaches of the Yellow River was 0.643, slightly higher than the national average. The regional TEE reduced first and then rose again, with the lowest point in 2011.

In terms of provincial differences, the TEE of Henan was relatively high, which first increased and then decreased, with the lowest point in 2011. The TEE of Henan reached the forefront of production at the end of the study period (2014–2016). The

Fig. 11.11 The mean value of TEE in Northeast from 2009 to 2016. *Data source* Edited by the authors

TEE of Inner Mongolia was second only to that of Henan in the region, ranging from 0.634 to 0.769. The TEE of Shanxi and Shaanxi was at a low level; the TEE of Shanxi first rose and then declined from 2009 to 2013, and it saw a sharp rise after 2013. The TEE of Shaanxi showed little change in the initial stage of the study period, and showed significant improvement after 2013 (Figs. 11.12 and 11.13).

11.3.2.6 Middle Reaches of the Yangtze River

The average TEE in Middle reaches of the Yangtze River during the study period was 0.610, which was very close to the average level in the whole country, and it first declined and then rose, with the lowest point in 2011.

In terms of provincial differences, Jiangxi had the highest level of TEE in the region, which was significantly higher than the average value of the whole country in the study period; the TEE of Jiangxi first decreased and then rose, with the lowest point in 2011.

The level of TEE in Anhui was lower than that in Jiangxi and slightly higher than the national average. The TEE in Anhui first decreased, then increased, and then decreased again, with the lowest point in 2011 and the highest point in 2014.

The annual average values of TEE in Hubei and Hunan were lower than the national average level. In the initial stage of the study period, the change trends of TEE in the two provinces were similar: the TEE first increased and then decreased

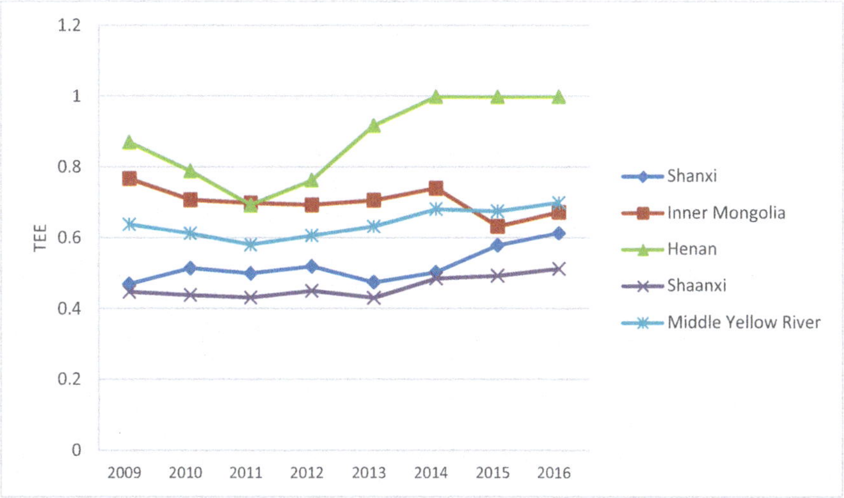

Fig. 11.12 The trend of TEE in Middle reaches of the Yellow River from 2009 to 2016. *Data source* Edited by the authors

Fig. 11.13 The mean value of TEE in Middle reaches of the Yellow River from 2009 to 2016. *Data source* Edited by the authors

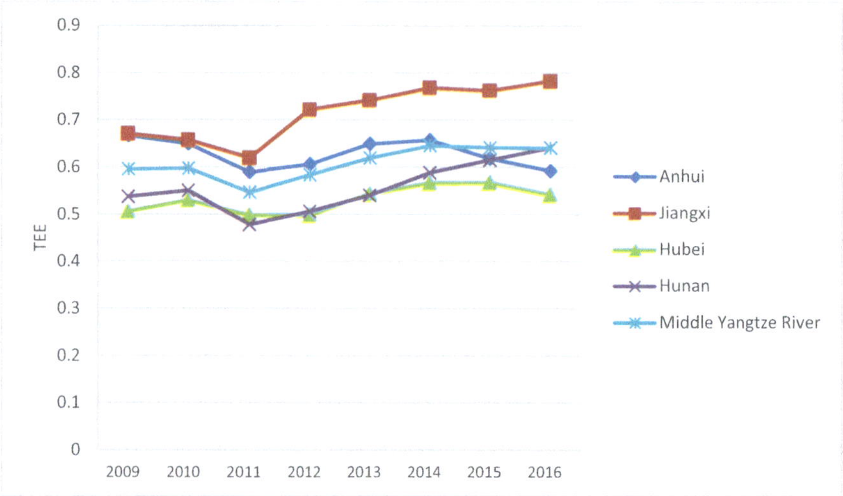

Fig. 11.14 The trend of TEE in Middle reaches of the Yangtze River from 2009 to 2016. *Data source* Edited by the authors

from 2009 to 2011; however, at the end of the study period, the TEE in Hunan presented a significant upward trend after 2011, while that in Hubei first increased and then fell, with the highest point in 2014 (Figs. 11.14 and 11.15).

11.3.2.7 Southwest

During the study period, the average value of provincial TEE in Southwest China was 0.456, which was the lowest among the eight regions, far lower than the national average. From 2009 to 2013, the average value of TEE in the region had little change, ranging from 0.425 to 0.429. After 2013, it showed significant improvement. The average TEE value in the region increased from 0.425 in 2013 to 0.534 in 2015.

In terms of provincial differences, the TEE in Guizhou was higher than that in other provinces in the region, which was above 0.7 over the study years. During the study period, the annual average values of TEE in Guangxi and Chongqing were 0.477 and 0.468, respectively, showing an upward trend from 2009 to 2016. The TEE of Sichuan was at a low level, with an annual average of 0.358 during the study period; it first decreased and then rose, with the lowest value in 2013. The TEE of Yunnan was the lowest not only in the region but also in the whole country. The average annual TEE in Yunnan was 0.2, and the TEE in Yunnan fluctuated at a low level during the study period (Figs. 11.16 and 11.17).

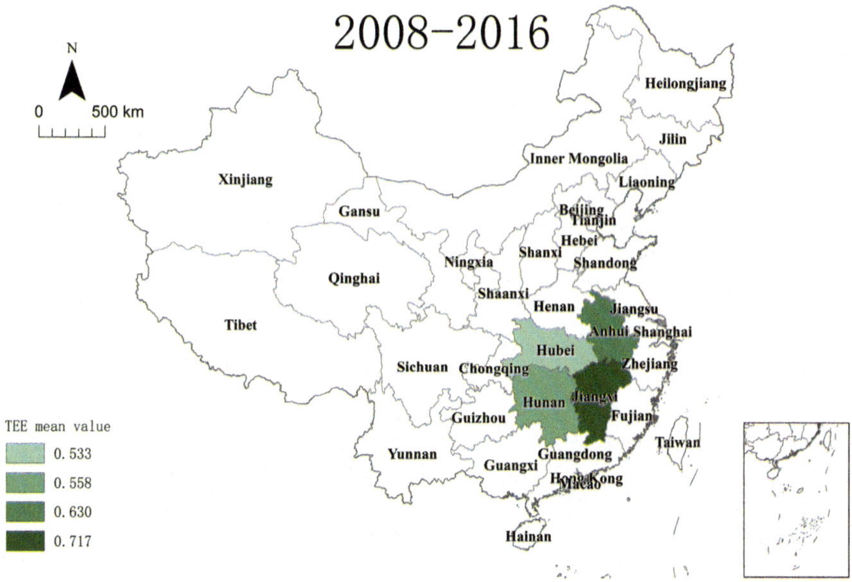

Fig. 11.15 The mean value of TEE in Middle reaches of the Yangtze River from 2009 to 2016. *Data source* Edited by the authors

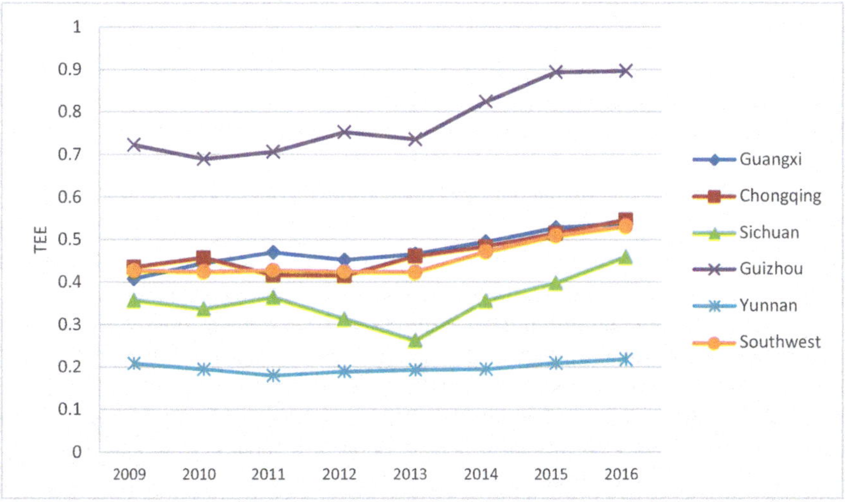

Fig. 11.16 The trend of TEE in Southwest from 2009 to 2016. *Data source* Edited by the authors

Fig. 11.17 The mean value of TEE in Southwest from 2009 to 2016. *Data source* Edited by the authors

11.3.2.8 Northwest

The average value of TEE in Northwest China was 0.541, which was at a low level among the eight regions. The average value of TEE in the region showed a downward trend.

The implementation of the Western Development Program can accelerate the investment in transport infrastructure and increase the stock of transport capital. However, on the whole, due to the relatively backward economic development in Northwest China, the transport volume has restrained the regional TEE.

In terms of provincial differences, Ningxia's TEE level was relatively high, and the average annual value was higher than the national average level. During the study period, Ningxia's TEE first increased and then decreased, which was at a high level from 2010 to 2013 and at a low level in 2009 and after 2013. At the beginning of the study period, the TEE of Gansu was at a high level, but it showed a downward trend at the later stage. At the end of the study period, it was at a low level; the TEE of Gansu was 0.371 in 2016, which was only 45.5% of that in 2009. Gansu is a key area of transport infrastructure construction in Western China, and the rapid rise of transport capital stock inhibited the TEE. During the study period, the TEE in Xinjiang was at a low level, with an average annual value of only 0.452, far lower than the average annual value of all provinces in China. The TEE in Xinjiang showed an ascent after an initial decline, and the TEE of Xinjiang in 2011 was the lowest. The overall level of TEE in Qinghai was the lowest in the region. The value of TEE in Qinghai fluctuated between 0.248 and 0.299 during the study period (Figs. 11.18 and 11.19).

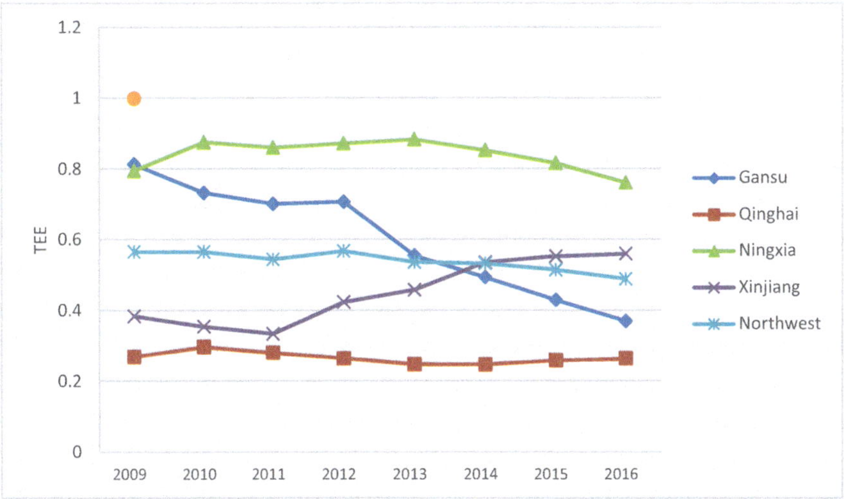

Fig. 11.18 The trend of TEE in Northwest from 2009 to 2016. *Data source* Edited by the authors

Fig. 11.19 The mean value of TEE in Northwest from 2009 to 2016. *Data source* Edited by the authors

Table 11.4 The value of Moran's *I* of provincial TEE in China from 2009 to 2016

Year	Moran's I	Z-score	*p*-value*
2009	0.209**	1.969	0.049
2010	0.281***	2.570	0.010
2011	0.247**	2.303	0.021
2012	0.217**	2.052	0.040
2013	0.234**	2.200	0.028
2014	0.250**	2.321	0.020
2015	0.263**	2.423	0.015
2016	0.286***	2.605	0.009

Note: ***, **, and * indicate significance at the 1%, 5%, and 10% levels, respectively
Data source Edited by the authors

11.4 The Spatial Autocorrelation of TEE

11.4.1 Global Spatial Autocorrelation Analysis

We use the Stata 13 software to analyze the spatial autocorrelation of TEE in China from 2009 to 2016. Table 11.4 lists the value of the Global Moran's I.

According to Table 11.4, the *p*-value of Moran's I index was lower than 0.05 during the study period, meaning that TEE displayed strong spatial autocorrelation in various provinces of China. The values of the Moran's I were greater than 0.2 at the significance level of 10%, which showed that the spatial distribution of the values of TEE was not random but had obvious spatial correlation; more specifically, the provinces with a high level of TEE gather together, and the provinces with a low level of TEE gather together. The value of Moran's I first rose and then declined from 2008 to 2012, which indicates that the degree of spatial autocorrelation strengthened first and then weakened. The value of Moran's I showed an upward trend after 2012, indicating the growth of spatial agglomeration effect of provincial TEE.

11.4.2 Local Spatial Autocorrelation Analysis

The global spatial autocorrelation test shows that the provincial TEE in China presented obvious positive agglomeration characteristics, while local spatial autocorrelation analysis can be used to measure the degree of local spatial clustering. The data of 2009, 2013 and 2016 are used for analysis and research.

Figures 11.20, 11.21, and 11.22 present the Moran scatterplot of TEE in China in 2009, 2012, and 2016. As shown in the figures, most of the provinces were distributed in the first and third quadrants. More specifically, Tianjin, Hebei, Liaoning, Jiangsu, Zhejiang, Fujian, Shandong and Henan remained in the first quadrant, which shows that the TEE of these eight provinces was very high. They had a diffusion effect on the surrounding areas, and they are mainly distributed in coastal areas. Guangxi, Sichuan, Chongqing, Yunnan, Qinghai and Xinjiang were always in the third quadrant in 2009, 2012 and 2016, which indicates that the TEE of these six provinces was at a low level. They are also surrounded by low-level regions, which are mainly distributed in western regions of China. Guangdong and Guizhou were always in the fourth quadrant, which shows that the TEE of these two provinces was at a high level, but these two provinces are surrounded by low-efficiency regions, which shows an obvious polarization effect.

Figures 11.23, 11.24, and 11.25 show the areas with significant (5%) locations color-coded by different types of spatial autocorrelation (local Moran's I) of four lifespan indicators, respectively. HH districts of TEE are mostly located in the Lower reaches of the Yellow River, whereas LL districts are mainly in southwestern China. More specifically, in 2009, the HH districts of TEE were composed by three obvious provinces: Hebei, Shandong, and Henan, which passed the test using a significance level of 5%; whereas the LL districts of TEE exhibited an obvious small cluster area, which only included two obvious provinces: Sichuan and Yunnan. By 2012, the HH districts of TEE decreased by one obvious province (Henan), while the LL districts of TEE increased by two obvious provinces (Guizhou and Qinghai). By 2016, the HH districts of TEE increased compared with 2012, which were composed by four obvious provinces: Hebei, Henan, Jiangsu, and Shandong, while the LL districts

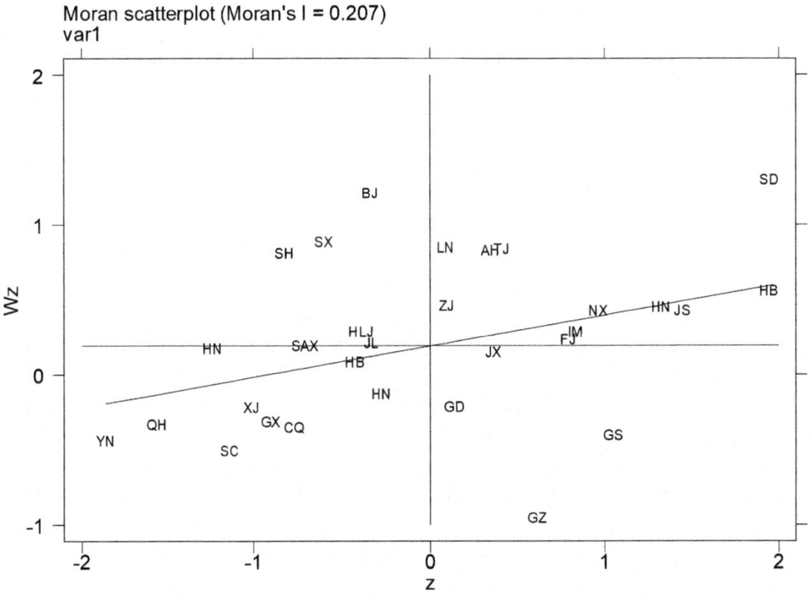

Fig. 11.20 The Moran's I scatterplot of TEE in 30 provinces in 2009. *Data source* Edited by the authors

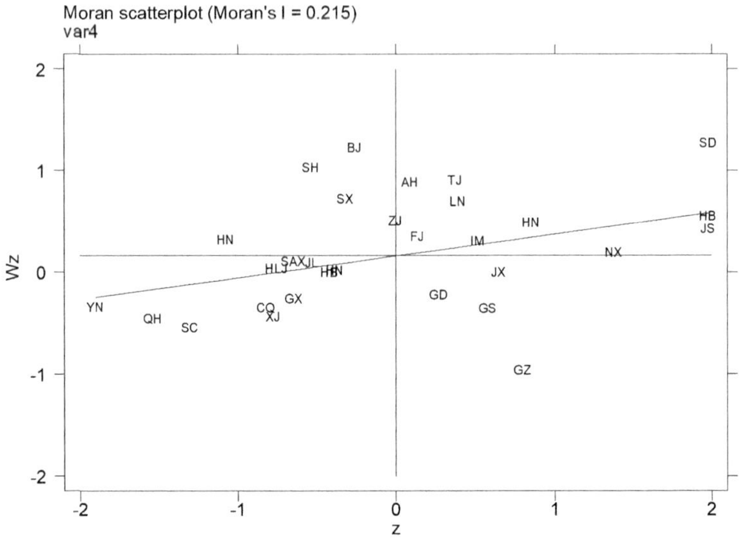

Fig. 11.21 The Moran's I scatterplot of TEE in 30 provinces in 2012. *Data source* Edited by the authors

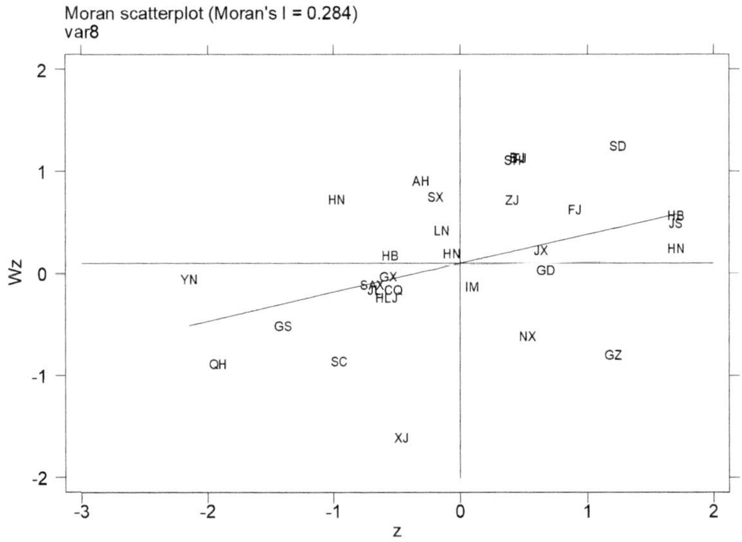

Fig. 11.22 The Moran's I scatterplot of TEE in 30 provinces in 2016. *Data source* Edited by the authors

Fig. 11.23 LISA diagram of TEE in 2009. *Data source* Edited by the authors

of TEE decreased compared with 2012, which included three obvious provinces: Sichuan, Gansu, and Qinghai.

11.5 Spatial Econometric Analysis of the Influencing Factors of TEE

11.5.1 Selection of Variables

Based on the spatial autocorrelation analysis of the TEE in China, this study uses the spatial Durbin model to analyze the influencing factors of TEE. On the basis of previous studies, this study selects economic development level, transport structure, transport infrastructure level, technological progress in transport energy utilization, urbanization level and urban population density as independent variables. The data are from the National Bureau of Statistics.

11.5.1.1 Economic Development Level

The higher the GDP per capita and income per capita, the greater the need for a higher quality environment; this can often lead to stricter environmental controls

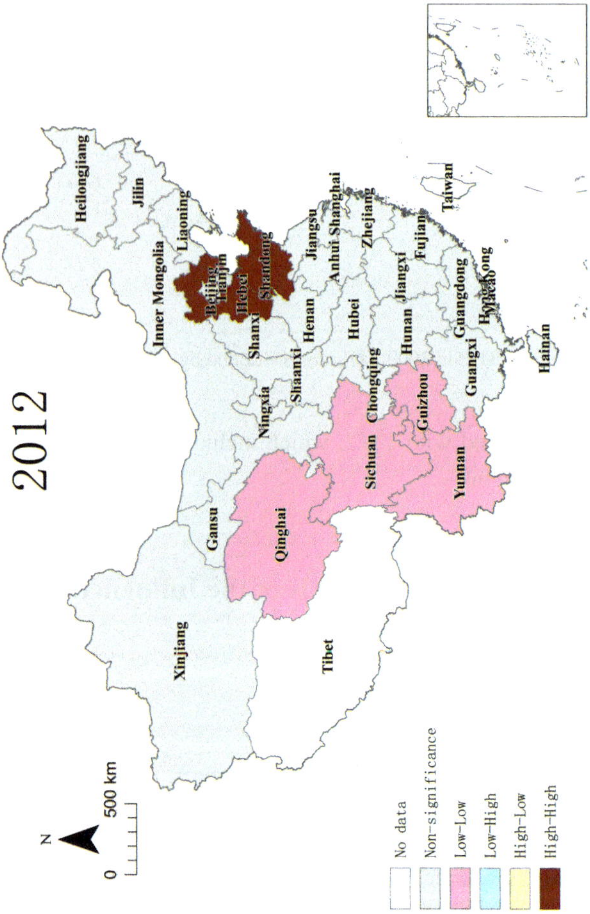

Fig. 11.24 LISA diagram of TEE in 2012. *Data source* Edited by the authors

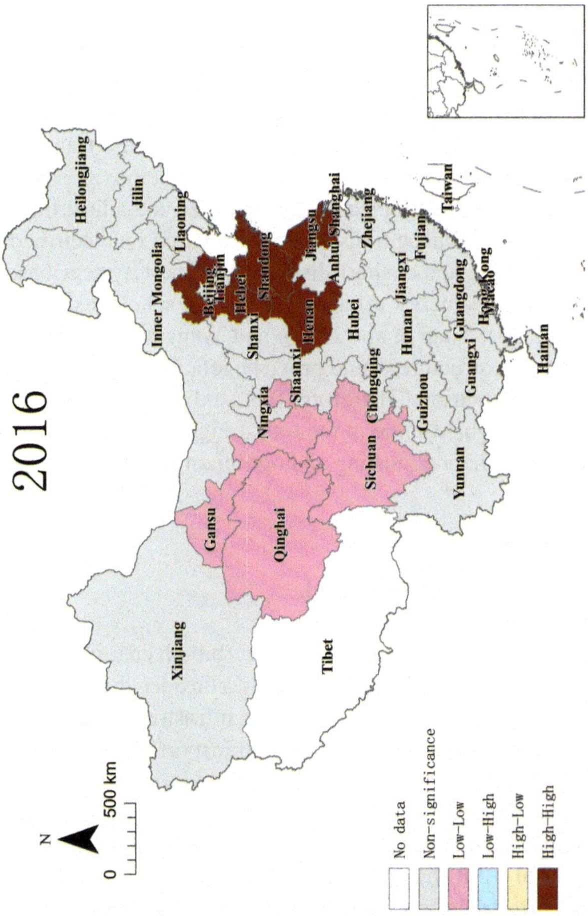

Fig. 11.25 LISA diagram of TEE in 2016. *Data source* Edited by the authors

enacted by local governments. Developed regions are better able to attract an excellent labor force, implement superior production techniques, develop advanced management experience, and access more investment capital. All these factors can lead to increases in TEE. Therefore, it is expected that there exists a positive influence between economic development level and TEE.

11.5.1.2 Transport Structure

According to the China Vehicle Environmental Management Annual Report 2018, the energy consumption and pollutant emissions per unit of highway freight are 7 times and 13 times those of railway transport, respectively. Waterway transport is characterized by economy, safety, low pollution, and low mass transport, so it has become a significant alternative for the transport of dangerous goods. The Report also points out that the idiographic measure to optimize and adjust transport structures is to establish a long-distance transport system based on electrified trains and environment-friendly ships, and a short-distance transport system based on low-emissions vehicles and new energy vehicles. Moreover, scholars such as Wei et al. (2013), Cui and Li (2015), Yuan et al. (2017), and Wang and He (2017) thought a more reasonable transport structure would prove beneficial in reducing carbon emissions from the transport industry. Therefore, this study uses the proportion of the railway and waterway freight volume in the total freight volume to represent transport structure. We predict that readjusting and optimizing the transport structure can improve TEE.

11.5.1.3 Transport Infrastructure Level

An excellent transport infrastructure system can curb both energy consumption and pollutant discharge, but transport planners should consider the level of per capita transport infrastructure. When per capita transport infrastructure is at a low level, it may lead to challenges resulting from increasing transport loads. Traffic congestion has been repeatedly considered as a main factor of road transport energy consumption and pollutant emissions (Oduyemi & Davidson, 1998). Therefore, an excellent transport system requires an adequate supply of transport infrastructure. Following the studies of Ma et al. (2018b) and Yang et al. (2019), this study measures the supply of transport infrastructure using per capita transport infrastructure level and offers the prediction that the improvement of per capita transport infrastructure will improve TEE.

11.5.1.4 Technological Progress

Technological progress in energy utilization is a key driving force for reducing carbon dioxide emissions (Bilgen, 2014; Dinda, 2004; Lee & Min, 2015; Wang et al., 2005), which has been widely considered and accepted. In general, lower energy intensity means lower environmental costs and better technology for transforming energy into economic output (Zeng et al., 2019). Referring to Scholl et al. (1996), Wei et al. (2013) and Yuan et al. (2017), we use the reciprocal of the energy intensity of the transport sector as an indicator of technological progress, and predict that technological progress will improve TEE.

11.5.1.5 Urbanization Level

Urbanization is thought to be correlated with transport carbon emissions. A large number of scholars, such as Wu et al. (2015), Xie et al. (2017), and Lv et al. (2019), have conducted in-depth studies on urbanization and transport carbon emissions. The ongoing urbanization has brought about large-scale construction of urban residential housing and infrastructure, and a notable increase of transport demand. However, the high transport demand from urbanization causes more energy consumption and gas pollutants.

11.5.1.6 Urban Population Density

Due to the existence of the scale effect and agglomeration effect, population growth can promote the improvement of urban production efficiency, resulting in economic benefits that exceed the external environmental costs of transport. Alford and Whiteman (2009) and Modarres (2013) found that urban commuters will consume relatively less energy and release less carbon dioxide in areas with high population density. However, when the urban population density reaches a certain level, the additional emission reduction effect brought by high population density may not be significant (Hong, 2015). Exorbitant urban population density may directly or indirectly have a negative impact on traffic flow, resulting in additional emissions of transport pollutants. Therefore, it is necessary to further empirically analyze the impact of urban population density on TEE (Table 11.5).

11.5.2 Results of Spatial Durbin Regression

The LR and Wald tests show that SDM is more suitable than SLM (LR chi2 (6) = 35.87, prob > chi2 = 0.0000; Wald chi2 (6) = 38.94, prob > chi2 = 0.0000), or SEM (LR chi2 (6) = 30.31, prob > chi2 = 0.0000; Wald chi2 (6) = 30.68, prob > chi2 = 0.0000). The results of the Hausmann test (chi2 (8) = 14.66, prob > chi2 = 0.0230) show that fixed-effect SDM is more suitable than random-effect SDM. Therefore, this study uses the SDM method to explore the influencing factors of TEE. The specific regression equation of the fixed-effect SDM model is as follows:

Table 11.5 Influencing factors of TEE

Explanatory variables	Definitions of variables	Pre-judgment
Economic development level (EDL)	GDP per capita (10^4 RMB)	Positive
Transport structure (TS)	Proportion of railway and waterway freight volumes to the total freight volume (%)	Positive
Transport infrastructure level (TIL)	Proportion of the length of railways and highways to the total area (km/km^2)	Positive
Technological progress (TP)	Ratio of GDP to energy consumption of the transport sector (tons of standard coal/10^4 RMB)	Positive
Urbanization level (UL)	Proportion of urban permanent resident population in the total permanent resident population (%)	Unknown
Urban population density (UPD)	Ratio of urban population to the urban area (persons/km^2)	Unknown

Data source The authors, edited from China Statistical Yearbook (2020) and China Energy Statistical Yearbook (2018)

$$\begin{aligned} TEE_{i,t} = {} & \rho \sum_{j=1}^{N} W_{i,j} TEE_{i,t} + \beta 1 DEL_{i,t} + \beta 2 TS_{i,t} + \beta 3 TIL_{i,t} + \beta 4 TP_{i,t} \\ & + \beta 5 UL_{i,t} + \beta 6 UPD_{i,t} + \theta 1 \sum_{j=1}^{N} W_{i,j} DEL_{i,t} + \theta 2 \sum_{j=1}^{N} W_{i,j} TS_{i,t} \\ & + \theta 3 \sum_{j=1}^{N} TIL_{i,t} + \theta 4 \sum_{j=1}^{N} TP_{i,t} + \theta 5 \sum_{j=1}^{N} W_{i,j} UL_{i,t} + \theta 6 \sum_{j=1}^{N} W_{i,j} UPD_{i,t} \\ & + \mu + \varepsilon_{i,t} \end{aligned} \tag{11.1}$$

Table 11.6 shows three regression results of SDM: spatial fixed-effects, time fixed-effects, and spatial and time fixed-effects. The LR test shows that the spatial and time fixed-effects model is better than the spatial fixed-effects model (chi2 (8) = 37.91; prob > chi2 = 0.0000) and the time fixed-effects model (chi2 (8) = 405.67; prob > chi2 = 0.0000). Therefore, the spatial and time fixed-effects model is selected to analyze the influencing factors of TEE.

The economic development level has a positive correlation with TEE, but it is not significant, which indicates that there is no significant effect of economic development level on TEE during the study period. Generally, the improvement of regional economic development level means more capital and technology investment in regional transport environment governance. However, the regression results show that the impact of economic development level on TEE is not significant. It means that there may be a rebound effect in transport energy utilization, which makes the improvement effect of economic development on TEE not significant.

Table 11.6 Spatial Durbin model regression results

	Spatial fixed-effects	Time fixed-effects	Spatial and time fixed-effects
EDL	0.0994488	0.43634442***	0.1054519
TS	0.0816585**	0.0169256	0.0569597*
TIL	0.0588662**	0.0460659	0.0428387
TP	0.5138017***	0.64497752***	0.5053315***
UL	−1.100005***	−0.8712***	−1.094156***
UPD	−0.2045165***	−0.049706*	−0.1938203***
*W*EDL*	−0.312968**	0.1016708	0.5845805**
*W*TS*	0.0260916	0.1630552***	0.0365516
*W*TIL*	0.2376842**	0.67383162***	0.0332762
*W*TP*	−0.2316394**	0.31413132***	−0.1486
*W*UL*	1.754223***	−0.175084	−0.0641437
W UPD*	0.2486216***	−0.21284242***	0.1426383*
R-squared	0.5829	0.3513	0.3705
Log-likelihood	298.4248	114.5434	317.3779

Note: ***, **, and * indicate significance at the 1%, 5%, and 10% levels, respectively
Data source Edited by the authors

The regression coefficient of transport structure is significantly positive, indicating that the higher the proportion of railway and waterway freight volumes in the total freight volume, the higher the TEE. This conclusion is consistent with the research conclusions of Wei et al. (2013), Cui and Li (2015), Yuan et al. (2017) and Wang et al. (2018). During the research period, the annual highway freight volume accounted for more than 75% of the total freight volume in China, and the proportion of railway and waterway freight volumes in the total freight volume was relatively low. Therefore, it is necessary to implement the transport structure adjustment policy and promote the use of railways or waterways, rather than highways, for long-distance bulk cargo transport (Liu et al., 2018).

The regression coefficient of transport infrastructure level is significantly positive, but it is not significant. Since the reform and opening-up, per capita transport infrastructure has improved significantly. However, there are still some differences between China and developed countries in per capita transport infrastructure, especially in small and medium-sized cities in western China. Therefore, the Chinese government should reasonably plan the construction of transport infrastructure.

There is a significant positive correlation between technological progress and TEE. This is consistent with the conclusions of Yang et al. (2017) and Cui and Li (2015). With a series of energy-saving policies and technologies formulated and widely implemented by the Chinese government, the energy intensity of the transport sector gradually decreased from 1.48 TSC/10^4 RMB in 2009 to 1.2 TSC/10^4 RMB in 2016 (NBSC, 2020), which reflects the continuous progress of energy consumption technology in China's transport sector.

The coefficient of urbanization level is negative at the significance test, which shows that urbanization is a significant negative factor affecting TEE during the study period. Although the improvement of urbanization level will promote technical development and the transformation and upgrading of the industrial structure, which is conducive to improving the energy utilization efficiency of the transport sector and reducing its pollution emissions to a certain extent, at this stage, the industrial and technological agglomerations brought by urbanization cannot offset the transport pollution emissions brought by rapid urbanization.

Urban population density has a significant negative impact on TEE. At present, the urban population in China continues to expand, which is an irreversible trend; the urban population density in China increased from 2209 people/km^2 in 2009 to 2408 people/km^2 in 2016 (NBSC, 2020). More attention should be paid to the areas with a rapid increase of urban population density. For example, the urban population density of Guizhou and Qinghai rocketed from 2009 to 2016, while the TEE in these two provinces decreased rapidly.

11.6 Review of Current Policies

11.6.1 Actions and Policies in 2017

11.6.1.1 Technical Policy on the Prevention and Control of Pollution by Motor Vehicle Emissions

In 2017, the Ministry of Ecology and Environment of China organized the revision of "Technical Policy on the Prevention and Control of Pollution by Motor Vehicle Emissions" (hereinafter referred to as "technical policy") to build a motor vehicle pollution prevention and control system with further improvement of environmental quality as the core, form a regional joint prevention and control mechanism, and promote the systematization, scientization, legalization, refinement and informatization of motor vehicle pollution prevention and control.

(1) **Building a prevention and control system and working mechanism on motor vehicle pollution**. Encouraging the development of transport products with ecological design such as light weight, modularization, no (low) harm and recycling of motor vehicles. The prevention policies and methods relating to various types of pollution (air pollution, noise pollution, water pollution, solid waste, electromagnetic radiation and so on) in the whole life cycle (design, production, use, recycling and so on) of motor vehicles are comprehensively considered.
(2) **Gradually tightening the emission limits on new motor vehicles**. The main limited pollutants are carbon monoxide (CO), total hydrocarbons (THC), nitrogen oxides (NOx) and particulate matter (PM). For newly produced motor vehicles, the Ministry of Ecology and Environment of China shall uniformly formulate national emission standards, and encourage local governments to

implement stricter pollutant emission standards and oil quality standards for new motor vehicles. The regulator will focus on strengthening the supervision of the production and sales of heavy-duty diesel vehicles, the inspection and maintenance of motor vehicles, and the supervision of high-emission vehicles and vehicles with high intensity of use, to ensure that the emissions of vehicles on the road meet the emission standard.

11.6.1.2 Optimization and Adjustment of the Structure of Motor Vehicles in Highway Transport

(1) **Accelerating the promotion of new energy vehicles and clean energy vehicles**. For encouraging the public to buy and use new energy vehicles, Beijing, Tianjin, Shanghai, Guangzhou, Shenzhen and other cities with serious motor vehicle pollution would directly grant the vehicle license, with no limit on the date or road. At the end of 2017, 32 cities in China had launched the plan of bus electrification.

(2) **Carrying out the rectification of high-emission vehicles**. In 2017, more than 3 million yellow label cars (the cars fail to meet exhaust emission standards) and old cars were eliminated in many regions in China, and the task of eliminating yellow label cars nationwide was basically completed based on the Air Pollution Prevention and Control Action Plan. Special rectification activities for high-emission vehicles were carried out across China to strictly check fuel vehicles for excessive emissions. Building a three-level (national, provincial and municipal) network monitoring platform for regular emission inspection of motor vehicles. More than 260 sets of remote sensing monitoring equipment have been built (including 126 sets in the s and its surrounding areas), and another 90 sets are under construction. The Ministry of Public Security of China has formulated a national unified code (6063) for the punishment for excessive discharge.

(3) **Taking transport management measures to reduce pollutant emissions from some trucks**. In 2007, the Beijing government issued "Notice on Taking transport Management Measures to Reduce Pollutant Emissions from Some Trucks", and adopted comprehensive economic, legal, technical and administrative measures to restrict the use of diesel vehicles complying with National I and II emission standards; the 38 road checkpoints on the way to Beijing would strictly check vehicles for excessive emissions.

The Tianjin government made stringent requirements on the installation of diesel particulate filters (DPF) in medium- and heavy-duty diesel trucks complying with the National I emission standard. From January 1, 2018, high-emission light-duty gasoline vehicles (National I and II emission standards) are prohibited to drive on urban outer ring roads and the roads within urban outer ring roads on weekdays.

The Shanghai government has built an environmental protection information platform for monitoring high-emission vehicles in the Yangtze River Delta region, strengthening the joint prevention and control of vehicle pollution in the region, and actively promoting the emission control of vehicles complying with the National III emission standard in the city.

11.6.1.3 Acceleration of Transport Structure Adjustment by Multiple Departments

(1) **Ministry of Transport of China**: distributing the "Opinions on Comprehensively and Deeply Promoting the Development of Green transport", and proposing to accelerate the increase of the proportion of railway passenger and freight transport in transport, speed up the improvement of railway logistics service, rationalize the formation mechanism of freight rate, and reduce the long-distance transport volume of diesel trucks.
(2) **Ministry of Ecology and Environment of China**: carrying out the research on transport structure adjustment, tracking the implementation of the policy of banning coal transport by automobiles in ports around the Bohai Sea in a timely manner, and inspecting the ineffective implementation of the policy in some ports, so as to promote the steady progress of transport structure adjustment.
(3) **China Railway Corporation**: actively implementing the central government's transport structure policies, and taking the improvement of the railway freight market share and the reduction of social logistics costs as important tasks in 2018.

11.6.2 Actions and Policies in 2018

11.6.2.1 Three-Year Action Plan for Winning the Blue Sky Defense Battle Issued by the State Council

In June 2018, the State Council of China issued the Three-Year Action Plan for Winning the Blue Sky Defense Battle, which defines the general idea, basic objectives, main tasks and safeguard measures of air pollution prevention and control, and puts forward the timetable and roadmap for fighting air pollution. The main measures are as follows: optimizing and adjusting the cargo transport structure, accelerating the upgrading of automobile, ship and oil products, and strengthening the prevention and control of pollution from mobile sources of various transport modes.

(1) **Optimizing and adjusting the cargo transport structure**. First, raising the proportion of railway cargo transport in China. By 2020, the national railway freight volume increased by 30% from the 2017 level; the railway freight volume in the Beijing–Tianjin–Hebei region and its surrounding areas, the Yangtze River Delta region, and the Fenwei Plain region increased by 40%, 10% and 25%, respectively. Second, increasing the construction of special railway lines, such as railway lines for ports and railway lines for industrial and mining enterprises. Third, promoting the construction of freight hubs with both combined transport

and a reasonable network of main lines and branch lines, and supporting multi-modal container transport.

(2) **Speeding up the upgrading of automobile and ship products**. First, promoting the use of new energy vehicles; increasing the purchase of new energy vehicles for public transport, sanitation, postal service, rental, commuting and light logistics in urban built-up areas, increasing the proportion of new energy vehicles in the total vehicles to above 80% in key areas, and speeding up the construction of centralized charging piles and fast charging piles. Second, vigorously promoting the elimination and renewal of operating diesel trucks in advance. By the end of 2020, the Beijing–Tianjin–Hebei region and its surrounding areas, and the Fenwei Plain region eliminated more than 1 million operating medium- and heavy-duty diesel trucks complying with the National VI emission standard or lower emission standards, respectively. From July 1, 2019, the key regions, the Pearl River Delta region, and the Chengdu–Chongqing region have implemented the National VI emission standard for motor vehicles in advance. Third, promoting the upgrading of ship types. From July 1, 2018, the National I emission standard for new marine engines has been fully implemented. Encouraging the elimination of inland ships that have been used for more than 20 years.

(3) **Strengthening the prevention and control of pollution from mobile sources of various modes of transport.** First, strictly supervising whether vehicles satisfy the requirements of environmental protection standards; severely cracking down on the illegal sale of new motor vehicles that fail to meet environmental protection standards; strengthening the supervision and management of motor vehicle emission inspection, and comprehensively promoting the sampling inspection on whether motor vehicles meet environmental protection standards. Second, strengthening the prevention and control of pollution from non-road mobile machinery and ships; **strengthening the delimitation of emission control areas**, delimiting areas where high-emission non-road mobile machinery are prohibited, and adjusting and expanding the scope of ship emission control areas, which covers key coastal ports and some inland river areas. Third, encouraging aircrafts and ships to use shore power.

11.6.2.2 Joint Action of Multiple Departments for Diesel Truck Pollution Control

In December 2018, the Ministry of Ecology and Environment, the National Development and Reform Commission, the Ministry of Industry and Information Technology, the Ministry of Transport and other 11 departments of China jointly issued the Action Plan of Diesel Truck Pollution Control, which mainly includes four aspects: clean diesel vehicles, clean diesel engines, clean transport and clean oil.

(1) **Clean diesel vehicles**. Strictly supervising law enforcement regarding the use of diesel vehicles; strengthening the emission inspection and maintenance management of diesel vehicles. Speeding up the elimination and in-depth management of old diesel vehicles. Promoting the construction and application of a monitoring system for diesel vehicles. Promoting the elimination of backward production capacity in transport and facilitating the centralized development of the diesel truck manufacturing industry;

(2) **Clean diesel engines for vehicles**. First, carrying out the strict management of new vehicle engines, non-road mobile machinery and ships; by the end of 2020, China implemented the National IV emission standard for non-road mobile machinery, as well as the National I emission standard and the National II emission standard ahead of schedule for marine engines. Second, strengthening the delimitation of emission control areas. By the end of 2019, the key regions and cities completed the zoning of areas where the use of high-emission non-road mobile machinery is banned, and other regions and cities completed the zoning work by the end of June 2020. Third, accelerating the management and elimination of non-road mobile machinery. Fourth, strengthening comprehensive supervision and management of transport construction projects. Fifth, promoting the development of electric energy in ports, and giving priority to shore power when providing ships at berth with electricity.

(3) **Clean freight transport**. First, increasing the railway freight volume; promoting the use of railways, rather than highways, for medium- and long-distance bulk cargo and container transport, and increasing the construction of special railway lines, such as rail lines for ports and rail lines for industrial and mining enterprises. Second, promoting the development of green freight transport. If there is a large demand for bulk cargo transport in the region, the new transport line construction should put priority on the development of railway, waterway or pipeline transport; promoting the construction of freight hubs with both combined transport and a reasonable network of main lines and branch lines, and supporting multi-modal container transport. Third, optimizing the structure of the transport fleet; increasing the purchase of new energy vehicles for public transport and some vehicles for sanitation, postal service, rental, commuting and light logistics in urban built-up areas, increasing the proportion of new energy vehicles in the total vehicles to above 80% in key areas, and speeding up the construction of centralized charging piles and fast charging piles.

(4) **Clean oil**. First, speeding up the improvement of oil and gas quality standards. Starting from January 1, 2019, the whole country fully supplied automobile gasoline and diesel that meet the National VI emission standard, stopped selling regular diesel and the automobile gasoline and diesel failing to meet the National VI emission standard, abandoned the standard for regular diesel, and used the same standard for automobile diesel, regular diesel and some marine oil. Second, improving the management system of fuel, detergent synergists, and diesel exhaust fluid. Third, promoting the recovery and treatment of oil and

gas. Fourth, strengthening the supervision of oil and gas production, storage, transport and use.

11.7 Policy Recommendations for Improving TEE

11.7.1 The Government Should Reasonably Plan Transport Investment, Establish and Improve the Legal System of Energy Conservation, Emission Reduction and Low-Carbon Economic Policies in the Transport Sector

During the 13th Five-Year Plan period, the total investment in the transport sector reached 15 trillion RMB. Therefore, the local government needs to plan transport construction, optimize the investment structure and allocate transport resources effectively. Local governments must continue to increase investment in energy conservation, emission reduction, information construction and other fields. The regional transport infrastructure plan shall be consistent with the regional integrated plan, the urban economic development plan and the existing transport plan. The government should implement the corresponding financial subsidy policy, encourage the research on and application of new energy, and promote the use of new energy vehicles. Tax incentives or reliefs shall be implemented for individuals and enterprises who purchase new energy vehicles.

11.7.2 According to the Regional Economic Development Level, Different Transport Emission Reduction Schemes Should Be Implemented in Different Regions

The coastal areas with a relatively developed economy and high TEE should perfect the energy management system, and improve the energy utilization technology and pollutant treatment technology of the transport sector. At the same time, coastal areas should take more responsibility for transport energy conservation and emission reduction. Coastal areas should carry out stricter restrictions on pollutant emissions from the transport sector and higher environmental standards. Coastal areas should also accelerate the transformation of the growth mode, and establish a low-carbon transport system. The inland areas with an underdeveloped economy and low TEE should learn the advanced technology from eastern regions. It's important to note that the growth of transport in inland areas must be balanced with minimizing environmental impact.

11.7.3 Optimizing Regional Cooperation in Energy Conservation and Emission Reduction of the Transport Sector

The eastern regions in China should pass on the advanced technology of energy conservation and emission reduction to central and western regions. Strengthening the technical cooperation and exchanges between coastal regions and inland regions can also promote the balanced development of TEE in all regions. Moreover, the central government of China and coastal provinces should assist central and western regions. The latter can utilize the advanced technology, management theory, capital and related supporting facilities of the transport industry in coastal areas.

11.7.4 Optimizing the Transport Structure

In terms of freight structure, the government should provide financial subsidies and tax reliefs for the transport enterprises that follow the policy of turning from highways to railways in freight transport. China should adjust the structure of passenger and freight transport, and promote the construction of expressways, high-speed rail and rail transit to increase the proportion of public passenger transport to total passenger transport. In urban transport planning, the public transport system should be prioritized.

11.7.5 Technological Innovation Needs to Be Strengthened

China should increase government investment in the research on emission reduction technology, and actively provide financial subsidies, tax incentives and other policies to transport enterprises and scientific research institutions for further research on low-carbon and energy-saving transport technology. In the aspect of transport management, China should innovate in transport management, such as simplifying the working process and optimizing the cumbersome procedures so as to improve the organization and service level of the transport sector.

11.7.6 Internal Structural Adjustment of the Transport Industry

The outdated production mode of the transport sector should be eliminated. China should accelerate the cultivation and development of the emerging strategic transport industry. For example, the Internet industry is booming, and the deep integration between transport, internet, big data, artificial intelligence and the real economy will drive the development of transport in the direction of digitization, networking and intelligence. The above policies will promote the modernization of transport.

11.8 Conclusions

11.8.1 The National TEE First Decreased and then Increased, and the National Policy Plays a Significant Role in TEE

The national average TEE showed different trends in two phases. Phase one lasted from 2009 to 2011, and the national average TEE exhibited a general downward trend and fell from 0.596 in 2009 to 0.578 in 2011. In the second phase, lasting from 2011 to 2016, the national average TEE showed an upward trend, reaching a value of 0.655 in 2016 and thereby exceeding the initial TEE level.

In order to confront the global economic crisis of 2008, China executed a positive financial policy and permissive monetary policy. This allowed various provinces to increase their investments in transport infrastructure, resulting in over-investment in transport (Li et al., 2015; Lu, 2012; Zhang & Zhou, 2020). Therefore, the national TEE exhibited a downward trend from 2010 to 2012. In 2011, the Ministry of Transport (2011) developed many policies intended to curb the release of transport pollutants. Notably, in 2013, the Communist Party of China (CPC) (2013) promoted the building of ecological civilization, and various provinces in China increased efforts to achieve energy conservation and emissions reduction. Many new energy-saving technologies were popularized and applied in the transport sector. These were conducive to improving TEE and can explain the upward trend in national TEE after 2011.

11.8.2 Coastal Regions Had the Highest Level of TEE, Followed by Central and Western Regions

A comparison of regions indicates that the Northern coast had the highest level of TEE, followed by the Eastern coast, Middle reaches of the Yellow River, Middle reaches of the Yangtze River, Southern coast, Northeast, Northwest, and Southwest. Coastal provinces enjoy geographical advantages and high levels of economic development. Middle reaches of the Yellow River and Middle reaches of the Yangtze River closely follow coastal areas and have been opened up to the outside world. Northeast, Northwest and Southwest regions are less able to acquire and introduce advanced technologies. After the reform and opening-up, a large number of talented workers moved from the Northeast, Northwest, and Southwest to coastal areas (Cheng & Lu, 2017). The resulting loss of talent has made it difficult to improve production technologies in various industries, especially in transport, so TEE levels remain low in these regions.

11.8.3 Transport Structure and Technological Progress Have a Positive Impact on TEE, While Urbanization Level and Urban Population Density Have a Significant Negative Impact on TEE

The global spatial autocorrelation test shows that the provincial TEE in China presented obvious positive agglomeration characteristics, meaning that the spatial panel data model fits better than the non-spatial panel data model. We apply the SDM method to analyze the influencing factors of TEE, and have found that transport structure and technological progress have a positive impact on TEE, while urbanization level and urban population density have a significant negative impact on TEE. What does seem fairly clear is that the overall level of TEE in China was low and total transport CO_2 emissions grew at the fastest rates seen for years. In the early stage of development, factors including backward technology, insufficient environmental protection consciousness, and extensive development had adverse impacts on the energy saving and emission reduction efforts in the transport sector. Although China has issued a series of policies to improve this situation, the effects of these policies have not emerged for a period of time due to the capital-intensive and long-cycle nature of the transport industry. To achieve the goals of the Paris Agreement, the Chinese government needs to focus on more effective and sustainable transport development policies.

References

Alford, G., & Whiteman, J. (2009). Macro-urban form and transport energy outcomes: Investigations for Melbourne. *Road & Transport Research, 18*(1), 53–67. https://www.mendeley.com/catalogue/609553a2-7b47-35ba-a296-c9ea79423b60/. Accessed April 24, 2020.

All Provincial Statistical Yearbooks. (2011–2017). All province bureaus of statistics of China, http://tongji.cnki.net/kns55/Navi/NaviDefault.aspx. Accessed April 24, 2020.

Banker, R. D., Charnes, A., & Cooper, W. W. (1984). Some models for estimating technical and scale inefficiencies in data envelopment analysis. *Management Science, 30*, 1078–1092. http://www.jstor.org/stable/2631725

Bian, L. H., & Ji, M. Q. (2019). Research on influencing factors and prediction of transportation carbon emissions in Qinghai. *Ecological Economy, 35*(2), 35–39+100. http://en.cnki.com.cn/Article_en/CJFDTotal-STJJ201902009.htm

Bilgen, S. (2014). Structure and environmental impact of global energy consumption. *Renewable Sustainable Energy Reviews, 38*, 890–902. https://doi.org/10.1016/j.rser.2014.07.004

Chang, Y. T., Chang, N., Danao, D., & Zhang, N. (2013). Environmental efficiency analysis of transportation system in China: A non-radial DEA approach. *Energy Policy, 58*, 277–283. https://doi.org/10.1016/j.enpol.2013.03.011

Charnes, A., Cooper, W. W., & Rhodes, E. (1978). Measuring the efficiency of decision making units. *European Journal of Operational Research, 2*(6), 429–444. https://doi.org/10.1016/0377-2217(78)90138-8

Chen, L., Chi, W. Y., Li, X. R., & He, T. (2019). Factor analysis of regional transportation carbon dioxide emissions based on high-quality transportation development. *Highway, 6*, 172–176. http://www.cnki.com.cn/Article/CJFDTotal-GLGL201906035.htm. Accessed December 7, 2019.

Cheng, M., & Lu, Y. (2017). Investment efficiency of urban infrastructure systems: Empirical measurement and implications for China. *Habitat International, 70*, 91–102. https://doi.org/10.1016/j.habitatint.2017.10.008

China Energy Statistical Yearbook (CESY). (2011–2017). China Statistical Publishing House, Beijing. http://tongji.oversea.cnki.net/oversea/engnavi/HomePage.aspx?id=N2018070147&name=YCXME&floor=1. Accessed April 24, 2020.

China Energy Statistical Yearbook (CESY). (2017). China Statistical Publishing House, Beijing. http://tongji.oversea.cnki.net/oversea/engnavi/YearBook.aspx?id=N2018070147&floor=1. Accessed April 23, 2020.

China Statistical Yearbooks (CSY). (2011–2017). China Statistical Publishing House, Beijing. http://tongji.oversea.cnki.net/oversea/engnavi/HomePage.aspx?id=N2017100312&name=YINFN&floor=1. Accessed April 24, 2020.

Cui, Q., & Li, Y. (2015). An empirical study on the influencing factors of transportation carbon efficiency: Evidences from fifteen countries. *Applied Energy, 141*, 209–217. https://doi.org/10.1016/j.apenergy.2014.12.040

Cui, Q., & Li, Y. (2017). Airline efficiency measures using a dynamic epsilon-based measure model. *Transportation Research Part A, 100*, 121–134. https://doi.org/10.1016/j.tra.2017.04.013

Darido, G., Torres-Montoya, M., & Mehndiratta, S. (2014). Urban transport and CO_2 emissions: some evidence from Chinese cities. *Wiley Interdisciplinary Reviews: Energy and Environment, 3*(2), 122–155. https://wires.onlinelibrary.wiley.com/doi/full/https://doi.org/10.1002/wene.71

Dinda, S. (2004). Environmental Kuznets curve hypothesis: A survey. *Ecological Economics, 49*(4), 431–455. https://doi.org/10.1016/j.ecolecon.2004.02.011

Lawrence, D. F., Brian, S. J., & William, B. (2000). Linking land use with household vehicle emissions in the central Puget sound: Methodological framework and findings. *Transportation Research Part D: Transport and Environment, 5*(3): 173–196. https://www.sciencedirect.com/science/article/pii/S1361920999000322

Feng, C., & Wang, M. (2018). Analysis of energy efficiency in China's transportation sector. *Renewable and Sustainable Energy Reviews, 94*, 565–575. https://doi.org/10.1016/j.rser.2018.06.037

Hong, J. (2015). Non-linear influences of the built environment on transportation emissions: Focusing on densities. *Journal of Transport and Land Use, 10*(1), 229–240. https://doi.org/10.5198/jtlu.2015.815

Ji, X. F., Bai, S. M., Chen, F., & Pu, Y. M. (2022). Allocation of provincial transportation carbon emission quota from the perspective of efficiency. *Journal of Arid Land Resources and Environment, 36*(4), 1–7. https://d.wanfangdata.com.cn/periodical/ghqzyyhj202204001

Lan, Z. R., & Zhang, H. G. (2014). Study on inter-provincial difference in carbon dioxide emissions efficiency of traffic and transportation sector in China. *Logistics Technology, 33*(4), 631–636. http://en.cnki.com.cn/Article_en/CJFDTotal-WLJS201407043.htm

Lee, S., & Lee, B. (2014). The influence of urban form on GHG emissions in the U. S. household sector. *Energy Policy, 68*, 534–549. https://www.sciencedirect.com/science/article/pii/S0301421514000299

Lee, K. H., & Min, B. (2015). Green R&D for eco-innovation and its impact on carbon dioxide emissions and firm performance. *Journal of Cleaner Production, 108*(Part A), 534–542. https://doi.org/10.1016/j.jclepro.2015.05.114

Lesage, J. P., & Pace, R. K. (2009). *Introduction to spatial econometrics*. Chapman and Hall Press. https://link.springer.com/article/https://doi.org/10.1007/BF03354894

Li, J. W., & Zhang, G. Q. (2016). Estimation of capital stock and capital return rate of China's transportation infrastructure. *Contemporary Finance & Economics, 6*, 3–14. http://en.cnki.com.cn/Article_en/CJFDTotal-DDCJ201606002.htm. Accessed December 7, 2019.

Li, T., Yang, W. Y., Zhang, H. R., & Cao, X. S. (2015). Evaluating the impact of transport investment on the efficiency of regional integrated transport systems in China. *Transport Policy., 45*, 66–76. https://doi.org/10.1016/j.tranpol.2015.09.005

Liu, Y. S., Tian, Y. H., & Luo, Y. (2018). Upgrading of industrial structure, energy efficiency, green total factor productivity. *The Theory and Practice of Finance and Economics,* 1, 118–126. http://en.cnki.com.cn/Article_en/CJFDTotal-CLSJ201801018.htm. Accessed April 25, 2020.

Lu, T. D. (2012). The proposition to avoid the over advance and inappropriate construction of China's transport infrastructures. *Scientia Geographica Sinica, 32*(1) , 2–111. http://geoscien.neigae.ac.cn/EN/Y2012/V32/I1/2. Accessed December 7, 2019.

Lv, Q., Liu, H., Yang, D., & Liu, H. (2019). Effects of urbanization on freight transport carbon dioxide emissions in China: Common characteristics and regional disparity. *Journal of Cleaner Production, 211*, 481–489. https://doi.org/10.1016/j.jclepro.2018.11.182

Ma, F., Wang, W. L., Sun, Q. P., Liu, F., & Li, X. D. (2018a). Integrated transport efficiency and its spatial convergence in China's Provinces: A Super-SBM DEA model considering undesirable outputs. *Applied Sciences, 8*(9), 1698–1720. https://doi.org/10.3390/app8091698

Ma, J. L., Liu, Y. H., Li, L., Yao, K., Zeng, X. L., & Dai, J. J. (2018b). Research on the influencing factors of carbon dioxide emissions in urban passenger transport based on principal component analysis. *Environmental Pollution & Control, 40*(10), 1188–1202. http://en.cnki.com.cn/Article_en/CJFDTotal-HJWR201810021.htm. Accessed December 7, 2019.

Ministry of Ecology and Environment of China (MEEC). (2018). China Vehicle Environmental Management Annual Report. Beijing, http://www.gov.cn/guoqing/2019-04/09/content_5380744.htm. Accessed December 7, 2019.

Ministry of Transport of China (MOT). (2011). The twelfth five-year plan of energy conservation and emission reduction of highway and waterway transportation. Beijing. http://www.gov.cn/zwgk/2011-07/08/content_1902139.htm. Accessed December 7, 2019.

Modarres, A. (2013). Commuting and energy consumption: Toward an equitable transportation policy. *Journal of Transport Geography, 33*, 240–249. https://doi.org/10.1016/j.jtrangeo.2013.09.005

National Bureau of Statistics of China (NBSC). (2020). http://data.stats.gov.cn/easyquery.htm?cn=E0103. Accessed April 29, 2020.

National Development and Reform Commission (NDRC). (2017). Baseline emissions factors for regional power grids in China. Beijing. http://m.tanpaifang.com/article/62660.html. Accessed April 24, 2020.

Oduyemi, K. O. K., & Davidson, B. (1998). The impacts of road traffic management on urban air quality. *The Science of the Total Environment, 218*(1), 59–66. https://doi.org/10.1016/S0048-9697(98)00201-0

Scholl, L., Schipper, L., & Kiang, N. (1996). CO_2 emissions from passenger transport: A comparison of international trends from 1973 to 1992. *Energy Policy, 24*(1), 17–30. https://doi.org/10.1016/0301-4215(95)00148-4

Song, M. L., Zhang, G. J., Zeng, W. X., Liu, J. H., & Fang, K. G. (2016). Railway transportation and environmental efficiency in China. *Transportation Research Part D: Transport and Environment, 48*, 488–489. https://doi.org/10.1016/j.trd.2015.07.003

Statistical Yearbook of the Chinese Investment in Fixed Assets. (2010–2013, 2015–2017). China Statistical Publishing House, Beijing. https://data.cnki.net/yearBook/single?id=N2019030174

The Communist Party of China. (2013). Decision of the Central Committee of the Communist Party of China on Some Major Issues Concerning Comprehensively Deepening the Reform, Beijing. http://www.china.org.cn/china/third_plenary_session/2014-01/16/content_31212602.htm. Accessed December 7, 2019.

Timilsina, G. R., & Shrestha, A. (2009). Factors affecting transport sector CO_2 emissions growth in Latin American and Caribbean Countries: An LMDI decomposition analysis. *International*

Journal of Energy Research, 33(4), 396–414. https://onlinelibrary.wiley.com/doi/abs/https://doi.org/10.1002/er.1486

Tobin, J. (1958). Estimation of relationships for limited dependent variables. *Econometrica, 26*(1), 24–36. https://doi.org/10.2307/1907382

Tone, K. (2001). Slacks-based measure of efficiency in data envelopment analysis. *European Journal of Operational Research, 130*(3), 498–509. https://doi.org/10.1016/S0377-2217(99)00407-5

Tone, K. (2004). Dealing with undesirable outputs in DEA: A slacks-based measure (SBM) approach. In: North American Productivity Workshop. Toronto. https://doc.guandang.net/s8615e0bc793c1a059a3db9f3.html

Tone, K., & Tsutsui, M. (2010). An epsilon-based measure of efficiency in DEA: A third pole of technical efficiency. *European Journal of Operational Research, 207*, 1554–1563. https://doi.org/10.1016/j.ejor.2010.07.014

Wang, B., Sun, Y. F., Chen, Q. X., & Wang, Z. H. (2018). Determinants analysis of carbon dioxide emissions in passenger and freight transportation sectors in China. *Structural Change and Economics D, 47*, 127–132. https://doi.org/10.1016/j.strueco.2018.08.003

Wang, C., Chen, J. N., & Zou, J. (2005). Decomposition of energy-related CO_2 emission in China: 1957–2000. *Energy, 30*, 73–83. https://doi.org/10.1016/j.energy.2004.04.002

Wang, Z. H., & He, W. J. (2017). CO_2 emissions efficiency and marginal abatement costs of the regional transportation sectors in China. *Transportation Research Part D: Transport and Environment, 50*, 83–97. https://doi.org/10.1016/j.trd.2016.10.004

Wei, Q. Q., Zhao, S. Z., & Xiao, W. (2013). A quantitative analysis of carbon dioxide emissions reduction ability of transportation structure optimization in China. *Journal of Transportation Systems Engineering and Information Technology, 13*(3), 10–17. https://doi.org/10.1016/S1570-6672(13)60109-9

Wu, C. F., Xiong, J. H., Wu, W. C., Gao, W. J., & Liu, X. B. (2015). Calculation and effect factor analysis of transport carbon dioxide emissions in Gansu Province based on STIRPAT Model. *Journal of Glaciology and Geocryology, 37*(3), 826–834. http://en.cnki.com.cn/Article_en/CJFDTotal-BCDT201503030.htm. Accessed December 7, 2019.

Wu, P., Wang, Y. Q., Chui, Y. H., Li, Y., & Lin, T. Y. (2019). Production efficiency and geographical location of Chinese coal enterprises-undesirable EBM DEA. *Resources Policy, 64*, 101527. https://doi.org/10.1016/j.resourpol.2019.101527

Xie, R., Fang, J., & Liu, C. (2017). The effects of transportation infrastructure on urban carbon dioxide emissions. *Applied Energy, 196*, 199–207. https://doi.org/10.1016/j.apenergy.2017.01.020

Yang, W. Y., Li, T., & Cao, X. S. (2015). Examining the impacts of socio-economic factors, urban form and transportation development on CO_2 emissions from transportation in China: A panel data analysis of China's Provinces. *Habitat International, 49*,212–220. https://www.sciencedirect.com/science/article/pii/S0306261916314957

Yang, K. H., Xia, Z. Y., & He, P. et al. (2017). Correlation of fuel quality and emissions of motor vehicle with atmospheric pollution in Beijing. *Journal of University of Chinese Academy of Sciences, 34*(3), 304–317. http://journal.ucas.ac.cn/EN/abstract/abstract12467.shtml

Yang, L., Wang, K. L., & Geng, J. C. (2018). China's regional ecological energy efficiency and energy saving and pollution abatement potentials: An empirical analysis using epsilon-based measure model. *Journal of Cleaner Production, 194*, 300–308. https://doi.org/10.1016/j.jclepro.2018.05.129

Yang, W. Y., Wang, W. L., & Ouyang, S. S. (2019). The influencing factors and spatial spillover effects of CO_2 emissions from transportation in China. *Science of the Total Environment, 696*, 133900. https://doi.org/10.1016/j.scitotenv.2019.133900

Yuan, C. W., Zhang, S., Jiao, P., & Wu, D. Y. (2017). Temporal and spatial variation and influencing factors research on total factor efficiency for transportation carbon dioxide emissions in China. *Resources Science, 39*(4), 687–697. http://en.cnki.com.cn/Article_en/CJFDTOTAL-ZRZY201704010.htm. Accessed December 7, 2019.

Zeng, L. E., Lu, H. Y., Liu, Y. P., Zhou, Y., & Hu, H. Y. (2019). Analysis of regional differences and influencing factors on China's carbon dioxide emissions efficiency in 2005–2015. *Energies, 12*(16), 3081–4001. https://doi.org/10.3390/en12163081

Zhang, J., Wu, G. Y., & Zhang, J. P. (2004). The estimation of China's Provincial capital stock: 1952–2000. *Economic Research Journal, 10*, 35–44. http://en.cnki.com.cn/Article_en/CJFDTOTAL-JJYJ200410004.htm. Accessed December 7, 2019.

Zhang, N., & Zhou, M. L. (2020). The inequality of city-level energy efficiency for China. *Journal of Environmental Management, 255*, 109843. https://doi.org/10.1016/j.jenvman.2019.109843

Zhang, T. X. (2012). Research on China's urban road transport carbon emissions under urbanization process. *China Population, Resources and Environment, 22*(8), 3–9. http://www.en.cnki.com.cn/Article_en/CJFDTOTAL-ZGRZ201208002.htm. http://en.cnki.com.cn/Article_en/CJFDTotal-CSFY201907004.htm. Accessed December 7, 2019.

Zhao, P. J., Zeng, L. E., Lu, H. Y., Zhou, Y., Hu, H. Y., & Wei, X. Y. (2020). Green economic efficiency and its influencing factors in China from 2008 to 2017: Based on the super-SBM model with undesirable outputs and spatial Dubin model. *Science of the Total Environment*, 140026. https://doi.org/10.1016/j.scitotenv.2020.140026

Zhao, P., Zeng, L., Li, P., Lu, H., Hu, H., Li, C., Zheng, M., Li, H., Yu, Z., Yuan, D., et al. (2022). China's transportation sector carbon dioxide emissions efficiency and its influencing factors based on the EBM DEA model with undesirable outputs and spatial Durbin model. *Energy, 238*, 121934. https://doi.org/10.1016/j.energy.2021.121934

Zhou, Y., Kong, Y., Sha, J., & Wang, H. K. (2019). The role of industrial structure upgrades in eco-efficiency evolution: Spatial correlation and spillover effects. *Science of the Total Environment, 687*(15), 1327–1336. https://doi.org/10.1016/j.scitotenv.2019.06.182

Chapter 12
Transport Safety

12.1 Research Progress in Traffic Safety

The influencing factors of transport safety can be divided into micro-level influencing factors and macro socio-economic factors. The research on micro-level influencing factors aims to put forward targeted improvement measures for specific traffic scenarios, while the research on macro socio-economic factors aims to explore macro traffic safety policies based on the relation between the socio-economic environment and traffic safety.

12.1.1 Micro-Level Influencing Factors

Scholars embarked upon the research on micro-level influencing factors of transport safety earlier, and achieved very great success. Micro-level influencing factors mainly include traffic participants, vehicles, roads (Verma et al., 2021), climatic environment (Chang & Wang, 2006; Eboli et al., 2020; Kopelias et al., 2007), and traffic management laws and regulations.

Traffic participants include drivers, passengers, and pedestrians. Zhang et al. (2010) analyzed China's road fatalities and their influencing factors through descriptive statistical analysis, and found that motor vehicle driver factors, such as driving time, driving years, alcohol, fatigue, speeding, and use of seat belts and headlights, have obvious effects on road safety. Petridou and Moustaki (2000) found that driver-related behavioral factors are the main causes of motor vehicle accidents. Zhuang and Wu (2011) revealed that the behaviors of pedestrians on roadways in China increase the odds of accidents and fatalities. Based on the data analysis of traffic accidents in China, Wang et al. (2019) found that speeding is the primary cause of traffic accidents. Based on the data of 2014, Bucsuházy et al. (2020) found that the most common factor contributing to accidents is driver inattention according to the Czech In-depth Accident Study database. Chand et al. (2021) found that fatigue and

P. Zhao and L. Zeng, *Transport Efficiency and Safety in China*, Population, Regional Development and Transport, https://doi.org/10.1007/978-981-99-1055-7_12

distracted driving are two factors leading to road traffic accidents. Yu and Tsai (2021) found that road safety education for motor vehicle drivers has a moderate effect in reducing adverse traffic outcomes based on the analysis of population-based data of Taiwan.

Vehicle factors include the vehicle type, design performance, and physical conditions. Rechnitzer et al. (2007) held that vehicle defects are a contributing factor to over 6% of road traffic accidents. Goel (2017) found that two-wheelers, heavy vehicles and passenger cars are related to a higher risk of road fatalities in India. Wang et al. (2019) found that the number of traffic injuries and deaths in traffic accidents related to freight vehicles is slightly higher than that in traffic accidents related to private cars in China. Shaik and Hossain held that buses and trucks are more involved in road traffic accidents in Bangladesh. Ganji et al. (2018) pointed out that the most contributing factor affecting vehicle safety is the braking system, followed by fuel supply and electrical systems.

Road factors mainly include road conditions, road geometry, and road class. Pei and Ma (2003) thought that road characteristics and facilities also have an impact on the behavior of traffic participants. Wang et al. (2019) noted that crashes tend to occur on the segments with curves, dense access points, and a high percentage of heavy vehicles, according to the studies on 161 road segments of eight suburban arterials in Shanghai. Macioszek (2011) found that motorcyclists and cyclists trajectories on small one-lane roundabouts have strongly influenced other traffic participants' behavior.

Climate factors, such as visibility, temperature, wind speed, rainfall and moistness, have effects on the likelihood and severity of vehicle collisions (Bergel-Hayat et al., 2013). Current studies show that there exists a significant positive correlation between rainfall and road accidents (Bergel-Hayat et al., 2013). Tu et al. (2015) found that the impacts of the haze weather condition on driving behavior are significant. Sangkharat et al. (2021) found that high rainfall levels could significantly increase road accidents in both the Southern and Northern provinces of Thailand from 2012 to 2018. Selecting 43 main roads in Finland from 2014 to 2016 as the research object, Malin et al. (2019) held that the risk in poor weather and road conditions is higher on motorways compared to two-lane and multiple-lane roads.

12.1.2 Macro Socio-Economic Factors

Macro socio-economic factors mainly include economics, urban level, population, length of highways, and so on, which are analyzed mainly through descriptive statistical analysis and regression analysis.

Western scholars have studied the relationship between traffic accidents and socio-economic activities since the 1940s. Smeed (1949) analyzed the relationship between traffic accidents and economic development based on the data of 20 European countries, and pointed out that traffic fatalities will first increase and then decrease with economic growth, which is called Smeed's Law. Scuffham (2003) held that these variables, such as real GDP per capita, unemployment rate, proportion of motorcycles to total motor vehicles, proportion of males to total population, alcohol consumption per capita, and speed limit, are important factors affecting traffic safety in New Zealand. Kopits and Cropper (2005) found an inverted U-shaped relationship between the road traffic fatality rate per population and GDP per capita. La Torre et al. (2007) applied multiple linear regression models to identify determinants of regional differences in traffic accident mortality in Italy from a socio-demographic perspective, and they found that the employment rate and alcohol use are two factors affecting traffic accident mortality. Gaygisiz (2009) found a positive link between favorable economic conditions (high per capita income, high employment rate, and low income inequality) and high traffic safety in more than 30 member countries of the Organization for Economic Co-operation and Development (OECD). Castillo-Manzano et al. (2014) developed a fixed-effects model to study the influencing factors of road traffic fatality rates of the EU-27 countries from 1999 to 2009, and they found that the density of hospital beds contributes to the fall in traffic-related fatalities, and that the quality of general medical facilities and technology associated with increases in health expenditure may also be a relevant factor in reducing road traffic fatalities. Pu et al. (2020) applied a panel data regression model to explore the impact of urbanization factors on traffic mortality in China, and their findings show that there exist spatial differences in the impact of urbanization on traffic mortality in the three major regions in China. Okui and Park (2021) analyzed geographic differences in road accident mortality and its associated factors using the vital statistics in Japan, and found that socio-economic characteristics are associated with geographic differences in road traffic mortality, particularly in men.

With the development of society and the economy, the number of vehicles in China is ever-increasing. Some scholars have paid a lot of academic attention to the relationship between road traffic safety and socio-economic factors. Shi et al. (2008) applied logistic regression analysis to explore the impact factors of expressway traffic accidents, and pointed out that gender, accident pattern, driver type, and responsibility reasons are the main influencing factors of expressway traffic accidents. In order to study the traffic accidents in highway tunnels, Hu et al. (2008) applied Pearson correlation coefficients to analyze the influencing factors of road traffic deaths per 100,000 population, and found strong positive correlations between road traffic death rates and the growing economy. Sun et al. (2019) used a fixed-effect model to analyze the effect of various economic factors on the casualties in traffic accidents in China, and held that economic development has a positive impact on improving traffic conditions, but new health institutions have no obvious effect on the casualties in traffic accidents. Li and Zhang (2021) applied a random-effects model to analyze the impact of GDP per capita, population, and vehicle- and road-related factors on traffic accidents, injuries, and fatalities in China from 2007 to 2010, and used the Shapley value decomposition method to determinate the relative contributions of these factors and their dynamic trends.

12.1.3 General Comment

By reviewing the literature, we've found that the empirical research on road traffic safety in China mainly focuses on micro-level influencing factors, and a few studies involve macro socio-economic factors. Although some scholars found that macro socio-economic factors have impacts on traffic accident casualties, there is no effective empirical analysis of the influencing factors of RTMR in China from a socio-economic perspective. Compared with the number of road traffic accident casualties, the RTMR can better reflect the actual transport safety level (Luoma & Sivak, 2014).

Based on the panel data of 31 Chinese provinces from 2003 to 2018, we analyze the main influencing factors of the RTMR from a socio-economic view using a Tobit panel model. Facing that the value of RTMR is limited in a certain range, the use of traditional regressions, such as ordinary least squares, could not lead to consistent estimators, due to the truncated data. Based on the principle of maximum likelihood estimation, the Tobit model can effectively avoid inconsistency and bias in parameter estimation (Yang et al., 2018), and generate more reasonable parameter estimation results.

12.2 The Characteristics of Road Traffic Accidents and Casualties in China

12.2.1 Traffic Accidents, Injuries and Deaths Have Dropped Sharply Since 2003

There has been a drop in the number of traffic accidents, injuries, and deaths in China since 2003. More specifically, the RTMR decreased sharply from 2003 to 2008 and the pace of decline eased after 2009. Since 2016, the number of traffic accidents, injuries, and deaths has rebounded slightly.

Compared with 2003, the number of traffic accidents, injuries and deaths decreased by 63.3%, 47.7% and 39.5% in 2018, respectively (Fig. 12.1). In the context of the rapid increase of the economy and automobiles, these achievements have not come easily. Since the beginning of the twenty-first century, the Chinese government has actively taken various measures to deal with transport safety issues (Wang et al., 2019), such as establishing and improving the legal system for transport safety, strengthening the construction of transport safety culture, continuously increasing investment in traffic construction and the network management system, and improving the safety performance of motor vehicles. These measures are important reasons for the overall improvement of transport safety.

The investment plan worth RMB 4 trillion contributed to a slowing rate of RTMR. In the wake of the 2008 financial crisis, China launched a massive stimulus plan and largely averted a recession. It is called the "4 trillion-yuan economic stimulus

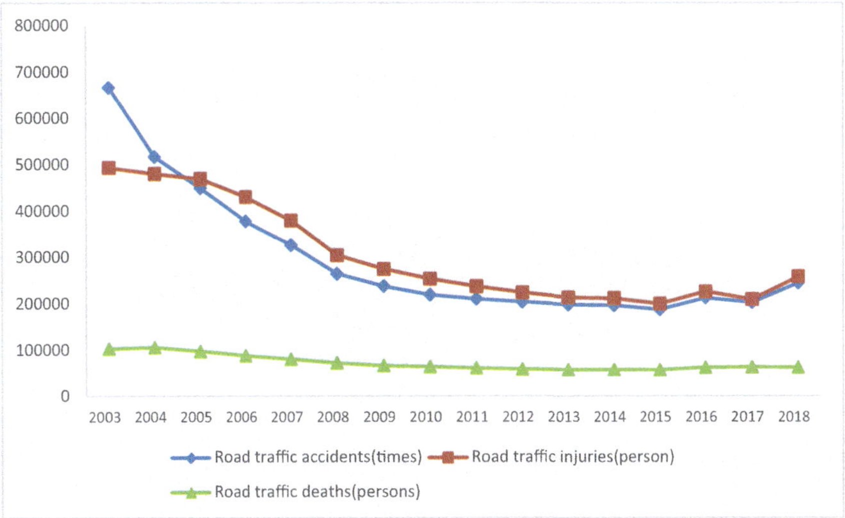

Fig. 12.1 The trend of road traffic accidents, injuries, and deaths in China from 2003 to 2018 Data source: The authors, edited from the National Bureau of Statistics of China (NBSC) (2022)

package", which sustained economic growth, promoted urbanization and brought a lot of traffic demand. With sustained economic growth, the number of cars and new car drivers showed rapid growth, which had a negative impact on the decline of road traffic accidents and casualties to a certain extent.

Since China promulgated Road Traffic Safety Law in 2004, the number of road accidents, fatalities, and injuries showed a trend of decline, but increased in 2016. This deterioration is likely to come from the implementation of the "Division of Work Safety Responsibilities of the State Council Work Safety Committee" in 2015 and the "Measures for the Administration of Statistics on Work Accidents" in 2016, which enhanced the ability of transport departments to collect road accident statistics (Zhang et al., 2019) (Fig. 12.1).

12.2.2 *The Rates of Road Traffic Accidents and Fatalities are High*

Compared with developed countries, the road traffic safety level in China is relatively low, which is mainly manifested in two aspects. One is that the road traffic accident casualty rate is high. In 2016, there were 63,093 deaths in road traffic accidents in China. In 2016, the mortality rate per 10,000 vehicles in China was 3.4, which were higher than those in the United States, Japan and Germany (Fig. 12.2). Wang et al. (2019) believed that the factors, such as special patterns of population mobility that involve more pedestrians and bicycle and motorcycle travel, poorer road

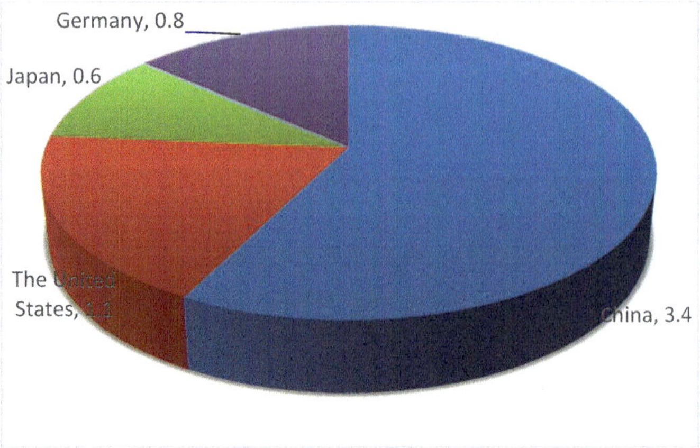

Fig. 12.2 The mortality rate per 10,000 vehicles, the United States, Japan, and Germany in 2016. *Data source* Yan et al. (2019)

infrastructure, more lax traffic control, and inferior emergency service and long-term trauma care in some regions of China, are the main causes attributing to this gap. The other one is that very serious road traffic accidents with more than 10 deaths occur frequently. Statistics show that there were many accidents of this kind recorded from 2001 to 2011 in China, with an average of 36.5 times per year.

12.2.3 The Heavy Goods Vehicle is a Major Vehicle Type in Road Traffic Accidents

With the rapid development of the logistics industry, road transport has become an important means of freight transport. The heavy goods vehicle has the advantages of large volume and fast turnover. But it poses a serious threat to people's lives. By the end of 2016, the number of civilian vehicles in China was about 185.7454 million, only 3% of which were heavy goods vehicles. The ratio of heavy goods vehicles to total civilian vehicles in China is very low, but the mortality rate is very high. According to the statistics on motor vehicles, heavy goods vehicles were involved in 18,578 accidents, accounting for 47.08% of the total number of road traffic accidents. They caused 10,684 deaths, accounting for 56.18% of the total number of traffic deaths, and resulted in direct economic losses of RMB 24.8 million, accounting for 59.65% of total economic losses. Therefore, the heavy goods vehicle is a main vehicle type in road traffic accidents (Fig. 12.3).

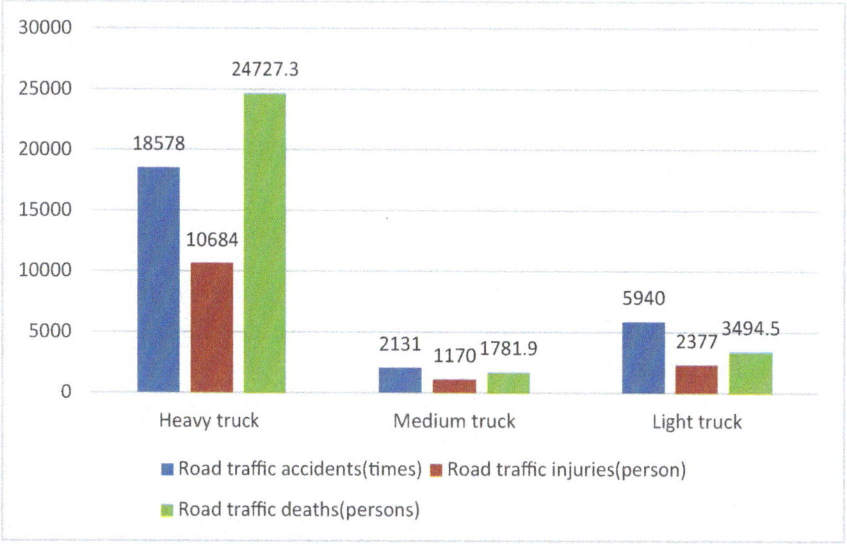

Fig. 12.3 The characteristics of different goods vehicles in road traffic accidents in 2016. *Data source* The authors, edited from Matthias et al. (2017)

12.2.4 Traffic Accidents on Expressways Should not be Ignored

In China, the expressway is becoming a major component of social transport for its high speed, great road network capacity, good transport conditions, safety and perfect road network coverage. In recent years, although the accident rate and fatality rate of expressway transport are declining, the overall traffic safety situation regarding expressways is still severe. During 2016, there happened 8934 expressway traffic accidents in China, resulting in 5947 deaths and 11,956 injuries. By the end of 2016, the expressway experienced rapid development in China, and the total length of national expressways reached 131,000 km, which accounted for 2.79% of the total length of highways. However, expressway traffic accidents accounted for 7.74% of the total highway traffic accidents and 13.68% of the total highway traffic deaths.

12.2.5 The Rural Road Safety Situation is Still Grim

The proportion of rural road traffic accidents to total road traffic accidents is high and increases year by year, which increased from 20% in 2005 to 24% in 2014. The proportion of rural road traffic casualties to total road traffic casualties increased from 22% in 2005 to 25% in 2014. Rural areas have a high incidence of extraordinarily serious traffic accidents. In 2014, among 12 extraordinarily serious traffic accidents, six took place in rural areas. The factors, such as poor conditions of transport infrastructure, lagging economic development, lower quality of prehospital trauma care and hospital treatment (Wang et al., 2019), farmers' weak consciousness of transport safety, many traffic tools with lagging performance, less stringent road traffic policies and law enforcement, and comparatively frequent illegal traffic activities, are important reasons for that (Huang et al., 2013).

12.2.6 The Number of Casualties in Urban Road Traffic Accidents is Relatively Small, but the Accident Frequency Rate is High

According to figures from the Ministry of Public Security of China (MPSC), for China, in 2016, because of urban road traffic accidents, 20,000 people lost their lives, 100,000 people were injured, and the total economic cost was about RMB 400 million. Compared with intercity highway transport, the average speed of urban road transport is relatively low, so urban road traffic accidents are less severe. However, due to the increasing number of vehicles and drivers, traffic accidents take place more frequently. At the end of 2016, the proportion of urban road length to the total road length was only 7.5%, but the proportion of urban road traffic accidents and casualties reached 45.8% and 38.8%, respectively. The number of traffic accidents on urban roads is four times that on expressways and ten times that on other roads.

12.2.7 Electric Bicycles Have Become a Major Killer on Roads

In recent years, the electric bicycle market has developed rapidly in China. According to statistics from the Ministry of Industry and Information Technology of China, the social stock of electric bicycles in China was 330 million in 2020, ranking first in the world. For Chinese residents in some small and medium cities, the electric bicycle is a main travel mode. Electric bicycles feature poor stability and high speed. Compared to bicycles, electric bicycles are heavier and faster, and may cause more serious injuries in collisions (Hu et al., 2014). In 2019, electric bicycles caused 8639 deaths and 44,677 injuries in road traffic accidents of China, accounting for nearly 70% of non-motor vehicle casualties. Compared with normal electric bicyclers, food couriers who ride electric bicycles are exposed to relatively long-time riding, so traffic accidents occur more frequently, and the casualty rate tends to be higher (Qin et al., 2021).

12.2.8 Costal and Southwestern Regions Face Serious Traffic Safety Issues

There are many traffic accidents in Guangdong, Zhejiang, Jiangsu, Shandong and Anhui. Except Anhui, the rest of these provinces lie in coastal areas. In 2016, the number of traffic accidents in these five provinces accounted for 37% of the total number of traffic accidents, and the number of traffic deaths in these five provinces accounted for 32.6% of the total number of traffic deaths. This shows that eastern coastal areas have a high number of traffic accidents and deaths. The increase in passenger and freight transport due to socio-economic development is an important reason for this.

The overall number of traffic accidents in central and southwestern regions is high, but the number of traffic deaths in central regions is significantly lower than that in southwestern regions. Southwestern regions occupy a narrow strip of land and their geological conditions are complex. There is heavy and concentrated rainfall, and geo-hazards such as collapse, landslides, and debris flows occur frequently, which will cause adverse effects on transport safety (Figs. 12.4 and 12.5).

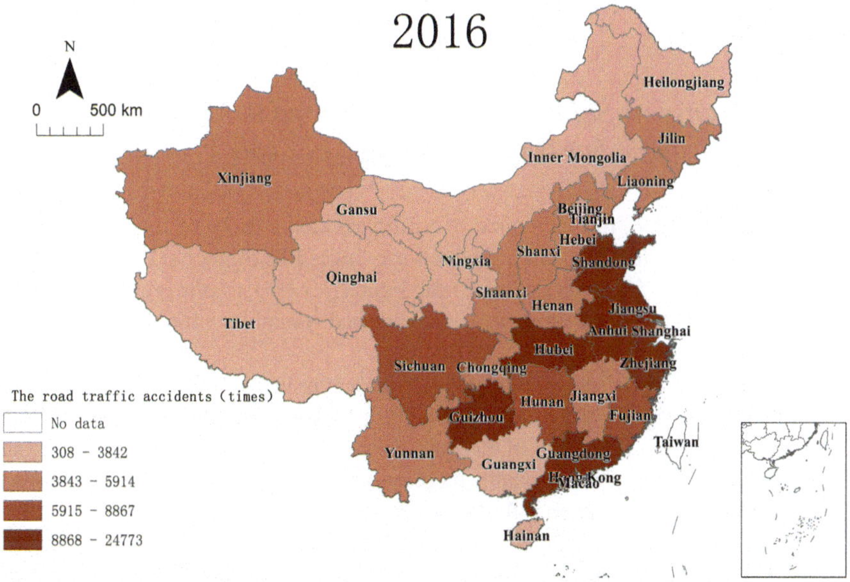

Fig. 12.4 The traffic accidents in China in 2016. *Data source* The authors, edited from China Statistical Yearbook (2017)

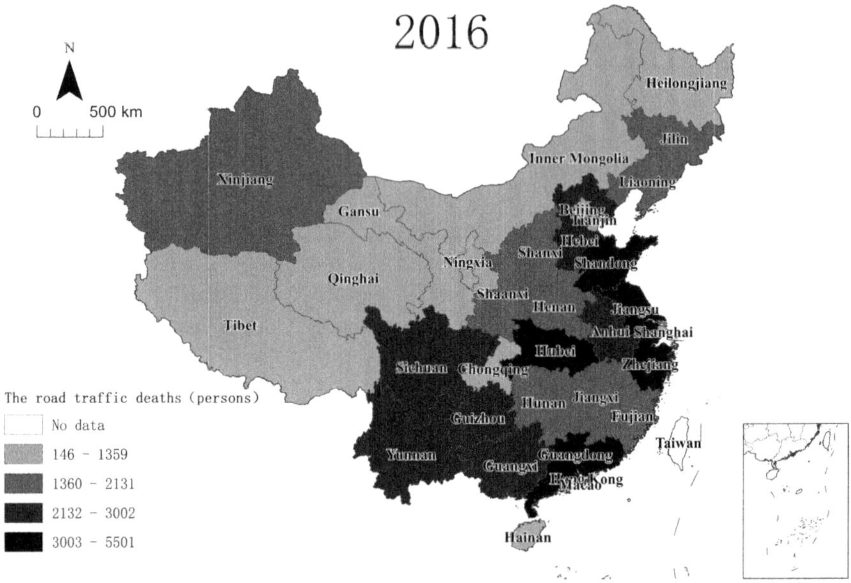

Fig. 12.5 The traffic deaths in China in 2016. *Data source* The authors, edited from China Statistical Yearbook (2017)

12.2.9 Transport Safety Has Been Improved in Most Provinces, but Has Deteriorated in Some Provinces

Compared with 2008, the incidence of traffic accidents in 26 of the 31 provinces in China decreased in 2016, among which provinces like Shanghai (71%), Sichuan (55.8%), Henan (49.5%), Tibet (48.7%) and Fujian (43.4%) saw the greatest decrease. However, the RTMR of five provinces increased. Guizhou (466.9%) and Hubei (120%) saw the largest increase.

From 2008 to 2016, the number of traffic deaths in 24 of the 31 provinces in China decreased significantly, among which provinces like Tibet (−62.0%), Inner Mongolia (−39.6%), Fujian (−38.6%) and Hunan (−37.9%) had the largest decline. However, the number of traffic deaths increased in seven provinces. Hubei (123.1%), Guizhou (68.7%) and Yunnan (44.2%) witnessed significant increases.

Generally speaking, the deterioration of transport safety is the most serious in Hubei, Guizhou and Yunnan. Against the backdrop of the rise of high-speed rail and civil aviation, railways and airways have been chosen over many highways for medium and long-distance passenger transport. The proportion of road passenger transport volume in the total passenger transport volume decreased from 93.5% in 2008 to 81.2% in 2016. However, only four provinces saw an increase in highway passenger transport volume from 2008 to 2016. These four provinces are Guizhou (128.2%), Yunnan (32.3%), Shanghai (16%), and Hubei (6.9%). The significant

Fig. 12.6 The change rate of traffic accidents in China from 2008 to 2016. *Data source* The authors, edited from China Statistical Yearbook (2017)

increase in road passenger transport, to a certain extent, increased road traffic accidents. Although Shanghai has an increase in road passenger transport volume, its high-quality population and stringent control on transport safety can reduce traffic accidents. At the same time, its advanced medical services can ensure more effective medical treatment and reduce traffic deaths. Therefore, in the context of the growing highway passenger transport volume, Shanghai has achieved an improvement in transport safety (Figs. 12.6 and 12.7).

12.3 Temporal and Spatial Characteristics of RTMR

12.3.1 Temporal and Spatial Characteristics of RTMR

See the Fig. 12.8.

12.3.2 Changes in RTMR

Most of the provinces with a higher RTMR are located in developed coastal areas or underdeveloped northwest regions, while most of the provinces with a lower RTMR

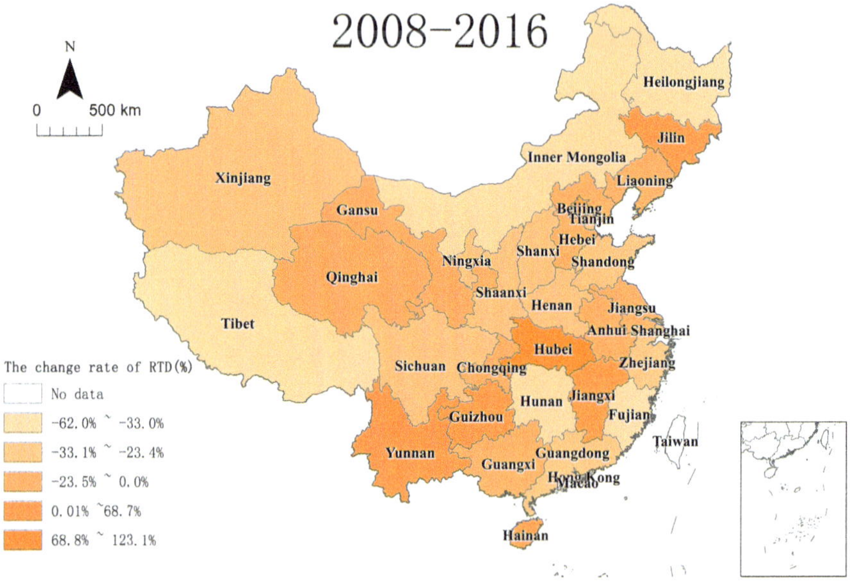

Fig. 12.7 The change rate of traffic deaths in China from 2008 to 2016. *Data source* The authors, edited from China Statistical Yearbook (2017)

Fig. 12.8 The mean value of RTMR (per 100,000 population) intervals in 31 provinces of China from 2003 to 2016. *Data source* The authors, edited from China Statistical Yearbook (2017)

are located in central regions. Such regional differences are caused by multiple factors. Developed coastal areas have a higher urbanization rate and higher per capita car ownership, resulting in frequent traffic accidents. Although underdeveloped western areas have a lower urbanization rate and lower per capita car ownership, the medical and health conditions are poor, resulting in a higher RTMR (Fig. 12.8).

The RTMR decreased sharply from 2004 to 2009, and the pace of decline obviously eased after 2009. For one thing, at the end of 2003, China promulgated Road Traffic Safety Law and then implemented it in an all-round way. Therefore, the number of road accidents in the whole society is decreasing year by year. For another, China has increased financial expenditure on medical care, and strengthened publicity work on transport safety. These policies are conducive to the sharp decline of RTMR.

The investment plan worth RMB 4 trillion contributed to a slowing rate of RTMR. In the wake of the 2008 financial crisis, China launched a massive stimulus plan and largely averted a recession. It is called the "4 trillion-yuan economic stimulus package", which sustained economic growth, promoted urbanization and brought a lot of traffic demand. With sustained economic growth, the number of cars and new car drivers showed rapid growth, which had a negative impact on the decline of RTMR to a certain extent.

12.3.3 Spatial Distribution Characteristics of RTMR

With the application of ArcGIS 10.2 software, the spatial distribution maps of RTMR of China in 2003, 2009 and 2016 are drawn. The classification of RTMR is based on natural breaks (Jenks, 1967). RTMR is divided into five classes (from high to low).

It can be found from Fig. 12.9 that in 2003, the first class only included Tibet, the second class included Ningxia, Zhejiang, Qinghai, Xinjiang, Guangdong, Fujian, Beijing, Tianjin and Shanxi, and the third class included Shandong, Jiangsu, Inner Mongolia, Liaoning, Gansu, Jilin, Shanghai, Guangxi, Yunnan and Hebei. The remaining 11 provinces were in the fourth class.

In 2009, Tibet was still in the first class, while the members of the second class changed greatly. The number of provinces belonging to the second class decreased from nine in 2003 to six in 2009. Zhejiang, Qinghai and Xinjiang entered the first class. The number of first-class provinces increased from one in 2003 to four in 2009. Beijing dropped from the second class to the third class. Shanxi, Fujian, Tianjin, Ningxia and Guangdong remained in the second class. The members of the third class also changed greatly. Jiangsu rose from the third class to the second class, while Yunnan and Hebei fell from the third class to the fourth class. Gansu, Inner Mongolia, Jilin, Guangxi, Liaoning, Shandong and Shanghai remained in the third class. As for the fourth class, Hainan, Shaanxi and Anhui were upgraded from the fourth class in 2003 to the third class in 2009, and the remaining provinces of the fourth class remained unchanged. The fourth class had 10 members in 2009 (Fig. 12.10).

In 2016, there were great changes in the members of the fourth class compared with 2009. Qinghai and Zhejiang were still in the first class; Hubei was upgraded from

Fig. 12.9 RTMR (per 100,000 population) intervals in 31 provinces of China in 2003. *Data source* The authors, edited from China Statistical Yearbook (2017)

the fourth class to the first class; the number of first-class provinces was three. There were nine provinces in the second class, which included Hainan, Jilin, Xinjiang, Guizhou, Yunnan, Beijing, Shanxi, Jiangsu and Ningxia. It is worth noting that Xinjiang fell to the second class from the first class, Beijing, Hainan and Jilin rose to the second class from the third class, and Guizhou and Yunnan were upgraded from the fourth class to the second class. Shanxi, Jiangsu and Ningxia remained in the second class. The number of third-class provinces was still nine and remained unchanged from 2009 to 2016. Tibet dropped from the first class to the third class. Tianjin, Guangdong and Fujian dropped from the second class to the third class. Jiangxi rose from the fourth class to the third class. Gansu, Guangxi, Liaoning, Anhui, Shaanxi and Inner Mongolia were the original third-class provinces. The fourth class included Shandong, Hebei, Shanghai, Chongqing, Heilongjiang, Sichuan, Hunan and Henan. Among them, Shandong and Shanghai dropped from the third class to the fourth class, meaning traffic safety was improved significantly, while the other provinces remained in the fourth class (Fig. 12.11).

Fig. 12.10 RTMR (per 100,000 population) intervals in 31 provinces of China in 2009. *Data source* The authors, edited from China Statistical Yearbook (2017)

Fig. 12.11 RTMR (per 100,000 population) intervals in 31 provinces of China in 2016. *Data source* The authors, edited from China Statistical Yearbook (2017)

12.4 Empirical Analysis the Influencing Factors of Traffic Safety

In this book, we choose the provincial RTMR as an dependent variable. The research period is from 2003 to 2016. Considering the complex macro socio-economic characteristics of China, the economic development level, urbanization level, vehicle density, medical assistance level, and government regulation are selected as influencing factors. The data is taken from the National Bureau of Statistics.

12.4.1 Selection of Macro Influencing Factors

12.4.1.1 Economic Development Level

In general, economic growth leads to increased traffic volume, traffic congestion, young male drivers, speeding, drunk driving, and ultimately an increase in fatal traffic accidents (Gulzar et al., 2012). Song and Zhang (2016) held that economic development will increase the burden of road traffic, resulting in an increase in road traffic injuries. But some authors believe that sustained economic growth enables road users in high-income countries to use safer transport tools, but not users in low-income countries (Bishai et al., 2006). With national economic growth, the government will be able to increase traffic safety investment, such as improving medical care standards, vehicle safety features, and road infrastructure. Based on the previous studies of Bishai et al. (2006) and Gaygisiz (2009), this study selects the economic development level as a major dependent variable for empirical research, and the relationship between the economic development level and RTMR needs to be tested empirically.

12.4.1.2 Urbanization Level

With the acceleration and deepening of the urbanization process, a huge population of rural residents has poured into the city, resulting in road traffic congestion, and pedestrians and non-motor vehicles arbitrarily pass through motor vehicle lanes, causing a large number of traffic accidents (Verma et al., 2021). But some scholars hold that due to higher levels of density of health resources and road maintenance and management (Nakamura, 1984), and the higher rate of public transport modes in urban areas, traffic crashes tend to be at a lower speed (Okui & Park, 2021). Based on the previous studies of Atubi and Gbadamosi (2015), Jadaan (1990), La Torre et al. (2007), and Li and Zhang (2021), the urbanization level is selected as a major dependent variable for empirical research.

Table 12.1 Influencing factors of RTMR

Explanatory Variables	Definitions of Variables	Pre-Judgment
Economic development level (DEL)	Per capita GDP (Dollar 10,000)	Unknown
Urbanization level (UL)	Proportion of city population in total inhabitants (%)	Unknown
Motorization level (ML)	Motor vehicles per 100,000 inhabitants (persons)	Positive
Medical assistance level (MAL)	Medical personnel in health care institutions per 10,0000 inhabitants (persons)	Negative
Government regulation (GR)	Government expenditure on health (RMB 100,000)	Negative

Data source Edited by the authors

12.4.1.3 Vehicle Density

Motorization level refers to the number of motor vehicles per unit population in a certain period. The increase in the number of vehicles and road length has enabled closer and more frequent regional exchanges, brought more convenience to people's life, and made great contributions to the economic development of all regions. Furthermore, it has led to the rapid expansion of road users and drivers, especially novice drivers (Fu et al., 2021). In 1949, when Smeed studied the relationship between traffic accidents and economic growth, he used the number of motor vehicles per thousand people as an explanatory variable. Based on the previous studies of Ali et al. (2019), He et al. (2015) and Li and Zhang (2021), the motorization level is added to the regression model in this research.

12.4.1.4 Medical Assistance Level

Research has indicated that RTMR may be reduced by first aid, specialist transport and emergency treatment of victims (Bishai et al., 2006; Mock et al., 1998). An effective medical emergency system based on the rescue chain makes it easier for aid agencies to save victims of road accidents. Based on the previous studies of La Torre et al. (2007), the medical assistance level is selected as a major dependent variable for empirical research, and the number of medical personnel in health care institutions per 10,000 inhabitants as a proxy variable of medical assistance level.

12.4.1.5 Government Regulation

Government expenditure on health, such as improving the healthcare services system, transferring payments to less advanced areas in medical and health services, and providing assistance to victims in road traffic accidents, would help to reduce traffic deaths. Therefore, this paper selects government expenditure on health as an important variable for regression analysis (Table 12.1).

12.4.2 Tobit Regression Analysis

This study evaluates the impact of the above influencing factors on the RTMR in China. The Tobit regression model assumes:

$$RTMR_{it} = \beta_0 + \beta_1 DEL_{it} + \beta_2 UL_{it} + \beta_3 ML_{it} + \beta_4 MAL_{it} + \beta_5 GHS_{it} + u_{it} \tag{12.1}$$

where $RTMR_{i,t}$ represents the RTMR value of province i in year t, β_0, β_1, β_2,..., β_5 stands for the coefficient of the independent variables, and $u_{i,t}$ is a random disturbance term. The parameters are estimated using Stata12.0 software (Table 12.2).

The economic development level has a significant negative impact on RTMR. With the strengthening of the comprehensive economic power of China, the expenditure on transport safety management, health and road maintenance is increasing, which will offset the increase in the number of traffic deaths caused by the rapid development of urbanization and motorization. For example, the Chinese government has formulated a national reform of health care and implemented a series of policies benefiting people since 2009 (Wang et al., 2013). Moreover, motorists and pedestrians in economically developed regions have a stronger awareness of traffic safety, the traffic laws are forcefully implemented, and the mandatory use of seatbelts and helmets reduces the likelihood of road traffic injuries (Li & Zhang, 2021).

The urbanization level has a significant positive impact on RTMR. China has been experiencing the largest urbanization, which has long-lasting and far-reaching effects on local and national public health (Li et al., 2016). The unplanned mode of urbanization development in some cities in China makes the problem of "urban diseases" increasingly outstanding. The increase of population density and the imperfection of traffic facilities lead to serious traffic congestion, the load capacity of urban roads exceeds the limit seriously, and traffic accidents occur frequently.

Table 12.2 Tobit regression results

Variable	Coefficient	Std. Err	*t*	P > |z|
EDL	−1.906579**	0.9030149	−2.11	0.035
UL	1.005031***	0.2782298	3.61	0.000
ML	0.0022332**	0.0006307	3.54	0.025
MAL	−0.0302844***	0.013503	−2.24	0.000
GHS	−0.0561096***	0.0057578	−9.74	0.000

Note: *** and ** indicate significance at the 1%, and 5%, respectively
Data source Edited by the authors

The motorization level has a significant positive impact on RTMR. Since the 2000s, the number of civilian vehicles in China has been growing at a rapid rate, which rose from 23.83 million in 2003 to 232.31 million in 2018, increasing the traffic burden and the risk of traffic accidents. Although a string of transport safety policies implemented by China from the late 1990s may have resulted in lower speeds and more traffic congestion, the rapid growth of the number of automobiles might increase the risk of exposure to road traffic accidents and the expansion of urban population might result in an increase in road traffic mortality (Castillo-Manzano et al., 2014; Wang & Chan, 2016).

The regression coefficients of the medical assistance level are significantly negative. Providing a better geographical coverage of health care for road accidents can help reduce the likelihood of traffic-related deaths (Buchmueller et al., 2004). At present, there are obvious differences between different regions in the allocation of public medical resources in China (Chou & Wang, 2009). The level of medical and healthcare services in some western areas and many remote villages lags behind other regions, and the government needs to adjust policy to rationally allocate medical resources.

The regression coefficient of government expenditure on health is significantly negative, which is in line with expectations. Although the proportion of fiscal expenditure on medical care to GDP increased from 4% to 7.1% from 2003 to 2018 (NBSC, 2022), it was still lower than the average level in OECD countries. According to the data of WHO, in 2018, the proportion of government expenditure on health care to GDP in the United Kingdom, the United States, and Japan was 19%, 22% and 24%, respectively (WHO, 2021), which was much higher than that of China in the same period. In order to curb RTMR, the Chinese government needs to increase fiscal expenditure on health.

12.5 Policy Review and Recommendations

12.5.1 Review of Transport Safety Policy in China

12.5.1.1 The Traffic Safety Policy of the Central Government

In 2010, the 12th Five-Year Plan clearly required that the RTMR per 10,000 vehicles should not exceed 2.2, and the number of serious traffic accidents with more than 10 deaths should decrease by more than 15% in 2015. The 13th Five-Year Plan for Road Traffic Safety issued in 2017 aims to reduce the RTMR by more than 4%, to reduce the death rate of operating vehicles by 6%, and to reduce the number of big road traffic accidents by more than 8% by 2020.

12.5.1.2 Formulation of Laws and Regulations on Traffic Safety

Criminal Law

With the energetic efforts of the MPSC, Amendment VIII to the Criminal Law added the provisions on the dangerous driving offence. Drunk driving has become a general violation of the law. In order to curb serious violations of key vehicles, the MPSC pushed through Amendment IX to the Criminal Law, which has been implemented since November 1st, 2015. In this Amendment, the serious overloading of school buses and highway passenger vehicles, speeding, and the illegal transport of dangerous chemicals and other dangerous goods are included.

Laws and Regulations on Road Traffic Safety

Law of the People's Republic of China on Road Traffic Safety was passed in 2003 and it came into force on May 1st, 2004. The latest revision can be found in the 2011 edition. In 2012, the State Council of China issued the "Opinions of the State Council on Strengthening Road Traffic Safety", which require an enhanced organizational guarantee of road traffic safety and also the incorporation of road traffic safety work into local governments' economic and social development plan.

Laws and Regulations on Railway Traffic Safety

In 1989, the State Council of China promulgated the "Regulations on the Safety and Protection of Railway transport", and the regulations were comprehensively revised in 2004, which played an important role in ensuring railway transport safety. In 2014, they were revised again. The 2014 edition is more comprehensive and pays more attention to the risk source control in the fields of equipment quality, construction and operation.

Laws and Regulations on Water Traffic Safety

The "Regulations of the People's Republic of China on Administration of Traffic Safety in Inland Waters" were adopted by the State Council of China in 2002 and implemented on August 1st, 2002. The revised versions were promulgated in 2011, 2017 and 2019, respectively. To ensure life and property safety in water transport, prevent pollution caused by ships, and strengthen the supervision and management of ship safety, the Ministry of Transport of China issued the "Regulations of the People's Republic of China on Ship Safety Supervision" in 2017.

Laws and Regulations on Air Traffic Safety

For promoting the safe, healthy and orderly development of international air transport, on May 17th, 2017, the provisions on the management of regular international air transport were adopted by the 8th Ministerial Meeting of the Ministry of Transport of China. The provisions on the management of civil aviation safety were promulgated by the Ministry of Transport of China on February 13th, 2018 and came into effect on March 16th, 2018.

12.5.1.3 Special Action on Traffic Safety

Pushing Forward Thorough Implementation of the "Civilized Transport Action Plan" and Promoting Collaborative Traffic Safety Governance

In 2010, the Central Commission for Guiding Cultural and Ethical Progress, MPSC, and Ministry of Transport of China jointly implemented the "Civilized transport Action Plan"; the three parties require local traffic law enforcement departments to take the rule of law as the guide and the people's livelihood as the ultimate objective, and adhere to the combination of public legal education, strict law enforcement and solicitous service, so as to gradually deepen people's understanding of traffic law. At present, the plan is still being implemented.

Deepening the Reform of Transport Management and Continuously Providing Convenient Services to People

In 2015, the MPSC began to build a public-oriented comprehensive online service management platform for transport safety nationwide. The platform has a national unified service telephone number "12,123", and provides more than 130 online services in 10 categories through webpages, mobile apps, SMS, voice calls and other means. The platform benefits 260 million motor vehicle owners and more than 300 million drivers across the country.

12.5.2 The Traffic Safety Policy of Local Governments

12.5.2.1 Beijing: Regulations on the Rapid Handling of Motor Vehicle Traffic Accidents

In order to ensure orderly, safe and smooth road traffic, alleviate congestion in the traffic network caused by traffic accidents, and improve road traffic efficiency, in 2018, Beijing formulated the "Regulations on the Rapid Handling of Motor Vehicle Traffic Accidents", which contain 14 rules.

12.5.2.2 Fujian: The "Three-Year Action Plan" for Road Traffic Safety Has Significantly Improved Transport Safety

The traffic deaths in Fujian decreased by 38.6% from 2008 to 2016, and the level of improvement in transport safety ranked among the top in China. In 2013, the transport mortality decreased by 13.5% compared with 2012. In that year, Fujian issued the "Three-Year Action Plan" for road traffic safety, which played an important role in improving transport safety.

The main content of the "Three-Year Action Plan" is as follows: (1) Establishing a "driver blacklist database", and realizing cross-sector sharing of information about employment, traffic violations and traffic accidents of drivers. (2) Fully implementing the school bus safety project in primary and secondary schools. (3) Strengthening the supervision of end-of-life vehicles and recycling vehicles, and eliminating illegal modification of vehicles. (4) Transport safety controls should be set up in particular places and periods, such as historic sites, temple fairs, and rush hours. (5) Effectively cracking down on illegal passenger transport services.

12.5.2.3 Inner Mongolia: Transport Safety Has Been Significantly Improved by Strengthening Road Traffic Safety Work

From 2008 to 2016, the traffic deaths in Inner Mongolia decreased by 39.6%, which was the biggest drop in China. In May 2013, Inner Mongolia issued new road transport safety policies: (1) Strengthening the safety management of road transport enterprises. (2) Implementing strict training, examination and management of drivers. (3) Enhancing vehicle safety supervision. (4) Reinforcing the management of road traffic safety in rural and pastoral areas. (5) Strengthening the enforcement of laws on road traffic safety. (6) Carrying out extensive publicity and education on transport safety.

12.5.2.4 Gansu: Establishing a Joint Supervision System for Major Road Traffic Accidents

In 2013, Gansu proposed the requirement for strict accountability measures for road traffic accidents, and an improved comprehensive supervision mechanism featuring "on-the-spot joint supervision, coordinated investigation, issuing of warning notices, key interviews and inspections, and follow-ups" for major road traffic accidents.

12.5.2.5 Shandong: Issuing the "Opinions on the Liability Investigation and Handling of Major Road Traffic Accidents (for Trial Implementation)"

In 2009, Shandong issued the "Opinions on the Liability Investigation and Handling of Major Road Traffic Accidents (for Trial Implementation)", which stipulate the scope, principles, composition and responsibilities of the accident investigation, the rights and obligations of the accident investigation team, and the investigation procedures for major road traffic accidents.

12.5.2.6 Jiangxi: Carrying Out Special Rectification Action on Truck Safety Devices

From June 2010 until the end of the year, Jiangxi initiated special rectification action on truck safety devices. For newly registered trucks and trailers with a total mass of more than 3500 kg, it is necessary to install safety devices that meet national standards. Special inspections and rectification of the installation of protective devices for registered trucks and trailers with a total weight of more than 3500 kg shall be carried out.

12.5.3 General Comment on Transport Safety Policy in China

Although the Chinese government has formulated many effective transport safety policies, there is still a large gap between current transport safety policies and the demand for safe travel due to the rapid socio-economic development. The existing deficiencies are as follows:

(1) The laws and regulations on corporate responsibility are not in place. The content of traffic policy still has the problem of insufficient safety education and training for employees, and difficulty in improving safety awareness, so there are still many violations.

(2) There is room for development in existing policies and regulations on the supervision of transport safety law enforcement. During the reform of the traffic law enforcement system, there are some problems, such as unclear definition of safety supervision responsibilities and insufficient allocation of law enforcement forces, which cannot meet the increasingly complex safety supervision needs of the current transport system.

(3) Current policies and regulations on traffic infrastructure are inadequate. The management and maintenance funds of some rural roads are not fundamentally and effectively guaranteed, resulting in problems such as inadequate management and maintenance and poor road conditions, and affecting driving safety. It is necessary to add the content of capital and technical support for traffic safety guarantees regarding traffic infrastructure to traffic policy.

(4) There is a lack of policies and regulations on scientific and technological support for traffic safety. Traditional means such as manual patrols are still widely used in China's safety and emergency management, and the application of big data and intelligent technology lags behind other countries. It is necessary to add the content of information technology to traffic safety policies and regulations.

(5) There are no sufficient policies and regulations on the institutionalization of traffic safety publicity. The transport education activities that have been held in China are neither systematic nor sustainable, and publicity and education efforts are insufficient.

12.5.4 Policy Recommendations

12.5.4.1 Vigorously Disseminating Knowledge of Traffic Safety

From the regression results, we can see that the level of economic development is significantly positively correlated with RTMR. With the development of the economy, the number of drivers and vehicles is growing rapidly in China, and many novice drivers lack transport safety awareness, resulting in many traffic accidents. Therefore, it is necessary to strengthen the dissemination of transport safety knowledge. It can not only enhance the awareness of public transport safety, but also improve the ability of the whole society to deal with transport safety. It is mandatory to implement the following measures: (1) facilitating the free dissemination of transport safety knowledge through TV, newspapers, the Internet and other media with legal obligations; (2) promoting the development of transport safety education in school; (3) popularizing transport safety knowledge in rural areas in various forms.

12.5.4.2 Speeding up the Informatization and Intelligentization of Traffic Safety Facilities Management

In addition to the construction of transport infrastructure and a better supply of transport means, the new path to urbanization requires the use of modern information technology to improve transport efficiency, to reduce the traffic burden and environmental pollution, and to ensure transport convenience and safety. Thus the informatization and intelligentization of transport facilities management becomes an inevitable tendency.

(1) Innovating traffic safety facilities management. The transport sector should make use of computer and information technology to promote the web-based, dynamic management of transport safety facilities. For example, information management software for transport safety facilities can be developed, and the electronic transport map can display the status of transport safety facilities in real time, so that we can accurately obtain the update, maintenance, performance attributes and other information of transport safety facilities. Furthermore, the transport sector should build a management service platform for transport safety facilities by integrating the social resources concerning traffic safety facilities, which can release the information about traffic safety facilities to the public in a timely manner.

(2) Speeding up the informatization of transport safety facilities. Formulating the information management standard for traffic safety facilities, and accelerating the research, application and popularization of the software and hardware of the information management system. In accordance with the principles of unified standards and hierarchical management, and the needs of departments for traffic safety facilities construction management, the government should set informatization management standards for transport safety facilities based on step-by-step implementation to accelerate the informatization of transport safety facilities. All regions should unite the whole construction management sector and other social forces, and actively strive for more government financial support as well as seek diversified social investment, providing funding for the informatization of transport safety facilities.
(3) Implementing the national strategy of intelligent development of transport safety facilities. The central government should design the national intelligent development layout of transport safety facilities and provide national standards for the Internet of things for transport safety facilities in collaboration with management departments and research-oriented companies in the safety facilities industry. Meanwhile, multiple departments should jointly issue incentive policies for stimulating the research vitality of enterprises with the production of intelligent products to guide the intelligent development of the transport safety facilities industry.
(4) Introducing advanced transport tools, technical devices, and safety facilities. The development and application of the Internet of things, cloud computing and other technologies provides new possibilities for the construction and management of transport safety facilities. To improve the intelligent level of transport safety facilities, it is necessary to learn and absorb new technologies to develop new safety facilities. For example, we can make full use of solar energy and other new energy sources to realize automatic detection and control of the transport safety system by replacing the traditional static mode. It is also necessary to pay attention to the intelligent investment in safety facilities materials, such as new thermoplastic materials, color mixtures, slurry and synthetic materials. They have better performance and are applied in the manufacturing of transport safety facilities.

12.5.4.3 Strengthening the Supervision of the Manufacturing and Use of Motor Vehicles and Improving the Safety Performance of Motor Vehicles

Since the density of motor vehicles has a positive impact on transport mortality, we make a series of policy recommendations: (1) Further perfecting the national standard system for vehicle safety, and encouraging and guiding the development of vehicles with better safety performance. In particular, the product quality and safety performance of three-wheelers, low-speed trucks, and minibuses need to be improved and perfected. We should also reduce and ban the vehicles with poor safety performance and high pollutant emissions. (2) Improving the safety standard of vehicles for highway transport, and raising the requirements for safety configuration of vehicles for highway transport. (3) Strengthening the daily management of vehicles, and carrying out strict vehicle safety inspection. Information about key vehicles, such as large and medium-sized buses, large and medium-sized lorries, vehicles carrying hazardous chemicals, and school buses, needs to be archived.

12.5.4.4 Improving the Rescue and First Aid System for Traffic Accidents to Reduce the Accident Fatality Rate

The regression results show that the density values of medical institutions and health workers have a significant positive impact on transport safety. Here are the suggestions: (1) Hospital administrative departments should strengthen the training of medical personnel for providing relief to victims of traffic accidents. (2) In order to deal with traffic emergencies, we should perfect the hardware and software support system of the emergency rescue mechanism, accurately comprehend the principle of emergency rescue in crisis management, and establish and perfect the rescue mechanism for traffic emergencies. (3) It is of great necessity to build and improve the legal system of social assistance funds for road traffic accidents, such as making more efforts to claim advance funding for traffic accidents to make sure more people injured in road traffic accidents receive timely medical treatment.

12.5.4.5 Fulfilling the Responsibility for Production Safety of Transport Enterprises and Strengthening Safety Supervision of Them

Local governments and transport departments should launch more extensive campaigns to tackle the production safety problems of transport, urge and guide transport enterprises to establish an effective internal responsibility system for transport safety, and also assist transport enterprises in implementing responsibilities to transport passengers and hazardous chemicals. It is advisable to establish an emergency response system for the transport of dangerous chemicals based on the technologies of GPS (Global Positioning System), GIS (Geographical Information System), ES (Expert System) and automatic monitoring.

12.6 Conclusions

12.6.1 China Has Made Great Progress in Road Traffic Safety, but the Overall Level of Road Transport Safety is Low Compared with Developed Countries

Road safety has always been, and will remain, a top priority of the road transport industry (World Road Transport Organisation, 2021). As one of the fastest growing economies, China has made great progress in road traffic safety, and the number of traffic accidents, injuries, and deaths has dropped significantly since 2004 (NBSC, 2022). China plays an extremely crucial role in realizing the target of the Decade of Action for Road Safety 2011–2020. However, it is notable that China is one of the countries with serious traffic accidents in the world, and the total number of road traffic accidents and casualties in China ranks second in the world. For achieving the goal of the Decade of Action for Road Safety 2021–2030, China needs to step up efforts to improve transport safety.

12.6.2 Coastal and Inland Regions Have Serious Road Traffic Safety Issues

Coastal and inland regions have serious road traffic safety issues. Such regional differences are caused by multiple factors. Developed coastal areas have a higher urbanization rate and higher per capita car ownership, resulting in frequent traffic accidents. Although underdeveloped western areas have a lower urbanization rate and lower per capita car ownership, the medical and health conditions are poor, resulting in a lower level of transport safety.

12.6.3 Influencing Factors of Transport Safety

The influencing factors of transport safety can be divided into micro-level influencing factors and macro socio-economic factors. Micro-level influencing factors mainly include traffic participants, vehicles roads, climatic environment (Chang & Wang, 2006; Eboli et al., 2020; Kopelias et al., 2007), and traffic management laws and regulations, and the research goal is to put forward targeted improvement measures for specific traffic scenarios. Macro socio-economic factors mainly include economics, urban level, population, length of highways, and so on. Descriptive statistical analysis and regression analysis are mainly applied to analyze these influencing factors.

References

Ali, Q., Yaseen, M. R., & Khan, M. T. I. (2019). Road traffic fatalities and its determinants in high-income countries: A continent-wise comparison. *Environmental Science and Pollution Research, 26*, 19915–19929. https://doi.org/10.1007/s11356-019-05410-9

Atubi, A. O., & Gbadamosi, K. T. (2015). Global positioning and socio-economic impact of road traffic accidents in Nigeria: matters arising. *American International Journal Contemporary Research, 5*,136–146. https://www.semanticscholar.org/paper/Global-Positioning-and-Socio-Economic-Impact-of-in-Ondo/531db4d45562e842a9e5ca9e9a899008181e0e7c?p2df

Bergel-Hayat, R., Debbarh, M., Antoniou, C., & Yannis, G. (2013). Explaining the road accident risk: Weather effects. *Accident Analysis and Prevention, 60*, 456–465. https://doi.org/10.1016/j.aap.2013.03.006

Bishai, D., Quresh, A., James, P., & Ghaffar, A. (2006). National road casualties and economic development. *Health Economics, 15*, 65–81. https://doi.org/10.1002/hec.1020

Buchmueller, T. C., Jacobson, M., & Wold, C. (2004). How far to the hospital? The effect of hospital closures on access to care. *Journal of Health Economics, 25*, 740–761. https://www.nber.org/system/files/working_papers/w10700/w10700.pdf

Bucsuhazy, K., Matuchov, E., Zůvala, R., Moravcov, P., Kostíkova, M., & Mikulec, R. (2020). Human factors contributing to the road traffic accident occurrence. *Transportation Research Procedia, 45*, 555–561. https://doi.org/10.1016/j.trpro.2020.03.057

Castillo-Manzano, J. I., Castro-Nuño, M., & Fageda, X. (2014). Can health public expenditure reduce the tragic consequences of road traffic accidents? The EU-27 experience. *The European Journal of Health Economics, 2014*(15), 645–652. https://doi.org/10.1007/s10198-013-0512-1

Chand, A., Jayesh, S., & Bhasi, A. B. (2021). Road traffic accidents: An overview of data sources, analysis techniques and contributing factors. *Materials Today: Proceedings*. https://doi.org/10.1016/j.matpr.2021.05.415

China Statistical Yearbooks (CSY). (2017). China Statistical Publishing House, Beijing. http://tongji.oversea.cnki.net/oversea/engnavi/HomePage.aspx?id=N2017100312&name=YINFN&floor=1

Chang, L. Y., & Wang, H. W. (2006). Analysis of traffic injury severity: An application of non-parametric classification tree techniques. *Accident Analysis and Prevention, 2006*(38), 1019–1027. https://doi.org/10.1016/j.aap.2006.04.009

Chou, W. L., & Wang, Z. (2009). Regional inequality in China's health care expenditures. *Journal of Health Economics, 18*, S137–S146. https://onlinelibrary.wiley.com/doi/pdf/https://doi.org/10.1002/hec.1511

Eboli, L., Forciniti, C., & Mazzulla, G. (2020). Factors influencing accident severity: An analysis by road accident type. *Transportation Research Procedia, 47*, 449–456. https://doi.org/10.1016/j.trpro.2020.03.120

Fu, Y., Zhong, Q., & Yang, Q. (2021). Economic development and road traffic safety in China: A status Quo analysis. In G. Zhang, Q. Zhong (Eds.), *Road safety in China*. Springer. https://doi.org/10.1007/978-981-16-0701-1_3

Ganji, S. R. S., Rassafi, A. A., & Kordani, A. A. (2018). Vehicle safety analysis based on a hybrid approach integrating DEMATEL, ANP and ER. *KSCE Journal of Civil Engineering, 2018*(22), 4580–4592. https://doi.org/10.1007/s12205-018-1720-0

Gaygisiz, E. (2009). Economic and cultural correlates of road-traffic accident fatality rates in OECD countries. *Perceptual and Motor Skills, 109*(2), 531–545, https://doi.org/10.2466/pms.109.2. https://xueshu.baidu.com/usercenter/paper/show?paperid=48c4a331abaf769dfd03be110c2044c6&site=xueshu_se

Goel, R. (2017). Modelling of road traffic fatalities in India. *Accident Analysis & Prevention, 112*, 105–115. https://doi.org/10.1016/j.aap.2017.12.019

Gulzar, S., Yahya, F., Mir, Z., & Zafar, R. (2012). Provincial analysis of traffic accidents in Pakistan. *Social Science Human, 3*(3), 365–374. https://www.oalib.com/paper/2072288

He, H., Paichadze, N., Hyder, A. A., et al. (2015). Economic development and road traffic fatalities in Russia: Analysis of federal regions 2004–2011. *Injury Epidemiology, 2*, 19. https://doi.org/10.1186/s40621-015-0051-6

Hu, F., Lv, D., Zhu, J., & Fang, J. (2014). Related risk factors for injury severity of e-bike and bicycle crashes in Hefei. *Traffic Injury Prevention 15*(3), 319–23. https://doi.org/10.1080/15389588.2013.817669

Hu, G., Wen, M., Baker, T.D., Baker, S.P. (2008). Road-traffic deaths in China, 1985–2005: *Threat and opportunity. Inj. Prev.14*, 149–153. https://injuryprevention.bmj.com/content/14/3/149

Huang, Y., Tian, D., Gao, L., Li, L., Deng, X., Mamady, K., & Hu, G. Q. (2013). Neglected increases in rural road traffic mortality in China: Findings based on health data from 2005 to 2010. *BMC Public Health, 13*, 1111. https://link.springer.com/article/ https://doi.org/10.1186/1471-2458-13-1111

Jadaan, K. S. (1990). Traffic accidents in Kuwait: An economic dimension. *Accident Analysis & Prevention, 22*(4), 399–401. https://pubmed.ncbi.nlm.nih.gov/2222704/

Jenks, G. F. (1967). The data model concept in statistical mapping. *International Yearbook of Cartography, 7*,186–190. https://xueshu.baidu.com/usercenter/paper/show?paperid=3a917c675547e8306d705d50ee3c73a6

Kopelias, P., Papadimitriou, F., Papandreou, K., & Prevedouros, P. (2007). Urban freeway crash analysis: Geometric, operational, and weather effects on crash number and severity. *Transportation Research Record*, 123–131. https://doi.org/10.3141/2015-14

Kopits, E., & Cropper, N. (2005). Traffic fatalities and economic growth. *Accident Analysis & Prevention, 37*(1), 169–178. https://doi.org/10.1016/j.aap.2004.04.006

La Torre, G., Van Beeck, E., Quaranta, G., Mannocci, A., & Ricciardil, W. (2007). Determinants of within-country variation in traffic accident mortality in Italy: A geographical analysis. *International Journal of Health Geographics, 6*, 49. https://doi.org/10.1186/1476-072X-6-49

Li, Y., & Zhang, G. (2021). Road traffic safety and differences in regional economic development: evidence from random effects model and Shapley value decomposition. In G. Zhang, & Q. Zhong (Eds.), *Road safety in China*. Springer. https://doi.org/10.1007/978-981-16-0701-1_4

Li, X., Song, J., & Lin, T. (2016). Urbanization and health in China, thinking at the national, local and individual levels. *Environmental Health, 15*, S32. https://doi.org/10.1186/s12940-016-0104-5

Luoma, J., & Sivak, M. (2014). Why is road safety in the U.S. not on par with Sweden, the U.K., and the Netherlands? Lessons to be learned. *European Transport Research Review, 6*, 295–302. https://doi.org/10.1007/s12544-014-0131-7

Macioszek, E. (2011). The influence of motorcycling and cycling on small one-lane roundabouts capacity. In: J. Mikulski (Eds.), *Modern transport telematics. TST 2011. Communications in computer and information science*, vol 239. Springer. https://doi.org/10.1007/978-3-642-24660-9_33

Malin, F., Norros, I., & Innamaa, N. (2019). Accident risk of road and weather conditions on different road types. *Accident Analysis & Prevention., 2019*(122), 181–188. https://doi.org/10.1016/j.aap.2018.10.014

Matthias, G., Hu, Y., Zhou, X., et al, (2017). 2017 Report on the Development of Road Traffic Safety. *China Emergency Management, No.134*(02):48-58. https://d.wanfangdata.com.cn/periodical/ChlQZXJpb2RpY2FsQ0hJTmV3UzIwMjMwNDE5EhdRS0MyMDE4MjAxODA0MjcwMDE5MDAyMRoIemY0NDNkeHM%3D

Mock, C. N, Jurkovich, G. J,. Arreola-Risa, C. & Maier, R. V. (1998). Trauma mortality patterns in three nations at different economic levels: implications for global trauma system development. *Journal of Trauma and Acute Care Surgery, 44*(5), 804–814. https://pubmed.ncbi.nlm.nih.gov/9603081/

Nakamura, M. (1984). A study of differences in the incidence rate of traffic accidents (first report)—Factors affecting incidence of traffic accidents in the prefectures of Japan. *Japanese Journal of Health and Human Ecology, 50*, 168–77 (In Japanese) https://xueshu.baidu.com/usercenter/paper/show?paperid=0579563d1adeff1ffed1f8114a717934

National Bureau of Statistics of China (NBSC). (2022). http://data.stats.gov.cn/easyquery.htm?cn=E0103

National Bureau of Statistics of China (NBSC). (2017). http://data.stats.gov.cn/easyquery.htm?cn=E0103 (accessed April 12, 2017).

Okui, T., & Park, J. (2021). Analysis of the regional distribution of road traffic mortality and associated factors in Japan. *Injury Epidemiology, 8*, 60. https://doi.org/10.1186/s40621-021-00356-4

Pei, Y. L. & Ma, J. (2003). Research on countermeasures for road condition causes of traffic accidents. *China Journal of Highway and Transport, 16*(4). 77–82. http://zgglxb.chd.edu.cn/en/y2003/v16/i4/77

Petridou, E., & Moustaki, M. (2000). Human factors in the causation of road traffic crashes. *European Journal of Epidemiology, 16*(9), 819–826. https://xueshu.baidu.com/usercenter/paper/show?paperid=1k4q04t0sv6404w0vj620vh06w156309&site=xueshu_se

Pu, H., Li, B., Luo, D., Wang, S., & Duan, S. (2020). Impact of urbanization factors on mortality due to unintentional injuries using panel data regression model and spatial-temporal analysis. *Environmental Science and Pollution Research, 27*, 2945–2954. https://doi.org/10.1007/s11356-019-07128-0

Qin, H., Wei, Y. H., Zhang, Q. D., & Ma, L. (2021). An observational study on the risk behaviors of electric bicycle riders performing meal delivery at urban intersections in China. *Transport Research F-Traffic, 79*, 107–117. https://doi.org/10.1016/j.trf.2021.04.010

Rechnitzer, G., Haworth, N., & Kowadlo, N. (2007). *The effect of vehicle roadworthiness on crash incidence and severity, Report No. 164*. Monash University, Accident Research Centre. http://www.monash.edu/__data/assets/pdf_file/0017/216710/muarc164.pdf

Sangkharat, K., Thornesa, J. E., Wachiradilok, P., & Pope, F. D. (2021). Determination of the impact of rainfall on road accidents in Thailand. *Heliyon., 7*(2), e06061. https://doi.org/10.1016/j.heliyon.2021.e06061

Scuffham, P. A. (2003). Economic factors and traffic crashes in New Zealand. *Applied Economics, 2003*(35), 179–188. https://doi.org/10.1080/0003684022000017566

Shaik, M. E., & Hossain, Q. S. (2021). Application of statistical models: Parameters estimation of road accident in Bangladesh. *SN Computer Science, 2021*(1), 261. https://doi.org/10.1007/s42979-020-00284-y

Shi, J. G., Yao, Y., Hou, Z. H., Zhou, J. H., & Wang, Z. G. (2008). Influencing factors of expressway traffic accidents. *Chinese Journal of Traumatology, 24*, 553–557. https://d.wanfangdata.com.cn/periodical/zhcs200807025

Smeed, R. J. (1949). Some statistical aspects of road safety research. *Journal of the Royal Statistical Society, 112*(1), 1–34. https://rss.onlinelibrary.wiley.com/doi/ https://doi.org/10.2307/2984177

Song, L., & Zhang, S. S. (2016). VAR model study on the impact of scale and speed of economic growth on traffic safety risk. *China Public Saf (Academic Edition), 3*, 83–85 https://d.wanfangdata.com.cn/periodical/zgggaq-xsb201603019

Sun, L. L., Liu, D., Chen, T., & He, M. T. (2019). Analysis on the accident casualties influenced by several economic factors based on the traffic-related data in China from 2004 to 2016. *Chinese Journal of Traumatology, 2019*(22), 75–79. https://doi.org/10.1016/j.cjtee.2019.02.002

Tu, H. Li, Z. Hao, L. Ke. Z. & Sun, L. (2015). Driving simulator fidelity and emergency driving behavior. Wwwtrborg Compendium of Papers Trb Meeting. 113–121. http://www.trb.org/main/blurbs/173968.aspx

Verma, A., Harsha, V., & Subramanian, G. H. (2021). Evolution of urban transportation policies in India: A review and analysis. *Transportation in Developing Economies, 7*, 25. https://doi.org/10.1007/s40890-021-00136-1

Wang, C. W., & Chan, C. L. W. (2016). Estimated trends and patterns of road traffic fatalities in China, 2002–2012. *Traffic Injury Prevention, 17*(2), 164–169. https://www.tandfonline.com/doi/pdf/https://doi.org/10.1080/15389588.2015.1045501?needAccess=true

Wang, C., Rao, K., Wu, S., & Liu, Q. (2013). Health care in China: Improvement, challenges, and reform. *Chest, 2013*(143), 524–531. https://doi.org/10.1378/chest.12-1839

Wang, D. Y., Liu, Q. Y., Ma, L., Zhang, Y. J., & Cong, H. Z. (2019). Road traffic accident severity analysis: A census-based study in China. *Journal of Safety Research, 70*, 135–147. https://doi.org/10.1016/j.jsr.2019.06.002

World Health Organization (WHO). (2021). *Global health expenditure database*. https://apps.who.int/nha/database/Select/Indicators/en

World Road Transport Organisation. (2021). *The industry's commitment*. Available online: https://www.iru.org/what-we-do/beingtrusted-voice-mobility-and-logistics/people/road-safety

Yan, X., Wu, B., He, Y., Zhong, M., Zhu, Y., Zhao. (2019). A Study of "Version Zero" Concept of Transportation Safety and Its Implementation Strategies in China. *Journal of Transport Information and Safety, 2019,37*(01):1–6. https://xueshu.baidu.com/usercenter/paper/show?paperid=190b0xm0643t0vr09t6w0ak00c482764

Yang, Z., Wang, D., Du, T., Zhang, A., & Zhou, Y. (2018). Total-factor energy efficiency in China's agricultural sector: Trends, disparities and potentials. *Energies, 11*(4), 853.

Yu, S. C., & Tsai, W. D. (2021). The effects of road safety education on the occurrence of motorcycle violations and accidents for novice riders: An analysis of population-based data. *Accident Analysis & Prevention, 163*, 106457. https://doi.org/10.1016/j.aap.2021.106457

Zhang, W., Tsimhoni, O., Sivak, M., & Flannagan, M. J. (2010). Road safety in China: Analysis of current challenges. *Journal of Safety Research, 41*(1), 25–30. https://doi.org/10.1016/j.jsr.2009.12.003

Zhang, J., Zhang, W., Xu, P., & Chen, N. (2019). Applicability of accident analysis methods to Chinese construction accidents. *Journal of Safety Research, 68*, 187–196. https://doi.org/10.1016/j.jsr.2018.11.006

Zhuang, X., & Wu, C. (2011). Pedestrians' crossing behaviors and safety at unmarked roadway in China. *Accident Analysis and Prevention, 43*(6), 1927–1936. https://doi.org/10.1016/j.aap.2011.05

Chapter 13
Summary

13.1 Main Findings

13.1.1 China's Railway Transport Features Low Technical Efficiency, Highway and Water Transport Features Moderate Technical Efficiency, and Air Transport Features High Technical Efficiency

The efficiency level of the railway passenger transport system is higher than that of the railway freight transport system, which indicates that the problem of investment waste is more serious in the railway freight transport system. This is partly due to the fact that freight transport is more resource-intensive. For example, cargo handling needs more human resources and facilities. In addition, the running speed of freight cars is generally lower than that of passenger cars, and the average occupation time of freight trains is longer than that of passenger trains (Yu & Lin, 2008).

The overall technical efficiency of China's highway system is at a moderate level. Highway transport is widely used for passenger and freight transport in China. The highway passenger volume and the highway freight volume account for more than 80% and 50% of the total passenger volume and freight volume, respectively. However, due to the large investment in highway transport infrastructure in recent years, HPTE (Highway passenger transport efficiency) and HFTE (Highway freight transport efficiency) are inhibited. In general, it is important to control the amount of investment and reduce the waste caused by excessive capital investment.

The overall level of WTE (Water transport efficiency) is also at a moderate level. The water freight volume accounts for a large proportion of the total freight volume, while the market share of water passenger transport is small. The proportion of water transport in the total fixed asset investment in transport over the years is small. In general, we need to fully exploit the advantages of river and ocean transport, and to further improve the market share of water freight transport.

P. Zhao and L. Zeng, *Transport Efficiency and Safety in China*, Population, Regional Development and Transport, https://doi.org/10.1007/978-981-99-1055-7_13

The overall level of ATE (Air transport efficiency) is high. From 2009 to 2016, air passenger and freight volumes grew at an average annual rate between 12.6% and 8.9%, which is much higher than the growth rate of passenger and freight volumes in the entire transport industry. The high-speed growth of air passenger volume contributes to ATE.

13.1.2 The HFTE, HPTE, RPTE, RFTE and TEE in China Show a Strong Spatial Correlation, While the ATE and WTE in China do not Have Any Spatial Correlation

From the perspective of global spatial autocorrelation, the HFTE, HPTE, RPTE (Railway freight transport efficiency), RFTE (Railway freight transport efficiency) and TEE (Transportation environment efficiency) in China have a highly positive spatial crowding effect rather than being random, and these values tend to be similar in provinces or regions that are close to each other, instead of a random spatial distribution in these years. However, the ATE and WTE do not show any spatial correlation. Based on local spatial autocorrelation analysis, there are mainly two positive spatial agglomerations for provincial HPTE, HFTE and RPTE: the HH agglomeration area in the first quadrant and the LL agglomeration area in the third quadrant. In contrast, there is mainly a positive spatial agglomeration for the provincial RFTE: the LL agglomeration area in the third quadrant, which further confirms that China's provincial HPTE, HFTE, RPTE, and RFTE present a significant positive spatial correlation.

13.1.3 The Efficiency of Various Transport Sub-Sectors is Mainly Affected by the Economic Development Level, Population Density, Urbanization Level, Transport Infrastructure Level, Industrial Structure and Other Factors

Based on the existence of spatial autocorrelation of HFTE, HPTE, RPTE, RFTE and TEE in China, this study applies SDM (spatial Dubin model) to perform a regression analysis on them. The results show that: (1) The level of economic development, highways and the tertiary industry can promote HPTE, while the development of high-speed railways can significantly inhibit HPTE. (2) The development of expressway construction and the secondary industry can promote HFTE, while population density significantly inhibits HFTE. (3) Population density and the urbanization level can significantly improve RPTE, while railway density and RPTE show a significant negative correlation with RFTE. (4) The urbanization level and population density have a significant negative correlation with RFTE. (5) The level of economic development, transport structure and infrastructure can significantly improve TEE, while the urbanization level can significantly inhibit TEE.

13.1.4 The TEE in China is Mainly Affected by the Economic Development Level, Urbanization Level, Transport Infrastructure Level and Industrial Structure

Based on the fact that there is no spatial effect for the ATE and WTE in China, and that the values of ATE and WTE of each province are between 0 and 1, which are truncated values, the Tobit model is applied to analyze the influencing factors of ATE and WTE. It is found that the economic development level, population density and the development of the tertiary industry show a positive correlation with ATE, while the development of high-speed rail inhibits ATE.

13.1.5 The Transport Sector is the Second Largest Energy Consuming Sector and the Third Largest Source of Carbon Emissions in China

In China, transport energy consumption and CO_2 emissions show an overall growth trend, but the energy intensity and carbon emission intensity show a downward trend. From 1990 to 2016, the end-use energy consumption grew rapidly, but the growth had slowed down since the beginning of the 12th Five-Year Plan. The end-use energy consumption of eastern and central provinces is higher than that of western provinces. However, the per capita transport energy consumption in Central China is relatively high. From 2009 to 2016, the spatial characteristics of transport CO_2 emissions in China presented transverse distribution, and the reduction trend from the coast to inland areas was not obvious. However, the CO_2 emission intensity of the transport sector in China obviously increased from the coast to inland areas. Compared with 2009, the growth rate of transport energy consumption in Central and Northwest China was higher than other regions in 2016.

13.1.6 The Number of Traffic Accidents and Injuries is on a Downward Trend, but the Overall Transport Safety is in a Severe Situation in China

After the peak in 2002, the number of traffic accidents and deaths in China decreased significantly. The road traffic accident rate and fatality rate in China are higher than developed countries. The number of traffic accidents is higher in developed coastal areas, but the mortality rate per 10,000 vehicles is higher in western areas. The heavy goods vehicle is a major vehicle type in road traffic accidents; traffic accidents on expressways cannot be ignored. The safety situation of rural roads is still grim.

13.1.7 During the Study Period, the RTMR Showed a Downward Trend

Most of the provinces with a higher RTMR (Road traffic mortality rate) are located in developed coastal areas or underdeveloped northwest regions, while most of the provinces with a lower RTMR are located in central regions. The reason is that developed coastal areas have a high level of urbanization and vehicle ownership, and western areas have poor health services. The RTMR generally showed a downward trend. China promulgated "Law of the People's Republic of China on Road Traffic Safety" and then implemented it in an all-round way. Meanwhile, China has increased financial expenditure on medical care, and strengthened publicity work on transport safety. These policies are conducive to the sharp decline of RTMR. The Tobit regression results show that the improvement of the economic development level, urbanization level and motorization level promotes RTMR, while the improvement of the medical level can inhibit RTMR.

13.2 Theoretical Contributions

13.2.1 Showing the Characteristics and Influence Mechanism of Transport Efficiency in China Through Systematic Studies

Since the end of the twentieth century, especially in recent years, many problems of the development of transport infrastructure in China have worsened, including over-expansion, duplicated construction, and unreasonable competition, among others (Lu, 2012). These problems are primarily the result of the ignorance of national conditions, the stage of development, and trends during transport planning. Against the background of the continuous intensification of resource and energy constraints in China, how to improve transport efficiency while maintaining socio-economic development is especially important for China. However, globalization has brought China closer than ever to the rest of the world, not only through its trade and transport networks, but also through the many transport-related issues that seem to be common among countries Therefore, our analysis on transport efficiency in China is meaningful and contributes to the literature on sustainable transport development. Furthermore, China is diverse in terms of geographical conditions, socio-economic levels, and development policies. Transport efficiency varies a lot from region to region. Hence, it is necessary to examine transport efficiency in China on the regional level.

Therefore, understanding transport efficiency of various transport sub-sectors (railway, highway, waterway and air transport) and its influencing factors in China is of particular importance because transport systems are a critical element of China's path towards rapid economic growth, urbanization, and sustainable development. The main objective of this study is to analyze the transport efficiency by province and to explore the determinants of transport efficiency.

13.2.2 Enriching the Research on Environmental Efficiency

In recent years, a lot of scholars have carried out fruitful research on the evaluation theory and evaluation method of environment efficiency and its influencing factors. But their research does not cover the transport sector. The development of transport infrastructure is characterized by significant capital input, high energy consumption, and heavy pollution emissions. For instance, according to the International Energy Agency (IEA), the transport sector accounted for about 22% of global CO_2 emissions in 2016, becoming the second-largest source of CO_2 emissions. In China, the transport sector contributed about 10.7% to national CO_2 emissions (IEA, 2018). Therefore, attempting to reduce resource consumption on the premise of ensuring output is a central theme in sustainable transport development worldwide.

On the basis of the actual situation in China, this study organically combines the frontier theories and research methods of environmental efficiency, and constructs a scientific theoretical system to calculate transport environmental efficiency of integrated transport systems, which is of great significance to improve the existing research on environmental efficiency. In addition, the evaluation of transport environmental efficiency under resource and environmental constraints will also provide a set of theoretical methods and a research framework for green GDP accounting.

13.2.3 Revealing the Impact Mechanism of Road Traffic Safety in China from a Macro Socio-Economic Perspective

Road traffic accidents are the main type of traffic accidents. At present, many researchers have analyzed the influencing factors of road traffic accident casualties and road traffic safety from various aspects. However, the existing research on the influencing factors of road traffic accident casualties mainly focuses on the impact of people, vehicles and roads on traffic accident casualties separately. There are few ways to consider economic, road and population factors, and there is a lack of research on the incremental impact of various variables (Sun et al., 2019). The policy recommendations are only for the optimization of rescue and management mechanisms. The existing research on road traffic safety is mainly conducted at a macro level, policy recommendations are only based on theoretical analysis, with no effective data support, and there is little research on traffic safety through data analysis related to traffic accident casualties.

This study conducts a regression analysis of the influencing factors of road transport safety from a macro socio-economic perspective. Through analyzing the influence of specific variables of each factor on road traffic accident casualties, we put forward corresponding suggestions and management measures for road traffic safety in China, so as to provide a theoretical basis and data support for improving road traffic safety in China.

13.3 Future Research Agenda

13.3.1 Future Changes in the World

(1) The world is witnessing a shift towards green transport

Since 2014, the new energy transport industry using electric energy and fuel cells as a replacement for fossil fuels has developed vigorously. In 2020, the number of electric vehicles in the world exceeded 10 million, with an annual growth rate of 43%, and the growth rate of fuel cell vehicles in the same period was 40%. Railway electrification, utilization of hydrogen energy in freight transport and intelligent upgrading of the transport system are becoming global trends.

(2) Global transport will become more integrated

The main modes of transport are water, highway, railway, air, and pipeline transport, each with its own advantages, disadvantages and application scope, and their development is uneven. In the future, various modes of transport will achieve in-depth coordinated and integrated development at a higher level and in a wider range with the progress of traffic technology. The trend of transport integration can reduce vicious competition among various transport modes and realize more efficient utilization of transport land and energy.

(3) The intelligent development of transport is an inevitable trend

The world has entered the ABC era (artificial intelligence + big data + cloud computing) in the twenty-first century. In the ABC era, the intelligent development of transport is inevitable. Intelligent transport strives to make efficient use of transport infrastructure and public resources. Big data-driven decision-making will provide a better travel experience, smoother goods circulation and more accurate government decision-making. In fact, many world-renowned companies, such as IBM, Google, Siemens, Alibaba and Baidu, are active in the field of intelligent transport, and new transport service modes emerge one after another. Looking forward to the future, transport will fully realize the digitization and networking of infrastructure and transport tools, as well as information-based and intelligent operation, by making full use of modern high technology.

(4) Technological progress and new energy development will contribute to a decline in global energy consumption and transport CO_2 emissions after they peaked in the twenty-first century

According to Vision 2050 issued by the World Business Council for Sustainable Development (WBCSD), low-carbon transport is expected to be popularized by 2050; WBCSD (2015) estimated that the use of light vehicles could reduce CO_2 emissions by 80%, and the CO_2 emissions of road freight, air and water transport would also be reduced by 50% (2010). In addition, solar energy will account for a large proportion in the global energy structure, and new energy vehicles such as electric vehicles are promising.

13.3.2 Prospects for China's Efficient Transport Network

(1) China's transport infrastructure construction aims to reach the world's advanced technological level

Currently, many technologies used in transport infrastructure construction, such as extra-large bridges, long tunnels, high-grade highways, high-speed railways, plateau railways, heavy-haul railways, deepwater ports, large airports and other projects, rank among the top in the world. The 14th Five-Year Plan involves more transport construction tasks than the 13th Five-Year Plan, which means that China will continue to pursue more advanced transport infrastructure construction technologies.

(2) China will make more breakthroughs in transport equipment technology

After years of effort, many technologies for transport equipment, such as high-speed trains, heavy-haul trains, and urban rail trains, have been ranked among the top in the world by technology import and innovation. In particular, high-speed trains have become a marquee name for China. Ocean-going ships, high-powered railway locomotives and marine machinery have reached the world's advanced level. During the 14th Five-Year Plan period, R&D expenditure will be increased to above 2.4% of GDP, which will continue to promote the progress of transport equipment technology in China.

(3) The transport industry in China will make full use of the new generation of information technology

The integration of the new generation of information technology and transport has been applied with surprising speed in full self-driving, online ride-hailing, delivery drones, and so on. For example, the full self-driving technology in China is gaining momentum, and transport development is included in the 14th Five-Year Plan. Baidu, Alibaba, Tencent, Huawei and other Internet giants have poured themselves into the full self-driving industry. JD.com has extensively used the unmanned aerial vehicle in the express industry.

(4) Key technologies of environmental protection lead the development of green transport

Significant progress has been made in the recycling of road materials and asphalt mixtures, operation and maintenance of transport infrastructure in ecologically sensitive areas, energy-saving lighting and intelligent control of tunnels, and emergency response technology for offshore oil spills. The development of new-energy vehicles is also accelerated, and pilot and demonstration projects such as green roads and green port stations are expedited.

(5) The traffic emergency response system has been greatly improved

China will continue to improve the traffic emergency response system, prevent and resolve major traffic safety risks in a timely manner, and effectively deal with all kinds of disasters and accidents. China will strengthen scientific research on major emergency support equipment. For example, the saturation deep diving has successfully realized the operation capacity of 320 m, and the maximum diving depth has exceeded 330 m.

13.3.3 Prospects for Research Topics

This study takes transport efficiency, transport environmental efficiency and transport safety in China as the research objects, and has made significant achievements in empirical research, but there are still some aspects to be further improved in the research process. The main deficiencies of this study and the main directions of future research are as follows:

(1) The big data analysis method can be applied in future research

The research on transport efficiency, transport environmental efficiency and transport safety in this book is based on the traditional econometric model, which has the characteristics of macro-thinking. The big data analysis method based on the new generation of information technology is a research method that has emerged in recent years; it has the characteristic of large amounts of micro data. In future research on transport efficiency, transport environmental efficiency and transport safety, we can try to apply the big data method to provide richer research results.

(2) Studying the impact of transport development on global climate change and extreme weather disasters

The transport sector is the second largest carbon emission source in the world. A in-depth analysis of the impact of transport on climate warming is helpful to global emission reduction decisions. The rapid development of the transport industry will inevitably occupy a large amount of land and affect the ecological environment, climate, and vegetation in the surrounding area. Therefore, it is very meaningful to analyze the impact of the expansion of transport land on the ecological environment or extreme weather disasters, so as to provide an effective reference for the formulation of global ecological policies.

(3) Pollutant emission factors can be considered in efficiency evaluation of specific transport modes

In the calculation of HPTE, HFTE, RPTE, RFTE, ATE, and WTE, pollutant emission factors are not considered, more specifically, A region of the expected output of technical efficiency measure impact, there is not fully consider undesirable outputs that environmental factors in previous studies, it will affect the accuracy of the results of environmental technical efficiency. In future research, we should try to incorporate energy and undesirable factors into the present technical efficiency evaluation system for different transport sub-sectors to improve its scientificity and authenticity.

(4) The efficiency evaluation of specific transport modes can be conducted at a municipal level

The research on HPTE, HFTE, RPTE, RFTE, ATE, WTE and TEE is carried out from a provincial perspective, and in the future we can further and expand the research at a micro level. Next, we can calculate the technical efficiency at the city, county, or enterprise level, and carry out a micro traffic behavior survey, which can reveal the influence mechanism of transport efficiency and transport environmental efficiency from a micro perspective. Then we can get more comprehensive and richer research results.

(5) The selection of the influencing factors of transport efficiency, transport environmental efficiency and transport safety from more angles

In the end, there are some deficiencies in the selection of factors affecting transport efficiency, transport environmental efficiency and transport safety. The reason is that

the factors affecting them; although we try to systematically select the influencing factors, owing to limited research ability and the accessibility of statistical data, some influencing factors are inevitably neglected. It is very difficult to include all the factors affecting both transport efficiency, transport environmental efficiency and transport safety. Therefore, in future studies, we can choose the influencing factors from more angles. For example, this study does not take into account related economic costs such as the price of the transport journey and the price of transport energy; in future research, we will introduce relevant price factors into the regression analysis, which can help us more comprehensively analyze the influencing factors of transport efficiency, transport environmental efficiency and transport safety.

References

International Energy Agency (IEA). (2018). World Energy Outlook (2018). https://www.iea.org/reports/world-energy-outlook-2018

Lu, T. D. (2012). The proposition to avoid the over advance and inappropriate construction of China's transport infrastructures. *Scientia Geographica Sinica, 32*(1), 2–111. http://geoscien.neigae.ac.cn/EN/Y2012/V32/I1/2. Accessed December 7, 2019.

Sun, L. L., Liu, D., Chen, T., & He, M. T. (2019). Analysis on the accident casualties influenced by several economic factors based on the traffic-related data in China from 2004 to 2016. *Chinese Journal of Traumatology, 22*, 75–79. https://doi.org/10.1016/j.cjtee.2019.02.002

World Business Council for Sustainable Development (WBCSD). (2015). Vision 2050. https://www.docin.com/p-1764452580.html

Yu, M. M., & Lin, E. T. J. (2008). Efficiency and effectiveness in railway performance using a multi-activity network DEA model. *Omega*, (36), 1005–1017. https://doi.org/10.1016/j.omega.2007.06.003

The manufacturer's authorised representative in the EU is Springer Nature Customer Service Centre GmbH, Europaplatz 3, 69115 Heidelberg, Germany. If you have any concerns regarding our products, please contact ProductSafety@springernature.com

Printed and bound by CPI Group (UK) Ltd, Croydon, CR0 4YY
01/10/2026
02225870-0010